ADAPTING TO DROUGHT

ADAPTING TO DROUGHT
FARMERS, FAMINES AND DESERTIFICATION IN WEST AFRICA

MICHAEL MORTIMORE

Formerly of Department of Geography, Bayero University, Kano

CAMBRIDGE UNIVERSITY PRESS

CAMBRIDGE

NEW YORK   NEW ROCHELLE   MELBOURNE   SYDNEY

Published by the Press Syndicate of the University of Cambridge
The Pitt Building, Trumpington Street, Cambridge CB2 1RP
32 East 57th Street, New York, NY 10022, USA
10 Stamford Road, Oakleigh, Melbourne 3166, Australia

© Cambridge University Press 1989

First published 1989

Printed in Great Britain at the University Press, Cambridge

British Library cataloguing in publication data
Mortimore, Michael
Adapting to drought: farmers, famines, desertification in West Africa.
1. Africa. West Africa. Environment.
Adaptation of. Man.
I. Title
304.2'0966

Library of Congress cataloguing in publication data
Mortimore, M.J., 1937–
Adapting to drought: farmers, famines, and desertification in
West Africa/Michael Mortimore.
    p.   cm.
Bibliography.
Includes index.
ISBN 0 521 32312 6
1. Famines – Nigeria, Northern. 2. Agriculture – Economic aspects – Nigeria, Northern. 3. Droughts – Nigeria, Northern.
4. Desertification – Nigeria, Northern. 5. Arid regions farming – Nigeria, Northern. 6. Nigeria, Northern – Rural conditions.
I. Title
HC 1055.Z7N6755 1988
333.73 – dc 19 88-2369 CIP

ISBN 0 521 32312 6

TM

For

Julia, Sally, Joanna, Nicholas and Rachel

CONTENTS

List of figures	page	ix
List of plates		xi
List of tables		xii
Preface		xv
Acknowledgements		xviii
A note on conventions and list of abbreviations		xxi

1	**Introduction**	**1**
	Crisis in rural tropical Africa	1
	Uncertainty and adaptation	3
	Science and ethno-science	4
	Aridity in Africa	6
	Aridity in northern Nigeria	9
	Drought	11
	Desertification: concept and controversy	12
	Desertification: definition and measurement	15
	Method	19
2	**From feast to famine?**	**23**
	The fat years	23
	Famine in Hausaland	36
	Postscript on the groundnut	40
3	**Drought in the 1970s**	**42**
	The droughts of 1972 and 1973	42
	631 villages	46
	Patterns of hardship	48
	Five villages	56
	Impact	57
	Response	63
	Evaluation	75
4	**Thirteen years in the life of a village**	**82**
	One village	82

	After Kakaduma: production and consumption	86
	A chronicle of adversity	89
	Household viability and change, 1975–86	97
	Adaptive response: the farming system	98
	Adaptive response: other systems	110
	Adaptive response: two families	113
	A balance sheet	116
5	**Wider horizons**	**117**
	Mobility in the city: Kano's strangers	117
	Mobility in the region	122
	Mobility in one village	130
	Territory and mobility	134
6	**Two dry decades**	**136**
	The persistence of meteorological drought	136
	Prediction	144
	Feedback	148
	Hydrological drought	151
	Land use and drought	155
7	**Shifting sands**	**157**
	The Manga Grasslands	157
	Ecological change, 1950–69	161
	Woodland	163
	Grassland	165
	Soils	168
	Surface materials	170
	Over-exploitation or under-precipitation?	185
8	**Interpretation**	**187**
	Adaptive capability	188
	Mobility	196
	Ecological degradation	199
9	**Policy directions**	**218**
	Insurance	219
	Diversification and mobility	221
	Intensification	225
	Conclusion	229
	Notes	231
	Bibliography	253
	Index	289

FIGURES

1.1	Bioclimatic zones and desertification in the subSaharan zone of northern Africa *page*	8
1.2	The Dry Zone of northern Nigeria	10
2.1	Northern Nigeria	24
2.2	The groundnut boom in northern Nigeria	25
2.3	Average crop purchases by the Northern Nigeria Marketing Board, 1963–65	26
2.4	Density of the rural population in northern Nigeria	31
2.5	Danbatta District in Kano State	37
3.1	Displacement of rainfall isohyets in 1972 and 1973, Kano State	43
3.2	(A) Regression lines for the start and end of the rains in 1973 and in an average year; (B) length of the rainy season and the southward shift of the northern limit of millet and groundnut production in 1973	45
3.3	Survey areas	47
3.4	Effects of drought on crop yields, 1973 (17 districts)	50
3.5	Grain and cattle prices, 1972–74	55
3.6	Evaluations of the 1972–74 famine, Kakaduma	78
4.1	The location of Dagaceri	82
4.2	Plan of Dagaceri	84
4.3	The settlement pattern in the vicinity of Dagaceri	85
4.4	Prices and rainfall, Kano State, 1970–85	91
4.5	Village and state-wide prices, 1980–81	93
4.6	M's farms at Dagaceri	114
5.1	Urban Kano, showing sample locations	118
5.2	First arrival in Kano: strangers interviewed in January–April 1974	118
5.3	The situation left at home by migrants to Kano	121
5.4	Out-migration: frequency of destinations cited by village heads in the sample area, 1973–74	123
5.5	In-migration: frequency of origins cited by village heads in the sample area, 1973–74	126

5.6 In-migration: percentage of village heads reporting, by district	127
5.7 Spatial preferences in seasonal circulation: five villages, 1973–74	129
6.1 Rainfall departures from the mean in 1973	139
6.2 Rainfall at Kano, 1906–85	140
6.3 Monthly distribution of rainfall at Kano Airport	142
6.4 Rainfall and mean weekly dust deposition at Kano, 1978–85	150
6.5 Trends in the depth to water table in pact of Kano State, 1973–83	154
7.1 The Manga Grasslands	158
7.2 Kaska dunes	174
7.3 Profile of the Kaska dune, 1979–86	175
7.4 Ligaridi Karami dunes	176
7.5 Ligaridi Babba dunes	177
7.6 Profiles of dunes: A, Ligaridi Babba; B, Gaptari; C, a rangeland dune	178
7.7 Bulari and Gaptari dunes	179
7.8 Ngelsandi and Lala'a dunes	179
7.9 Rangeland dunes near Kaska	181

PLATES

1 A farmer recovers the bird-damaged remnants of his millet crop, near Babura, north Kano State (October 1973)   *page* 49
2 Emaciated cattle attempting to graze regenerating dum (*Hyphaene thebaica*) in barren farmland near Babura, north Kano State (July 1973)   51
3 Street scene in Dagaceri (October 1986)   83
4 Manga woman hoeing a groundnut plot at Dagaceri (late July 1978)   101
5 The *ashasha* in use at Dagaceri (June 1978)   109
6 Rope- and mat-making gets under way after a bad harvest in Dagaceri (October 1986)   112
7 The Manga Grasslands in the early rains (late July 1981)   159
8 The leading edge of the Kaska dune proceeds slowly through the village market place (July 1986)   173
9 Rangeland dune developing on the crest of a thinly grassed stable slope, overlooking an intermediate depression with farmland and sparse woodland (July 1980)   182
10 Cattle grazing around a semi-nomadic Fuli'be camp (*ruga*): a major dune formation occupies the entire skyline (July 1986)   184

TABLES

		page	
1.1	Desertification in the ecosystem	page	19
2.1	Estimated area and average annual production of some major crops in Northern Nigeria, 1965–67		28
2.2	Estimated numbers of livestock in Northern Nigeria, 1965 and 1975		29
2.3	1973 and 1974 compared in Roni and Danbatta Districts		39
3.1	Latitude correlated with rainfall in West Africa		44
3.2	Reported crop yields in 1973 as a fraction of 'normal'		49
3.3	Reported livestock mortality, 1973		52
3.4	Prices of some major foodstuffs, 1968–70 and 1973–74		54
3.5	Effects of drought on the production of some main crops		58
3.6	The livestock account, 1972–74		59
3.7	Meals taken on the day before interview		60
3.8	Food supplies by village, from west to east		62
3.9	Number of families not owning livestock		62
3.10	Losses of cattle, sheep and goats		63
3.11	Carbohydrate and animal protein foods in use, May–July 1974		67
3.12	Plant foods used in famine, northern Kano State, 1972–74		68
3.13	Some names given to famines in northern Kano		76
3.14	Reasons given for evaluations of the famine		79
3.15	Reasons given for the calamity		80
4.1	Average crop yields		87
4.2	Grain yields and requirements		88
4.3	Indices of rainfall at Kano, Nguru, Zinder and Maine Soroa, 1970–85		89
4.4	Changes in residential households in Dagaceri, 1975–86		98
4.5	Manuring and distance from the village of Dagaceri		100
4.6	The effect of drought on manuring		100
4.7	Size of farm holdings in Dagaceri		103
4.8	Modes of acquisition of farmland		103
4.9	Distribution of farm holdings around Dagaceri		104
4.10	Changes in land use at Dagaceri, 1950–81		105

List of tables

4.11	Marketing Board purchases of groundnuts in northern Hadejia, 1963–75	106
5.1	Geographical and ethnic origins of strangers interviewed in Kano urban area in the dry season of 1973–74	119
5.2	Livestock ownership, Kano strangers in January–May 1974	121
5.3	Frequency of destinations cited in villages reporting out-migration, 1973–74	124
5.4	Frequency of origins cited in villages reporting in-migration, 1973–74	125
5.5	Duration of trips: seasonal circulation, five villages, 1973–74	128
5.6	Participation in circulation, five villages, 1973–74	128
5.7	Sources of income during seasonal circulation, five villages, 1973–74	129
5.8	Male family heads temporarily absent from Dagaceri village in March 1976	130
6.1	Monthly distribution of rainfall at Kano (June–September), 1906–85	143
6.2	August rainfall 1966–85: departure from the 1931–60 mean	144
6.3	Trends in the groundwater table in part of Kano State, 1973–83	154
7.1	Ecological change in the Manga Grasslands, 1950–69	162
7.2	Rainfall at Maine Soroa, grain yields and grazings in the Manga Grasslands	166
7.3	Soil properties	169
7.4	The development of moving sand formations, 1950–69	171
7.5	Sand mobilisation in selected localities	172

PREFACE

Late in May 1974, the suffering caused by two successive harvest failures was at its height, and the planting rains of earlier that month had not been followed by more, causing the early millet to sprout and die in all the fields across the north of Kano State. In an unlit government rest house on the outskirts of Gumel, Cecil Woodham-Smith's *The great hunger* provided pretty sombre reading against a mocking background of sporadic lightning in the night sky. There is no denying that famine may hold a morbid fascination for those not called to suffer it, a fascination the more powerful for the way the subject transcends the bounds of culture and of history. There is no use in pretending a scientific detachment from the tragedies and triumphs, the greed and generosity, the degradation and the dignity in a major social disaster. Yet to further understanding, something more than committed journalism is called for. Tools of observation and enquiry, knocked up in the workshops of the social and environmental sciences, ought to help the search for system and significance.

According to a minority view, the relations between society and its environment lie at the heart of geography and form its noblest theme. A famine following in the wake of a major drought ought therefore to offer quite an opportunity. Not many responses, however, are evident. There is good enough reason for this. Such is the reluctance to scale the 'Berlin Wall' (as Paul Richards calls it) separating the social from the natural sciences that no theoretical orientation seems to exist which can adequately support such an attempt. Consequently either one approaches from the angle of social structure and risks portraying rainfall failure merely as an aggravation of a crisis of distribution, or one follows the ramifications of environmental variability into social behaviour, and risks being accused of environmental determinism. Examples of both these forms of imbalance have been in evidence since the West African droughts of the 1970s.

How do village people look at it? Polly Hill, in her study *Rural Hausa: a village and a setting*, gave us all an example of allowing rural perceptions and behaviour to speak for themselves, through painstaking fieldwork encumbered with a minimum of theoretical preconceptions. It is in this spirit that the present study has been attempted. Villagers seem to see rainfall as a short-term variable calling for adaptive response within the limits set by fixed social constraints – among which may feature an entrenched ruling class, a government dominated by urban bias, or the extortions of taxation. It is this adaptive behaviour that first impressed me when Hausa villagers, whom I had learnt to respect during my researches in the Kano Close-Settled Zone in the sixties, had to live with the disasters of the seventies. Instead of the system cracking, it survived. How and why? What were the roots of such resilience, that a society seemed even to gain in the dignity of suffering as it lost in material well-being?

I have been angered and bemused in turn by some of the strictures laid down in the literature on desertification. Much of this literature has been inspired by the World Conference on Desertification (1977), which was the United Nations' own institutionalised adaptation to the Sahelian Drought. According to some, the very communities I was learning to admire were themselves the architects of degradation in their own environment. Furthermore, such views were being given hearing not only in international circles but even by the governments of the affected communities. The reason was clear. These were the judgements of the 'experts' in the natural sciences, and the new technology of remote sensing, whose basis in 'ground truth' does *not* include enquiries into the operations of indigenous land-use systems. So, hypotheses reign – even to the extent that West Africans have been accused of causing their own droughts. The deterministic flavour of such hypotheses, and the authoritarian solutions they suggested, clashed with my own observations and reflections, and stimulated me to search for an alternative.

It seemed to me significant that earlier path-breaking studies such as those of M.G. Smith and Guy Nicolas in Hausaland had given a relatively small place to environmental variability as a factor helping to explain social behaviour; neither was it addressed in the thorough studies that were undertaken by the Rural Economy Research Unit of Ahmadu Bello University in the 1960s, even though one set of villages selected for those studies was as far north as Sokoto. Possibly this situation reflected the comparatively stable rainfall regime which came to be taken for granted (by researchers) in the fifties and sixties. Again it was left for Polly Hill to rediscover the place of crop failures in contributing to patterns of poverty in rural Hausaland. But in socio-economic analyses, the natural environment tends to be treated as an independent variable. Even Michael Watts' *Silent violence. Food, famine and the peasantry in northern Nigeria* gives pre-eminence to social theoretical concerns; although the wealth of adaptive skills for dealing with an agricultural drought is very well documented for one village. The environment as a major focus – its short-term variability and the question of its degradation in linkage with land-use systems in the longer term – seems too broad and complex a question to be successfully encompassed by the theoretical frameworks thus far employed.

The present study does not succeed in fully remedying this situation. I see it as a preliminary skirmish, little more than an exercise in clearing the ground. Unlike the literature to which I have just referred, this book had not the benefit of intensive field research sustained for several months or more in a single locality; rather it contains the results of repeated, often hasty and ill-prepared forays that took place over a number of years. During that time I was employed by two Nigerian universities whose pedagogic and administrative requirements tended to preclude long absences. The extent of commitment to the interests of the rural poor among researchers in Third World institutions has recently been questioned by Robert Chambers (*Rural development: putting the last first*); they have been accused of being elitist, urban biased and infrequent visitors to the rural scene. It seems to me that such a facile view takes inadequate account of the assets of indigenous research. Many who work in universities or institutes themselves have families in rural areas, and are not unfamiliar with rural problems. The student population in teaching institutions ebbs and flows several times a year between town and village, forcing rural perspectives on the attention of the teacher. And there is the possibility, little appreciated by the short-term foreign visitor, of continuing reflective interaction with the flow of events – flood or famine, boom or depression, coup or reformation as perceived within the country. All this provides a critical context for rural research, rather different from the detached debate, the contending hypotheses, that inform the specialist abroad. Research in African countries is still too dependent on the training, literature and ideas provided in Western universities, and there is something to be said for exposure to a local perspective, together with repeated circulation at short intervals of time between the study and the field.

Under the green shade of its protective neem trees, the thatch and straw compounds of Dagaceri village give little hint of the recurring waves of misfortune that have afflicted communities throughout the semi-arid zone of subSaharan Africa during the past two decades. Struck by the reassuring way in which this community resurfaced after sinking under the famine of Kakaduma in 1972–74, and wanting to observe how it made out in subsequent years, I fell upon the longitudinal perspective, which is central to this study, largely by accident. A preoccupation with 'development by transformation' and its corollary, the downgrading of existing systems of production to a residual status, has for too long obscured the need to understand resilience as a way of life, in these environments of uncertainty. The subject has not lost its relevance, as, following widespread failure of the rains in 1987, Africa bows once more under the sombre threat of famine.

MICHAEL MORTIMORE
Milborne Port

ACKNOWLEDGEMENTS

The research that is reported in this book was made possible by grants from the research funds of Ahmadu Bello University, Zaria, and Bayero University, Kano; a period of six months' study leave in 1974; and a sabbatical year in 1984–85. The support provided by these institutions, under whose aegis I worked in Nigeria from 1962 to 1986, went beyond this, however. Although both universities were founded principally with a view to training indigenous graduates for professional service in their regions of origin, it is hard to underestimate the potential importance of social and environmental research in their catchment areas: which for Ahmadu Bello University, when it was founded in 1962, included the whole former Northern Region of Nigeria, and for Bayero University, which became an independent institution in 1976, the smaller but no less significant Kano hinterland. The departments of geography in which I served always wholeheartedly espoused the priority of field research, which ought to be the grist to the geographer's mill in places where such rich interactions of natural environments with human communities form a backloth to the teacher's calling; where new insights are as likely to spring from some student, fresh back from dissertation fieldwork, as from a teacher; and where, as students graduated into teachers themselves, indigenous experience met expatriate curiosity in a continually fertile mix. It is to H.A. Moisley, who inaugurated the Chair of Geography at Ahmadu Bello University (1963–65), that I owe my first research assignment in Kano, and to his successors at Zaria that I owe subsequent encouragement in my endeavours. Through the present heads of those departments – Professor K.O. Ologe at Zaria and Dr E.A. Olofin at Kano – I can best express my appreciation to all my former colleagues and students, whose contributions to this study are more important than they realise. These good years owed much to Professor Ishaya Audu, Vice-Chancellor of Ahmadu Bello University from 1967 to 1975, and Professor Ibrahim Umar, Vice-Chancellor of Bayero University from 1979 to 1986.

During the long history of this study I have benefited from short-term associations with four British universities. I enjoyed a link with the Department of Geography at the University of Leeds during a term's study leave in 1974, and the University of Liverpool provided computer facilities; here the African interests of Professor R.W. Steel, CBE, and Professor R. Mansell Prothero have for long provided me a distant source of encouragement. The hospitality of the

Acknowledgements

Department of Geography at the University of Durham was put at my disposal during a sabbatical year in 1984–85, a pleasure I owe to the initiative of Dr W.T.W. Morgan and the support of Professor J.I. Clarke. My stay at Durham was made possible by the award of a Royal Society Developing Country Fellowship, and of the Research Fellowship of St Mary's College, whose Common Room provides the visitor an extra bonus. The Principal, Miss Joan Kenworthy, worked hard to create a climate conducive to study. During the 1986–87 academic year, I enjoyed an Honorary Fellowship at the Centre of West African Studies in the University of Birmingham, which I owe to the efforts of the Director, Mr Douglas Rimmer.

Many people have contributed to my investigations in the field. Among these are A. Hamzat and J. Oklobia, one-time students at Ahmadu Bello University, who pioneered the first drought surveys in Danbatta district in 1973; I. Balami, I. Ruma, E. Shallangwa and D. Warip, who administered the extensive surveys in Kano state in the following year; Ismaila Daudu who, after accompanying me to Dagaceri, decided to stay there and write his own dedicated study, for which act he is still remembered well in the village; Levi Egbunu who returned again and again to the heat and the sand, and a back seat in an overloaded, antiquated Land Rover; Bamidele Samuel and Adamu Gombe who on several occasions assumed the toil of driving. The insights and companionship of several persons helped transform fieldwork from a mere challenge to a never-ending learning experience. John Schultz came with me on a brief but unforgettable tour of northern Kano at the height of the famine of 1972–74. Kingsley Ologe, with whom I worked in Dagaceri in 1975–76 and jointly led a memorable tour, in 1978, of the Working Group on Desertification of the International Geographical Union, was sadly missed on all my later visits, for his humour as well as his physical geographer's eye. Ken Brown, always ready to contribute generously in time and effort, brought a social scientist's perspective even when changing tyres, digging out, or running with tape and Abney across the dunes. My greatest debt is to Abigail Ruston, whose mastery of the Hausa tongue more than adequately repaired my own deficiencies. Dropping other commitments in order to sustain the regularity of my field visits, she added humanity as well as continuity to our relations with our informants and their families.

The illustrations for this book were prepared in the Cartographic Unit of the Department of Geography, Bayero University, by John F. Antwi, and I am grateful for his cheerful acceptance of my sometimes unreasonable demands. S.O. Taiwo drew figure 3.2. Parts of the draft manuscript were read by Ken Brown, Brian Giles, Joan Kenworthy and Ken Swindell, who all helped me with their criticism. The book was prepared amidst my own uprooting from African soil, and that it was accomplished at all is due to the help of Marguerite Batchelor and Julia Kirkpatrick, and to Daphne Mitchell's skill and perseverance in deciphering and typing the manuscript.

The co-operation of the governments of Kano and Borno States, and their representatives in local government, allowed the field investigations to proceed. In particular, Sarkin Arewa of Birniwa (who died in 1985), and the Lawan of Kumagunnam enthusiastically welcomed my interest in their districts. The Dagace of Dagaceri and the Bulama of Kaska went beyond the call of duty to welcome us several times to the hospitality of their homes. Finally, tribute must be paid to the unnamed people whose patient answers to innumerable questions provide the bedrock of this book. Their readiness to trust information about their personal affairs to an outsider with nothing to offer in return has always deeply impressed me, and imposed an obligation to make the most honest use of it I can. Their dignity under conditions of extreme hardship is sometimes little less than heroic, and deserves a better record than I am able to provide.

A NOTE ON CONVENTIONS AND LIST OF ABBREVIATIONS

The use of vernacular terms offers clear advantages where an English equivalent is either insufficiently specific or unwieldy. Hausa terms are widely used in the literature on Nigeria and her neighbouring countries and some (for example, the word *fadama*) are in process of acquiring scientific status in their own right. The Hausa language is the most widely spoken in that area of West Africa in which this study is located, either as a first or second language. For these reasons I use Hausa terms in this book in preference to those of other languages spoken by the subjects, except where the Manga, Kanuri or Fulfulde term is clearly preferable. Vernacular terms are shown in brackets in the text preceded by H (Hausa), M (Manga), K (Kanuri) or F (Fulfulde).

Plants are referred to by their scientific names only, except where their importance merits the additional use of their Hausa or common names in the text. Trees, shrubs and some grasses are listed in table 3.12, where the reader may find equivalents if required.

The name Fulani is employed to refer to the Fulani people in general, but in chapters 4 and 7, when dealing with specific groups, I have followed common practice in using the term Ful'be which is the name they give themselves.

Except when indicated otherwise, references to tons are in metric units. However, if a significant margin of error exists, or an order of magnitude only is required, round numbers in imperial tons may be retained.

The Nigerian Naira (₦) was worth S£0.85–0.95 for the greater part of the period of this study, but it fell in value to S£0.50 in 1986, when the introduction of the Second Tier Foreign Exchange Market further reduced its value for most commercial transactions to S£0.15–0.20.

The following abbreviations are used in the text:

APPER	African Priority Programme for Economic Recovery
BBC	British Broadcasting Corporation
CWR	Crop and Weather Report
ECOWAS	Economic Community of West African States
EEC	European Economic Community
F	Fulfulde
FAO	Food and Agriculture Organisation

FSR	farming systems research
H	Hausa
ILO	International Labour Organisation
ITD	Inter-Tropical Discontinuity
K	Kanuri
LGA	local government authority
M	Manga
MAB	Man and the Biosphere (UNESCO programme)
MSS	multi-spectrum analyser
OECD	Organisation for Economic Co-operation and Development
TFR	total fertility rate
UNDP	United Nations Development Programme
UNEP	United Nations Environment Programme
UNESCO	United Nations Educational, Scientific and Cultural Organisation
USA	United States of America
USAID	United States Agency for International Development
WMO	World Meteorological Organisation

1

INTRODUCTION

CRISIS IN RURAL TROPICAL AFRICA

The problem is not whether there is a crisis in rural Africa, but what its nature really is. Food shortages are occurring yearly in different parts of the continent. Since the Sahelian Drought of the 1970s, they have captured the able and compassionate attention of film-makers and provoked unprecedented international awareness, charitable giving and voluntary agency activity. But rural poverty is not new to Africa. It has persisted through half a century or more of colonialism, two decades of independence, and many ambitious programmes of development aid. Neither the drive towards industrialisation in the sixties nor a contemporary emphasis on the transformation of agriculture has yet succeeded in abolishing it. The blame has swung widely between the natural environment on the one hand, through various local economic and political constraints, to international indebtedness on the other, but a solution seems as elusive as ever. Failure to arrive at a consensus is embarrassing for applied science, but not altogether surprising in view of the high level of generality at which many prefer to discuss the affairs of this vastly diverse continent.

Diagnostic and prescriptive fashions succeed one another almost as rapidly as the ups and downs of the African rainfall. For the farming sector alone, a plethora of constraints has been identified. A typical list might include the following: low productivity caused by infertile soils, erosion, drought, flood, pests and diseases ('ecological constraints'); labour shortage, caused by rural–urban migration; technical backwardness, caused by scarcity of inputs; conservative farm management, caused by an ageing farm population, and by the low status given to agriculture in educating the young; high marketing costs, caused by inadequate rural infrastructure, and by restrictive trade practices; under-investment, caused by scarcity of credit, by low farm prices, unfavourable or vacillating price policies, and unfavourable terms of trade between the rural and urban sectors.

Behind such attributions lies a wide divergence of views about the policies and attitudes of colonial, post-colonial and donor governments, and the nature of the changes required (Lofchie, 1986). In irrigation, we are told, there has been too much dependence on public investments in large-scale projects, and insufficient emphasis on small-scale enterprise (World Bank, 1981). The proper role of government is a matter for controversy: interventions based on capital-intensive projects have been in and out of fashion in a couple of decades, leaving (in irrigation for example) a record of inadequate planning, and unforeseen social and ecological effects (Adams and Grove, 1985). Unquestioning acceptance of state economic interventions has been replaced by advocacy of social outlays on the provision of basic needs (ILO, 1981). The virtues of self-reliance are being rediscovered (Sandbrook, 1986).

Agricultural research and policy, it is now being argued, need to pay sufficient attention to the internal characteristics and rationale of indigenous farming systems (Richards, 1985). Too much research has been specialised and technically inappropriate; so 'farming systems research' is advocated, stressing the participation of small-scale farmers in research, and the design and implementation of new technologies (Norman, Simmons and Hays, 1982). According to Hyden (1983), however, the 'peasant mode of production' cannot make dramatic production increases as long as household units continue to co-operate for both productive and reproductive purposes, owing to such constraints as family labour, and high dependency ratios. In the 'economy of affection', kinship ties and patron–client relations still have an inhibiting effect on change. What is needed is a dynamic, capital-intensive large-scale sector, according to Hart (1982). Emphasis on export crop production has been in turn rejected (because of commodity dependence) and rehabilitated (because of the need for foreign exchange). But agricultural policies have ignored small farmers (e.g. Wallace, 1981); and improved input packages – 'integrated rural development' as interpreted by the World Bank – seem to favour some at the expense of others (for example, the Funtua Agricultural Development Project in Nigeria: Mabogunje and Gana, 1981). And for some scholars, the impoverishment of the peasantry is, anyway, an inevitable historical outcome of the penetration of capitalist forces into the countryside under colonial and postcolonial conditions (Watts, 1983a).[1] Agricultural – and rural – problems form a part of the larger development dilemma which continues to generate controversy. Africa's Priority Programme for Economic Recovery, 1986–1990 (APPER), which was first approved by the Organisation of African Unity in 1985, allocated 45 per cent of its proposed budget to agriculture with the primary objective of increasing food production.

This book does not offer another analysis of small-scale farming systems. The rural sector is, of course, much broader than farming, although many discussions leave the reader with the impression that they are synonymous. The problems of livestock production, forestry or arboriculture, fishing, collecting, hunting, marketing, services or rural manufacturing are not identical with those of farming. Specialised livestock production systems have received increased attention from researchers recently, more particularly in regions beyond the limits of farming. But the other subsectors are still being neglected. It often seems to be assumed that a decline in rural non-agricultural activity will accompany urbanisation and the transformation of agriculture. But such an assumption does violence to the multi-faceted structure of West African rural economies. Diversification is very much alive and, as I hope to show, is also intimately related to ecology.

The place given to 'ecological constraints' in economic analyses of African rural systems is usually quite inadequate, limited too often to a ritual obeisance to the restrictions imposed by adverse conditions of soil and water. In like manner, the place given to social and economic systems in ecological studies may be perfunctory or even dismissive. Thanks to the international media, one ecological factor – drought – has now a large place in the popular understanding of African poverty. Perhaps this is an exaggerated place. On the one hand, drought can be made a scapegoat for social ills. On the other, however, the role of political factors in causing food shortages may be emphasised almost to the exclusion of crop and pasture failures. The need for a balanced interpretation has not always been taken sufficiently seriously, perhaps because it is so difficult to achieve.

UNCERTAINTY AND ADAPTATION

There are certain parallels between the ways in which uncertainty has been conceived in the ecosystem and in the human system. Uncertainty in the ecosystem is founded on rainfall behaviour, which is conceptualised either as subject to a downward trend (diminution) or as intrinsically variable in time and space about a fluctuating 30-year mean (variation). The present and future status of ecosystems and their primary productivity depends on which of these views is correct. Is the system 'fragile' or 'resilient'? Uncertainty in the human system, and in particular with regard to its adaptive capability when faced with food shortage, centres similarly on a controversy between a view of a system that is disintegrating in the long term, and an alternative view of flexible, experimental and innovatory adaptation. Is the system 'vulnerable', or resilient?[2]

The title of this book should not be taken to imply that adaptation is being viewed in a deterministic light as a one-way response to drought and desertification, acting as exogenous variables on human systems. Such a viewpoint has been criticised, and rightly so, for its neglect of the economic, social and political parameters of hunger. Rather, adaptation is understood as a sequential process in which solutions to problems become in turn a part of the next problem. According to Bennett (1978), a chain of problems and solutions, each solution begetting another problem is called behavioural or social adaptation, the distinctive feature of human cognition being its anticipatory characteristic. Individuals, families or communities have to confront the situations in which they find themselves at a given time, with the resources of which they are aware under given constraints of land, labour, capital and mobility. Adaptation is being used not merely as a model of stimulus and response, but rather as an interpretive tool for pursuing the intricacies in space and time of human ecological behaviour. In choosing to place some emphasis on rainfall and the ecosystem, I am concerned to reflect the perceptions of the communities I have studied, for whom rainfall variability (in particular) occupies a large place in the spectrum of uncertainty. This is not to belittle other sources of uncertainty, whose significance will be apparent from the analysis.

I do not propose to observe the distinction suggested by Burton, Kates and White (1978: 34–48) between adaptation (biological and cultural) on the one hand and adjustment (incidental and purposeful) on the other. There is an implied difference in time-scale between these categories, the first long-term and the second short. While I shall be concerned mainly with short-term adaptation, and not at all with biological, such a difference is not intrinsic, for adaptations made on a one-year time-scale (or even shorter) will add up in time to a culture trait. And culture is not immutable but purposefully adaptive.

The recurrence of drought, associated with large-scale food shortages, during the last two decades has shown that the operation of ecological factors cannot be divorced from the perspective of time. On the one hand, a single year's drought administers a shock to the systems of primary production which – given present meterological capability – cannot be predicted. The weakness of disaster insurance systems – or the inadequacies of the food entitlement system (Sen, 1981) – bestows far-reaching social and economic repercussions on such an event, including the possibility of a regional famine. On the other hand, recurrent drought, independently or together with certain trends in land use, may bring about a process of degradation, or a loss of primary productivity, whose configuration and eventual outcome may be only dimly perceived by the participants.

It is now commonplace to point out the inadequacy of development programmes or projects based on 'average' ecological conditions, rainfall in particular. But huge public investments stand as memorials to the errors made.[3] Other programmes may respond more quickly to diminished rainfall. But government agencies clearly have difficulty in incorporating the probabilities – or uncertainties – of such extreme events into their short-term decision making. On the other hand, there is abundant evidence that the inhabitants of areas prone to ecological uncertainty have developed a flexibility in adaptive response that stands in marked contrast to the behaviour of governments, notwithstanding their greater resources. Furthermore they are able to use such skills against uncertainties of a kind originating within the political economy, as well as against ecological events. This contrast between bureaucratic and indigenous perceptions has inhibited effective assistance to the poor in times of disaster.

SCIENCE AND ETHNO-SCIENCE

In the outburst of concern over the 'African crisis', and in particular the linked phenomena of drought, famine, desertification and impoverishment, too little attention has been given to empirical investigation. Worse, relationships have been postulated without the intervening variables even being identified, or the nature of the linkages supposed. Such a case is the commonly asserted relationship between population growth and soil erosion. Consequently, practicable solutions are elusive. Meanwhile, the survival of millions of families, now as aforetime, depends on the successful exercise of self-reliance. Governments tend to intervene late, inadequately or ineffectively.

A huge divergence, indeed, has opened between a scientific orthodoxy (and the attitudes of governments based on it) and the behaviour, if not perceptions, of land users in regions prone to drought and desertification (Heathcote, 1983: 289–93). It may be characterised, for example, in such contradictions as the following: expert opinion says that grazing too many animals reduces the productivity of pastures – but livestock owners try to increase their herds; ploughing up land increases its susceptibility to erosion – but farmers cultivate more land; indiscriminate deforestation causes desert encroachment – but people go on cutting trees; population growth causes pressure on the land – but couples beget. Breakdowns of understanding between official and popular views of how the environment should be managed are as old as the modern colonial experience in Africa. One could cite the running battle between colonial opponents of bush burning and its ubiquitous practitioners.[4] That pejorative concept, 'indiscriminate deforestation', was already being employed in the Bechuanaland Protectorate as early as 1897.[5] Corrective measures that are advocated for the four malpractices enumerated above are destocking, 'rational land use', conservation forestry and population control. All tend to be as unacceptable socially as they seem necessary 'rationally'. Yet healthy animals, productive farmland, useful trees and prosperous families are valued in African societies. Why this dialogue of the deaf?

Much misunderstanding is due to plain ignorance of the rationale for indigenous land-use practices. Experts on whistle-stop tours, urban-biased bureaucrats, and a failure of grass-root participation in decision making conspire to ensure that plenty of mistakes are made. Fuelwood plantations are set out with trees whose wood is unsuitable for domestic use; crop varieties are recommended that lack desirable properties; families are compulsorily resettled on infertile land; and so on. Belatedly, the integrity, rationality and ecological wisdom in indigenous land-use systems are receiving recognition, even in the much maligned shifting cultivation systems (Richards, 1986).

Another part of the answer lies in conflicting perspectives of scale and time. The scale perspective of most land users is restricted, by experience and information, to a small area (or a series of linked areas) and they are oblivious of the statistical projections which generate such concern at national or global levels. Likewise, the priority of household survival, which recurs annually, may preclude the adoption of a longer perspective in decisions relating to land use. Such a viewpoint may be ignored by those impatient to impose change.

Ecological processes, such as desertification, may operate on a time-scale too long to be appreciated adequately and acted on either by the victims or their governments. Diagnosis, and the prescription of solutions, tend to be left to experts who are, not unnaturally, often specialists in ecology, pedology or geomorphology. The measurement, explanation and prediction of ecological degradation has therefore tended to proceed along lines of its own, closely associated with the development of remote sensing techniques. Since, in most hypotheses of degradation, anthropogenic factors assume major importance, the policy prescriptions that follow may be draconian in their social implications. Adverse criticism of such prescriptions may lead to their softening, in the face of the complexity, and intractability, of the social and economic problems. But we are still a long way from an effective integration of natural and social scientific viewpoints, still less of scientific and 'folk' perceptions of environmental management.

However, it is no longer unusual to hear calls for integrative or holistic approaches. Notwithstanding the difficulty of reconciling natural and social scientific viewpoints, such an approach seems essential if the complexities of these relationships linking societies and their natural environments are to be more adequately understood. As a disciplinary orientation for the present study, it is consistent with a view of 'geography as human ecology' (Jones and Eyre, 1967). And a link with ethno-science can be achieved only by means of a strong empirical emphasis. What is needed is an approach which can marry the objectivities of ecological change with the perceptions of the communities affected by, or themselves effecting, such change. Such empiricism must be specific in terms of time and place. There have been many calls for such studies (for example: Dyson-Hudson, 1972 and Baker, 1974a – in relation to pastoral societies: O'Keefe and Wisner, 1977: 230 – in relation to human ecologies; Leftwich and Harvie, 1986 – in relation to famine studies). In the arid and semiarid zones of Africa, pastoralists are receiving more attention than farmers, perhaps because of the mistaken notion that farming systems research in the subhumid zone has general applicability.

ARIDITY IN AFRICA

The need for a more specific locational reference for desertification studies is illustrated by Berry's (1984a, b) review of economic, demographic and political trends in the Sudano-Sahelian region. This study was based on information provided by national governments to the United Nations Environment Programme. National population estimates, for example, are listed for the Sudano-Sahelian region, yet not only do substantial parts of some of these countries fall outside this region, but it is well known that important demographic differences exist between nomadic and sedentary populations in the Sahel (Hill, 1985). Recourse to using national estimates and FAO statistics is also subject to the fundamental criticism that such data are only rarely based on systematic censuses, and are insensitive to the annual fluctuations that are an essential characteristic of dry areas. A prerequisite for a more accurate monitoring of problems related to drought and desertification is a regional framework giving full recognition to relevant ecological parameters. Such a framework is provided by the provisional *World Map of Desertification* (UNEP, 1977a). Following the World Meteorological Organisation, Africa is divided into four bioclimatic zones based on the ratio between precipitation and evaporation: hyperarid, arid, semi-arid and subhumid (figure 1.1). The *arid zone* has an aridity index between 0.03 and 0.20 and the *semi-arid zone* has an aridity index between 0.20 and 0.50. This regionalisation differs from those based on mean annual isohyets (cf. Rapp, 1978:44). It has two advantages over the commonly used ecological zones (Sudan, Sahel, etc.). Firstly, it is based directly on climatological variables instead of using climax vegetation as a proxy. Vegetation, moreover, is subject to widespread modification. (It may be added that the politicisation of the term Sahel has largely destroyed its value as an ecological term.) Secondly, it corresponds more closely to significant determinants of land use. The aridity index of 0.20, which separates the arid and semi-arid bioclimatic zones, approximately marks the northern limit of rain-fed farming, thereby distinguishing the semi-arid zone as one of both animal and rain-fed crop husbandry. The boundary corresponds roughly with the northern legal limit of rain-fed farming in Niger (the 350 mm isohyet), as well as to a minimum probability of receiving 75 per cent or more of normal rainfall in eight years out of ten (Davy et al., 1976:25, 42).[6] Both the southern Sahel and the Sudan zones are thus together designated semi-arid. The aridity index of 0.50 delimits the subhumid zone (equivalent at this point to the Northern Guinea), where a significant amelioration occurs in the length of the rainy season, diversity of cropping systems and rainfall reliability.[7]

The UNEP map (1977a) provides, as a second step in regionalising desertification, three classes of hazard: very high, high and moderate. These are also shown on figure 1.1. In attempting to take into account both physical processes (such as sand movement, sheet wash and salinisation) and pressures exerted by human and animal populations, these units are bound to be provisional in nature. They do not correspond well with the visible evidences of ecological status in northern Nigeria. It is expected that further work will improve their usefulness.

A third step, not yet attempted, needs to differentiate amongst production systems, and take into account the relations between farming and pastoral systems. Notwithstanding some similarities in soils, topography and drainage, and in some economic characteristics, across the semi-arid zone, there are differences at this level that have analytical importance. They include, for example, differences in material culture, population density, settlement pattern and the spatial mobility of people and herds. The individuality of African food production systems is as relevant to controlling desertification as to improving productivity, and prescriptions also need to be adaptable to the social and political values in community regions (Morgan and Pugh, 1969).

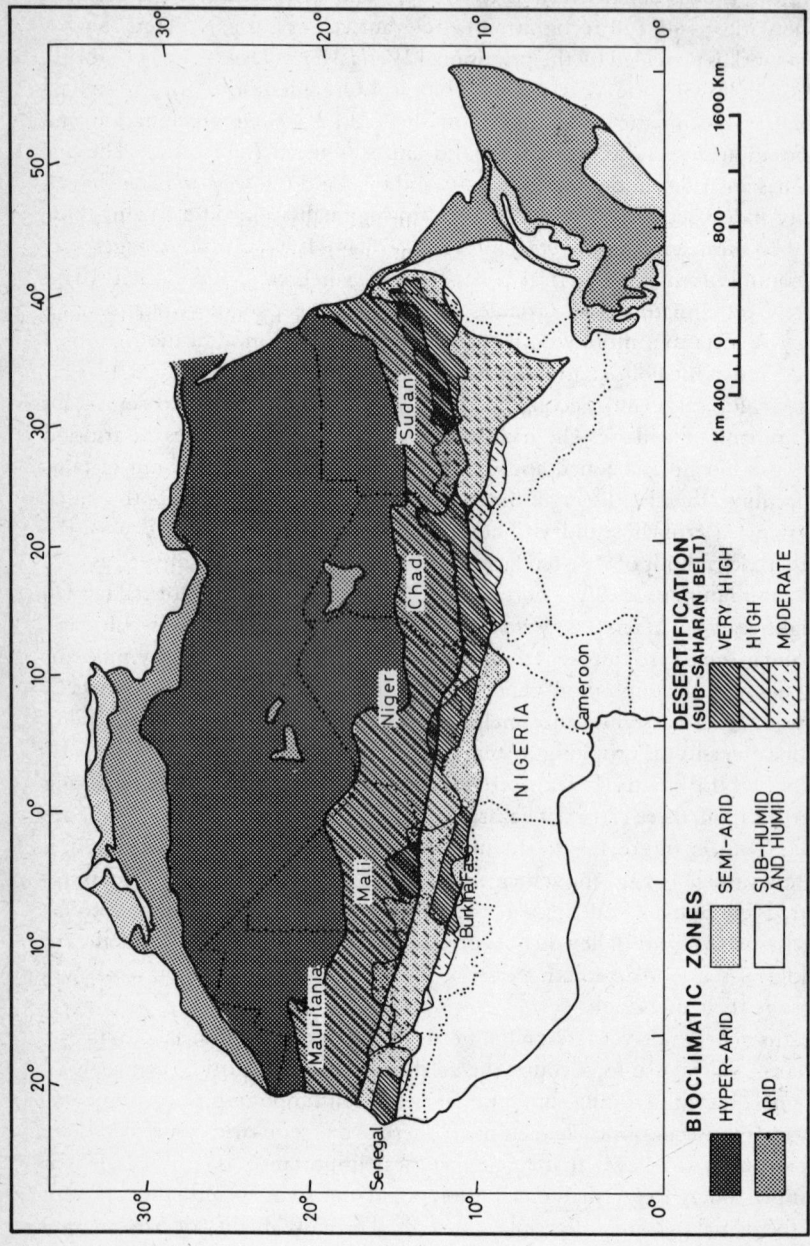

Figure 1.1 Bioclimatic zones and desertification in the subSaharan zone of northern Africa (after UNEP, 1977a)

Finally, national boundaries cut across the semi-arid zone, giving each state its unique political economy of aridity. Five major variables contribute. First is the distribution of semi-arid (and arid) territory in the national space, varying from totality (Mauritania) to a tiny segment (Cameroon). Second is the demographic ratio between semi-arid (and arid) areas and the state as a whole. Uneven distribution of the population may give the semi-arid areas greater significance than their spatial extent suggests (Nigeria). Expressed in political conflict, a demographic imbalance (or balance?) may seriously impede economic development efforts (Chad). Thirdly, continentality varies between coastal countries (Senegal, Nigeria) and interior countries (Niger, Mali) and, with it, access to some economic options. Fourthly, international ties which reflect past colonial affiliations, present global alignments, and national policies on foreign capital create variable patterns of assisted rural development programmes and emergency food aid. Finally, the possession of exploited resources of exportable minerals affects the degree of national self-sufficiency attained.

The existence of so many variables in the aridity equation suggests that the regionalisation of the problem should be taken more seriously than it has been so far.

ARIDITY IN NORTHERN NIGERIA

Those who have accepted an equation between drought and the Sahel, and between the Sahel and francophone West African countries, may be surprised to find the problem discussed in relation to Nigeria. Certainly, no attention was given by the international media to the millions of drought-affected Nigerians who, in the 1970s, out-numbered those of Senegal, Mauritania, Mali and Niger combined. The view is also commonly encountered that Nigeria's oil wealth somehow excluded her from the list of casualties, as if governments, rather than people, suffer from starvation. Certainly no appeal was made for international food aid, excluding Nigeria from the demeaning controversy surrounding the Sahelian relief operation (Sheets and Morris, 1974). It will be clear from the evidence presented in later chapters that drought occurred in northern Nigeria on a scale, and at an intensity, comparable with other areas of Africa; the significance of the processes of ecological degradation known as desertification (Sagua et al., 1987) will also be made apparent. The areas affected fall within the Dry Zone of northern Nigeria (figure 1.2) and are continuous with adjacent areas of Niger Republic.

The Dry (or semi-arid) Zone is delimited on the basis of a number of approximately coincident climatic parameters mapped by Kowal and Knabe (1972). These include the isohyets for 1,000 mm annual rainfall, and 40 days' growing season. Since actual rainfall is subject to wide annual variation, the Dry Zone may also be described in terms of 9:1 confidence limits. Its southern boundary has an upper limit of 1,270 mm (that is, rainfall may be expected to

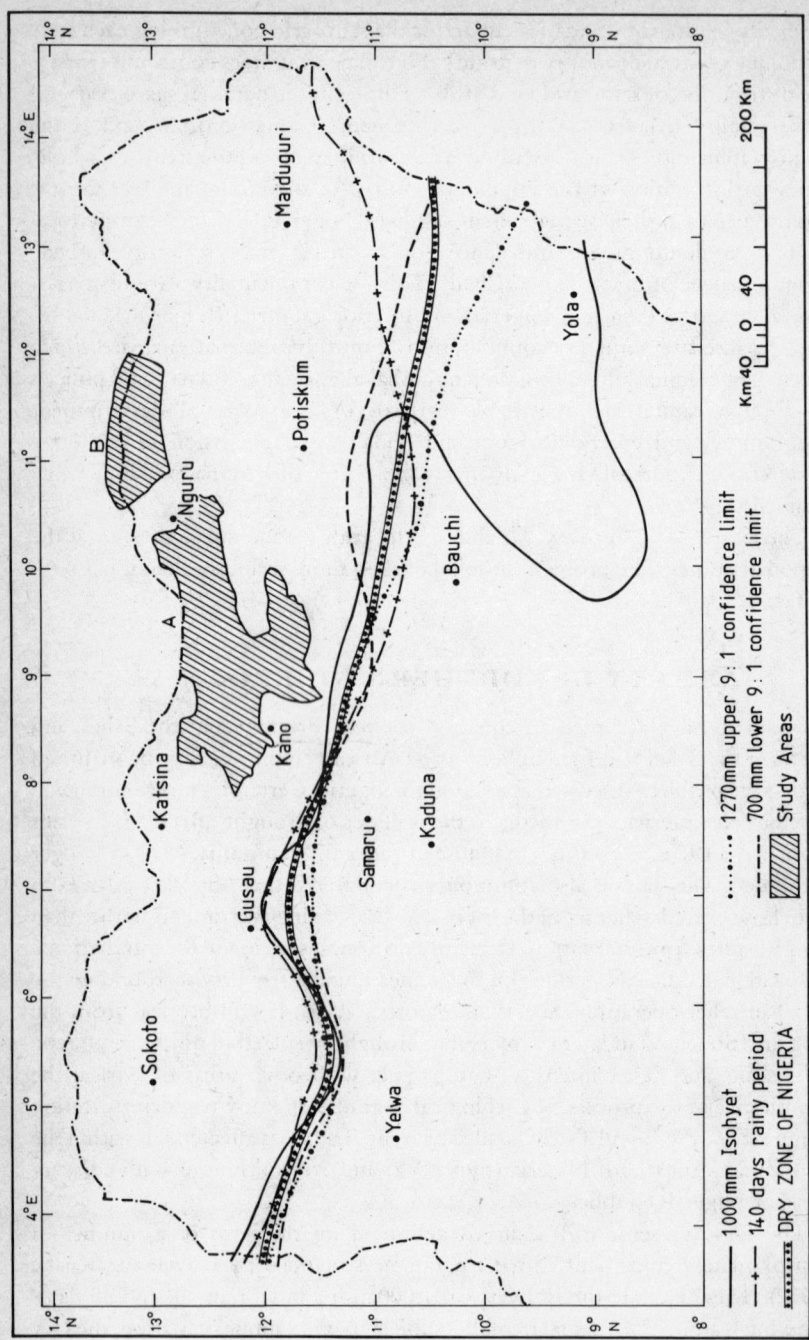

Figure 1.2 The Dry Zone of northern Nigeria

exceed this amount one year in ten), and a lower one of 760 mm. This boundary crosses Nigeria from Lat. 12°N on the western frontier to Lat. 10°30' on the eastern. Annual rainfall diminishes northwards to an average of less than 500 mm in a season of 60 days, where the upper and lower confidence limits are 635 and 389 mm.[8] The Dry Zone corresponds with the areas severely affected by drought in the 1970s; and the regional pattern of rainfall deficiency tends to be symmetrical with the mean distribution. It also includes, therefore, those areas affected by the droughts of 1913 (Grove, 1973) and 1926, the low rainfall of the early forties, and the drought of 1949 (Grove, 1952), as well as the recurrent droughts of the seventies and eighties. However, it should be noted that droughts of agricultural importance have also occurred further south (Oguntoyinbo and Richards, 1977).

Before proceeding further, some discussion is necessary on the scientific and social meaning of the terms drought and desertification.

DROUGHT

Drought is easily defined as a deficiency of rainfall significantly below the normal or expected amount for period in question (a year, a rainy season, a month or less). It is customary to distinguish between three types of drought (Farmer and Wigley, 1985: 24). *Meteorological drought* is defined in statistical terms, such as percentage departure from the mean rainfall, or in relation to the quartile, quintile and decile distributions. *Hydrological drought* occurs when surface or groundwater levels fall below average, and is affected not only by precipitation, but also by infiltration, evaporation and evapotranspiration. *Agricultural drought* is a shortage of water for crop growth, and may be defined as 'a consistently high soil moisture deficit over the growing season' (*ibid.*), although other indices have also been proposed. To these three types, it is necessary to add *ecological drought*, which occurs when the primary productivity of a natural or managed ecosystem (such as unimproved rangeland) falls significantly owing to reduced precipitation.

Rainfall totals for the year, or for the rainy season, are misleading indicators of drought. Hydrological drought is affected by groundwater storage over periods up to several years. Agricultural and ecological drought result not only from an overall shortage of rainfall, but equally from a maladjustment of the seasonal distribution of rainfall to the growth cycle of plants. Thus when total rainfall is below the mean, excellent crops may be obtained when the distribution is satisfactory, as in 1985. But what may be a satisfactory distribution of rainfall for crops may support only a poor growth of fodder grasses. A drought is necessarily perceived in terms of the needs of a given community (Hewitt and Burton, 1971), and it is the failure of primary productivity that has for long justified attempts at scientific explanation (e.g. Tannehill, 1947), analysis of the relations between 'hunger and politics' (De Castro, 1952), and administrative response (e.g. Union of South Africa, 1923).

In this book I shall be concerned primarily with agricultural and ecological drought, as they affect rural systems of primary production, and as they are perceived by the producers. For such, the most important time period is the rainy season as a whole (June–September), but within this period, the distribution of rainfall in periods as short as ten days ('decades' as defined by Kowal and Knabe (1972)) may also be critical.

There is no simple relationship between drought and hunger (or famine), since food production is governed by other factors besides the rainfall, and the efficacy of insurance, storage and distribution systems is variable, between places, between social groups, and at different times. It is the continuing importance of subsistence production to the great majority of farming families, and the weakness of their insurance and storage capabilities, that maintains the link between drought and hunger, a link that has been successfully broken in many other parts of the world.

DESERTIFICATION: CONCEPT AND CONTROVERSY

It is not so simple to arrive at a satisfactory definition of the concept of desertification. Generally credited to Aubréville (1949), this term, since its adoption by the United Nations Conference on Desertification (UNEP 1977c), is swallowing up a number of related terms found in earlier literature, including desert encroachment, 'the advancing Sahara', desiccation, desertisation, and the French *dessèchement*.

The history of the concept goes back at least as far as the second decade of colonial rule in West Africa, when the French Comité d'Etudes for West Africa commissioned a study of *dessèchement progressif* (Hubert, 1920). Drawing on 31 reports relating mainly to locations in modern Mali, Mauritania and Senegal, the study reviewed geological, prehistoric, historical and contemporary evidence. On the longer time-scale, evidence for *dessèchement* was considered to be irrefutable: fossils and archaeological remains in desert areas, and the isolation of some water-loving animals and fish, owing to a diminution of surface water. Other evidence for *dessèchement* included: the contemporary diminution of surface water; the destruction of vegetation (which was squarely blamed on both indigenous and European woodcutting); moving sand on vegetated or fixed dunes, especially in Mauritania and Senegal;[9] and diminishing rainfall, recorded over several decades, in Senegal.[10] Following Hubert, and also inspired by Huntingdon's theories of climatic change, Bovill (1921) cited diminishing rainfall and well water levels in Sokoto as evidence of the 'encroachment of the Sahara on the Sudan', but his inferences could not be substantiated from the rainfall records then available. He reported that some crops would no longer grow, pastures were depleted, and there was migration to the south, but alternative explanations were available to account for these observations (Raeburn, 1928; Prothero, 1962).

It seems probable that these early views were influenced by the low rainfall which was recorded in several years between 1905 and 1920, especially in 1913, when famine caused heavy human mortality in many areas of West Africa. It is striking that after a relatively short acquaintance with interior West Africa, a number of European observers were already convinced of the reality of ecological degradation in the medium and longer term. Generalising to the region as a whole, Renner (1926: 587) reported that 'there is a general belief that aridity is increasing in the Sudan'.

Stebbing (1935, 1937a, b, 1938a, b, 1953), a forester with Indian experience, was the most energetic advocate of the idea of an encroaching Sahara. Visiting West Africa in the dry season, he interpreted his observations of extensive burning, cutting and clearing as a process that was transforming the Sudan savanna woodland (a 'mixed deciduous forest') into a man-made savanna (a 'scattered thorn bush'). The savanna in turn was being downgraded into desert. Hubert (1920:422) had noted earlier that destroyed vegetation was usually succeeded by a type characteristic of a drier regime. Although Stebbing conceded a role in this process to 'intermittent rainfall' (increasing length and severity of dry periods), man with his 'misutilisation of the soils' was the chief culprit: 'over-utilisation of the vegetable covering of the soil... commences the reduction of water supplies and the lowering of water table in the soil' (1953:120); and erosion was bringing intermittent rainfall in its train. The mechanisms, however, were not clearly spelt out. Stebbing even extended himself to predicting the rate of advance of the Sahara (1935:510) – a prophecy which turned out to be false.[11]

Stebbing's views were criticised by Jones (1938) and Stamp (1940); and the Anglo-French Forestry Commission (1937) toured both sides of the border separating Nigeria from the French colony of Niger. Its main conclusion (p. 8) was that 'there is no obviously apparent danger of desiccation' (a natural increase of arid conditions), but

there is unquestionably an impoverishment of the sylvan conditions of the country. This impoverishment is due almost entirely to uncontrolled expansion of shifting cultivation as a result of the security afforded by European administration... Consequently a large part of the natural woodland which covered the country has been so seriously degraded in quality that it is of very little use and it has not been replaced, except in very few localities, by farmland of real value.

This hostility to shifting cultivation was tempered by two members of the Commission (Collier and Dundas, 1937) who pointed out that

shifting cultivation, wasteful though it may be, can do no permanent harm to the soil provided there is an adequate 'bush fallow' period for recuperation. But once the demand for land becomes such that the recuperative period allowable is insufficient, the only hope of maintaining fertility lies in replacing the bush fallow by manure; a requirement that is realized in the fine, well-timbered permanent farmlands of Kano. (p. 192)

Almost the same sentiments were repeated 40 years later by the National Committee on Arid Zone Afforestation (1978:14): 'the existing pressure on land in most of the arid zone areas is such that it could no longer support the expensive shifting cultivation practice without serious consequences on the vegetation and soil of those areas'.

Aubréville (1949), also a forester, used the term desertification to describe the degradation by burning, clearing and erosion in the entire forest and savanna zones of West Africa, and graphically illustrated his thesis with cross-sections of the latitudinal vegetational zones, before and after human impact. Like Stebbing, he was quick to convert an observed spatial ordering into a sequential hypothesis. With a broad brush, he painted a gloomy picture of tomorrow's Africa (p. 341):

The closed forests are shrinking and disappearing like evaporating spots. The trees of the open forests and savannas become more and more spaced out. On all sides the bare skin of Africa appears as its thin green veil of savanna burns releasing a grey fog of dust into the atmosphere. Arable land is carried away by the yellow waters of rivers in flood. Slabs of sterile truncated soil, bearing tufts of grass around uprooted bushes, recall a kind of leprosy that is spreading over the face of Africa; elsewhere great slabs of ferruginous brown or blackish rock abound. Great banks of sand are deposited in the beds of streams and rivers around which small threads of water meander in the dry season. Billions and billions of small red or grey particles multiply evenly over the soil; these are the great termitaria whose populations share out the debris of the wasting forest vegetation or search out the relative coolness of its sap. The mountains are magnificently bare, their erosion gullies, their fractures, etched in sharp lines, their structure clearly visible. Above, the atmosphere vibrates with an intense heat. Alluvial soils, sands and valleys are cultivated intensively; in the dry season winds carry away clouds of dust. In the thorn steppes, whose disappearance marks the beginning of the desert, the shrubs become more and more spaced out, no longer reproducing; the rains have ceased or appear only very irregularly, the winds that bring the summer rain are no longer humid enough. Elsewhere broad-leaved trees dry up one by one and die; they are not replaced by young individuals but, instead, thorn trees gather as if favoured by a dry season that has become longer and more arid... During the dry season, the whole of Africa burns, lines of fire running everywhere, chased by the dry winds, no portion left undamaged; this is a sign for great rejoicing among the people, because the time for hunting rats is come. Thus we see how tropical Africa would be transformed if the 'savanisation' towards which she is fast proceeding were some day to be accomplished.

Drawing on fifty years of personal observations in French West Africa, Chevalier (1950a–e, 1952) repeated the dreary litany, by then very familiar, of *dessèchement progressif*: the remobilisation of sand in the north, and the appearance of laterites in the south; soil erosion and degradation; the disappearance of surface water and falling well levels; the replacement of perennial by seasonal streams; and the disappearance of gallery woodlands.

The unanimity of the colonial forestry school on the nature of the West African problem is thus impressive. Independence brought only an intensification of the gloom. Not to be outdone by its colonial antecedents, the USAID

(1972:4) claimed that several studies of the Sahara (which were not cited)[12] had 'concluded that there had been a net advance in some places, along a 2,000 mile southern front of as much as 30 miles per year'. Such a conception of desertification as a linear encroachment of desert conditions (popularised by Eckholm and Brown, 1977: 9) may be highly misleading. Bernus (1977a: 92; 1979) demonstrated that in the arid zone of Niger, it 'is not a front whose advance can be calculated over the last 40 years. Desertification happens at particular points: it is patchy, not linear.' Mainguet (1980) proposed that distinctions should be made between linear extension, 'extension ponctuelle' at wells and 'extension annulaire' around villages.

On the other hand, in a rare attempt to measure the phenomenon on the ground, Lamprey (1975) estimated from transects in the Republic of Sudan that the boundary of the desert had shifted southward by 90–100 km in the 17 years since the ecological boundaries were previously mapped (Harrison and Jackson, 1958). A study of charcoal production in the hinterland of Khartoum showed that between 1960 and 1980 the northern limit of charcoal burning shifted south by an average of 15–20 km per year, owing to the cutting of suitable trees and the absence of regeneration (Berry, 1984a: 64–5). Such a trend is as likely due to Khartoum's appetite for fuel as to any climatic shift.

A large literature on desertification (see Leng, 1982) has grown out of a less pessimistic tradition of arid zone research in the 1950s (see UNESCO; White, 1955; Hills, 1966). A forerunner of the burgeoning official concern of the seventies was a report on desert encroachment in South Africa (Union of South Africa, 1951). Even before the Sahelian Drought, pessimistic evaluations were being made (Dregne, 1970), but it was that event which popularised the issue (Eckholm, 1976) and provoked a veritable sandstorm of literature surrounding the United Nations Conference of 1977. The human dimensions of desertification were the primary focus of interest (e.g. Johnson, 1977; Biswas and Biswas, 1980). Throughout the history of the debate on desertification, it is fair to say that the case has rested, to an undesirable extent, on fragmentary evidence, unsystematic field observations and hypothetical arguments; and scepticism has not been lacking. Prerequisites for the resolution of the matter must be a consensus on the definition of the concept and on measurable indicators.

DESERTIFICATION: DEFINITION AND MEASUREMENT

If desertification is now accepted as the most appropriate term, agreement on its exact meaning remains elusive. In a review of more than a hundred definitions taken from the literature, Glantz and Orlovsky (1983) suggest that the concept, as it is used by researchers, may mean either a process of change, or the end state of a process (or both, as implied in the term itself). If a process, it may be conceived either as a negative change, from a productive to a less productive state, or as a transfer of the unproductive characteristics of one area (such as a desert) to another, as implied in the word 'encroachment'.

Most contemporary definitions confine desertification to arid and semi-arid zones on the fringes of deserts. But the contributory processes are not so confined, having also been observed in humid and subhumid areas. Glantz has suggested a return to Aubréville's catholic usage, but the difficulties of field research argue rather for a narrowing of the concept.

According to the United Nations (UNEP, 1977c: 3),

Desertification is the diminution or destruction of the biological potential of the land, and can lead ultimately to desert-like conditions. It is an aspect of the widespread deterioration of ecosystems, and has diminished or destroyed the biological potential, i.e. plant and animal production, for multiple use purposes at a time when increased productivity is needed to support growing populations in quest of development. Important factors in contemporary society – the struggle for development and the effort to increase food production, and to adapt and apply modern technologies, set against a background of population growth and demographic change – interlock in a network of cause and effect. Progress in development, planned population growth and improvements in all types of biological production and relevant technologies must therefore be integrated. The deterioration of productive ecosystems is an obvious and serious threat to human progress. In general, the quest for ever greater productivity has intensified exploitation and has carried disturbance by man into less productive and more fragile lands. Overexploitation gives rise to degradation of vegetation, soil and water, the three elements which serve as the natural foundation for human existence. In exceptionally fragile ecosystems, such as those on the desert margins, the loss of biological productivity through the degradation of plant, animal, soil and water resources can easily become irreversible, and permanently reduce their capacity to support human life. Desertification is a self-accelerating process, feeding on itself, and as it advances, rehabilitation costs rise exponentially. Action to combat desertification is required urgently before the costs of rehabilitation rise beyond practical possibility or before the opportunity to act is lost forever.

Using the UNEP's definition, global estimates of the extent and rate of progression of desertification have been widely disseminated. According to UNEP (Tolba, 1986), six million hectares of land are lost 'totally' each year and, in a further 21 m ha, productive capacity is reduced 'to the point of zero economic productivity', which means that a farmer can obtain no net profit from its exploitation.[13] Between 1977 and 1983, the annual rate of loss rose by one million hectares. Desertification, according to these estimates, affects 80 per cent of the rangeland, 60 per cent of rain-fed cropland and 30 per cent of irrigated cropland in the world's drylands (arid, semi-arid and subhumid bioclimatic zones: Stiles, 1984). Many of these countries affected by desertification on a large scale increased their dependence on imported food during the seventies and eighties. They include some of the poorest countries in the world, measured by income per capita. Those in tropical Africa are experiencing rapid population growth, having total fertility rates of 5.5–8.3 (compared with China, whose TFR is 2.3: Milas, 1984). The latest assessment of the UNEP (1984) indicates a world population for the drylands of 850 million, of whom

280.5 million live in rural areas affected by 'moderate' or 'severe' desertification (Mabbutt, 1985).

In the Sudano-Sahelian region, 87 per cent of the rural population living on the rangelands is affected by desertification, 78 per cent of the population of rain-fed croplands, and 30 per cent of that on irrigated cropland – altogether, 49.5 million people. Accelerating desertification trends were assessed in 1984 for rangelands, rain-fed croplands and woodlands in Sahelian countries (Berry, 1984a: 45–6; 1984b). In subSahelian countries (including Nigeria), the deterioration in cropland was 'moderate', but in other land classes it was as bad as in the Sahel. Sand dune encroachment was 'severe' in the Sahel and 'moderate' in the subSahel. In both zones, the deterioration of groundwater was 'moderate' and of surface water, 'severe'. These trends were established from qualitative estimates provided by officials in the countries concerned, and it would be inadvisable to place much reliance on them.[14]

But the definition and interpretation of the concept of desertification embodied in the quotation above (from UNEP (1977c)) are not free from ambiguity.

Five fundamental questions arise. First is the question of irreversibility. The 'destruction of the biological potential' is an ambiguous phrase, because given time and favourable climatic conditions (or unlimited technical resources) virtually any desert is reversible. Linked to this is the undefined idea of 'exceptionally fragile ecosystems', carrying the connotation of easy destructibility. But arid and semi-arid ecosystems are biologically adapted to frequent climatic oscillations, which have occurred in the last few millennia. There is evidence that grassland and woodland can recover, given protection, even on sand dunes.

Secondly, the concept of biological potential can have meaning only in relation to a supposed prior state – whose estimation, however, is open to doubt. It is not even agreed whether the savannas of West Africa are a climatic climax or a fire climax (Anyadike, 1982), let alone what the potential may be in smaller areas with a long history of dense settlement. The operational difficulty of such a concept is admitted in the following revealing remark (FAO/UNEP, n.d.: 24): 'Obviously, only experienced range scientists can make reasonably reliable estimates of the productivity of the present plant community compared to what is possible using the best management practices.' For land under agricultural use, a valid estimate of biological potential seems even less attainable.

Thirdly, the meaning of 'desert-like conditions' is by no means clear since a desert can be variously conceived as a rainless, unvegetated, or uninhabited place, or as a landscape category. To define the process of desertification in terms of its expected end state – however conceived – is unsatisfactory, since the inevitability of such a progression is open to doubt, once the concept of irreversibility has been questioned.

Fourthly, to limit the concept of desertification to the land, rather than to the ecosystems as a whole, would be to overlook the fact that the human communities who inhabit the areas prone to desertification – and whose plight is the main justification for concern – exploit the vegetal resources of natural and managed ecosystems, as well as the soils and water resources. This is implicitly accepted in the subsequent use of the term, in the quotation above. But an explicitly systemic approach has many advantages for understanding, and ameliorating, desertification. One of these is that emphasis is placed on the circularities, such as feedback loops, instead of searching for 'cause and effect' models.

Finally, while a climatic cause for desertification is not overtly excluded, the emphasis is placed overwhelmingly on 'over-exploitation' by man. Such an emphasis (Dregne, 1983) prejudges the issue of natural versus anthropogenic causes. Like the concept of biological potential, over-exploitation (over-grazing, over-cultivation, excessive woodcutting) can be defined only in relation to an optimal level of exploitation – carrying capacity, critical population density, sustained yield – whose specification depends in turn on numerous assumptions concerning management practices, macro- and micro-economic determinants and the possibilities for change.

These objections can be met by simplifying the definition of desertification to *the degradation of ecosystems in arid or semi-arid regions*, where 'degradation' means the loss of primary productivity. Such a usage does not imply irreversibility, does not need to define potential productivity, does not depend on an end state, takes account of the whole ecosystem and does not prejudge the question of causation. Roughly equivalent to Rapp and Hellden's (1979: 115) succinct 'dryland degradation', such a definition is relevant to the objectives of the present study, in that productivity is the primary concern of the human communities. Losses in productivity, irrespective of the past or future state of an ecosystem, can be measured objectively over a period of time.

It is not difficult to identify indicators of ecological degradation, and their monitoring presents few technical problems. A suitable list might include (Reining, 1978; Dregne, 1983): soil characteristics (depth, organic matter, crusts, salt or alkali content, and dust deposits); water (depth to groundwater, quality, extent of surface water, status of drainage systems); and vegetation (percentage ground cover, reflectance, biomass above the ground, yield, species composition). The difficulties arise when social indicators are included, in recognition of the human significance of desertification. For such indicators as have been proposed (Reining, 1978) include certain characteristics of land and water use, settlement patterns, demography and social processes which may be cause *or* effect of degradation, or *both*, or *neither*. The selection of such an indicator again prejudges the explanatory hypothesis, risking self-fulfilling prophecy. Indeed, descriptive and explanatory modes are frequently confused in discussing desertification.[15]

The elaboration of indicators to be used in the mapping and measurement of desertification (FAO/UNEP, n.d.) is beyond the scope of this discussion; for a

Table 1.1 *Desertification in the ecosystem*

	Subsystem	Scope	Management regime	
			Exploitation	Conservation
1	Hydrological	Water	a Off-take b Catchment modification	a Storage b Catchment protection
2	Vegetational	Woodland	a Cutting b Burning	a Planting b Protection
3	Vegetational	Grassland	a Grazing b Burning	a Improvement b Protection
4	Pedological	Soils	a Cultivation b Irrigation	a Fertilisation b Drainage
5	Morphodynamic	Surface materials	a Exposure b Disturbance	a Protection b Restoration

study limited in space and time, a more rough and ready approach is called for. If desertification is defined in the restricted sense suggested above, indicators of trends in productivity will not be hard to specify, although their use will be constrained by the types of evidence available.

It will be argued in chapter 8 that a deterministic hierarchical model of cause and effect is inappropriate to the study of desertification, owing to the systemic nature of the linkages between society and its natural environment. There are no independent variables – except the rainfall (and even this has been questioned – see chapter 6). An alternative framework, which will be used as a basis for the empirical investigations reported in chapter 7, is shown in table 1.1. The ecosystem is divided into five subsystems which correspond to the most commonly used categories in the literature. Appropriate indicators are selected in relation to each subsystem. Management regimes – a term intended to include all physical relations between society and the land – emphasise exploitation and conservation objectives in variable proportions (cf. Heathcote, 1983:122f.) but some kind of balance is obviously necessary for social reproduction. The analysis of such regimes forms an integral part of investigating desertification. Monitoring ecological indicators is not enough.

METHOD

Most field studies conducted on drought in West Africa have been retrospective. Exceptions were Poncet's (1974) survey of the impact of drought in part of Niger, and Laya's (1975) interviews with farmers and livestock breeders. Concurrent studies necessarily tend to be designed in haste, short on methodological rigour and constrained by a need for quick information about as wide an area as possible. The present study had its origin in such a survey of the drought-affected areas of Kano State (area A in figure 1.2), which was carried out from May 1973 to July 1974 (Mortimore, 1973, 1975a, 1975b).[16] Accepting the limitations of the survey method, it was, nevertheless, conducted during the lifetime of the disaster. An analysis of the results of this survey is offered in chapter 3.

The famine of 1972–74 became known in Hausaland by the name Kakaduma.[17] Notwithstanding the severity of its effects, by the late seventies it was clear that the rural production systems, far from being destroyed, were continuing to function, and the aftermath promised to be as instructive as the disaster itself (Van Apeldoorn, 1978b). Field studies in rural areas usually have a short duration – eighteen months or less – and longer-term variability has been neglected.[18] Worse, such analyses have been assumed to describe 'normal' conditions, and in this guise have influenced both policy and conventional wisdom. Even the most thorough studies of small-scale agriculture yet conducted in northern Nigeria, whose influence in the literature has been enormous, depend exceedingly heavily on farm management studies carried out in three villages in the one year 1966–67 (see Norman, Simmons and Hays, 1982).

The repercussions of variability in the monthly and annual rainfall alone provide justification for the use of a longitudinal approach. The seasonal dimension affects rural production, consumption, welfare and health quite profoundly (Chambers, Longhurst and Pacey, 1981). Even more important, in the present context, is the need to identify longer-term trends (Prothero, 1974), both in socio-economic variables and in primary productivity. But longitudinal studies are expensive, if truly compatible data, reasonably free from ambiguity, are to be obtained from widely representative samples. The attempt that is made in chapter 4, using selective interviews on multi-round visits, is a low-cost reconnaissance profile of the fluctuating fortunes of one village community over a period of 13 years. It is intended to throw some light not only on the obvious – but neglected – fact that no two years are the same, but also on the reasons for the resilience of the rural system in a marginal (or high-risk) environment. The droughts of the seventies, and the effects of other external shocks, may be seen thereby in time perspective, at least in the medium term. A longer-term exploration of ecological trends is attempted in chapter 7. However, a thorough pursuit of historical evidence was not attempted; this calls for a separate study with adequate resources.

Famine hits the poor first and hardest (Hill, 1972, 1977, 1982). Social differentiation is arguably more urgent an issue than spatial and, unlike it, cannot be shrugged off as due to rainfall variations. An appropriate methodological response would be micro-scale studies of household welfare, including intra-household transactions, and of the position of women, the young and the aged. But not only would an intimate longitudinal study of selected (and perhaps reluctant) families be required, but the series would have to start before the onset of the first major drought. Such was beyond the resources available for the present study. Some use will be made of case studies at the family level, but these are primarily illustrative, and no rigid definition of the family will be attempted.[19]

In an analysis of spatial mobility as a form of response to drought and food shortage (chapter 5) reliance is again placed on survey method (supported by case studies), in order to obtain information from larger samples, but the longitudinal profile of chapter 4 provides a time perspective in which to place the spatial patterns. The ramifications of seasonal circulation, especially, extend throughout the rural economy, and it is difficult to devise adequate methods to pursue then thoroughly. Neither is it necessary, in order to gain basic insight into the policy implications.

The measurement of meteorological drought, distinct from its effect on food supply or welfare, is possible using standard rainfall data and methods (chapter 6); so is the persistence of drought and eventually, perhaps, its prediction. Theories of 'feedback', however, which link rainfall with land use at the regional scale, are extremely difficult to put to the test and lie beyond the scope of this study. A review of hydrological drought in northern Nigeria has, by contrast, to be based on very inadequate data which are not yet free from ambiguity.

Such data are not available at all for physical indicators of desertification. In order to answer the question, 'How far and how fast has ecological degradation proceeded?' in a defined area, methods have been devised (chapter 7) using a combination of air photo interpretation, published resource inventories, field measurements, and interviews with resident observers. The area used for this investigation is area B in figure 1.2. The analysis shows the possibility of arriving at firm conclusions about ecological degradation in the medium to long term using inexpensive methods. Earth satellite data, which at small scales offer greater speed at even less cost, are not yet available at a resolution adequate for micro-scale studies. The convenience of remote sensing methods is qualified by the need for 'ground truth' and for field measurements of other related variables, especially if the materials available are incomplete or dated.

What may appear as a rag-bag of methods – questionnaire surveys, multi-round visits, case studies, unstructured interviews, remote sensing, field measurements – is defended as a multiple methodology for a multivariate problem. The evidence from these several approaches does converge on conclusions of significance (chapter 8). On such a basis, more specifically targeted studies may be attempted in future. Monocausal, generalised and oversimplified interpretations are revealed as artifacts of inadequate method.

Cost-effectiveness was a major consideration, both of time and money. While rapid rural appraisal techniques are not appropriate to all research objectives, it is noticeable that beyond the reconnaissance level the marginal returns of extra methodological finesse may be unimpressive. The diminishing resources of many African universities and the increasing urgency of problems of social, economic and environmental change must compel a low-cost resolution of rural research priorities.[20]

2

FROM FEAST TO FAMINE?

I will give you rain at the proper time; the land shall yield its produce

Leviticus 26:4

THE FAT YEARS

On what kind of a world did the droughts of the early 1970s so rudely erupt? In the previous decade, the Northern Region of Nigeria (now the 'northern states' of the Federal Republic: figure 2.1), and also the adjacent areas of Niger Republic, reached levels of agricultural production probably unsurpassed before or since. Independence had come a decade previously, but it was an essentially unaltered colonial system of export agriculture that continued to sustain both economies. Although peasant producers, whose memories are long, were under no such illusion, delayed or inept bureaucratic responses to the drought crisis showed that stability had come to be taken for granted.

The crops and the animals

Any visitors to Kano in the 1960s could see with their own eyes that the fortunes of Northern Nigeria were built upon the groundnut. The famous storage pyramids portrayed the fruitfulness of the earth as vividly as the practical difficulties of evacuating the crop. Exports of the 'blessed groundnut'[1] had grown from 50,000 tons in 1916 (six years after the arrival of the railhead at Kano) to 872,000 in 1962–63, when Nigeria was Africa's leading producer (Hogendorn, 1978: 123, 133). The groundnut-producing area lay mainly in the Dry Zone (figure 1.2) where more than half the population of the Northern Region lived, and where more than half the farmers produced groundnuts for the market.

The years 1952–66 were prosperous years for the agricultural economy of Northern Nigeria. The fortunes of the groundnut provide a good indicator (figure 2.2). For 12 of these 15 years, annual rainfall was above the long-term mean. A conspicuous upward trend was recorded in the volume of groundnuts purchased for industrial use or export, by the Regional Marketing Board, reaching an all-time peak in the 1966–67 buying season of 1,042,854 imperial tons. In addition, substantial quantities passed into local trade, besides those consumed at home. Throughout the period groundnuts comprised 18–19 per cent of the value of Nigerian exports (Wells, 1974: 61). Yet the official price index actually fell by more than 20 points after 1960–61. Producers were

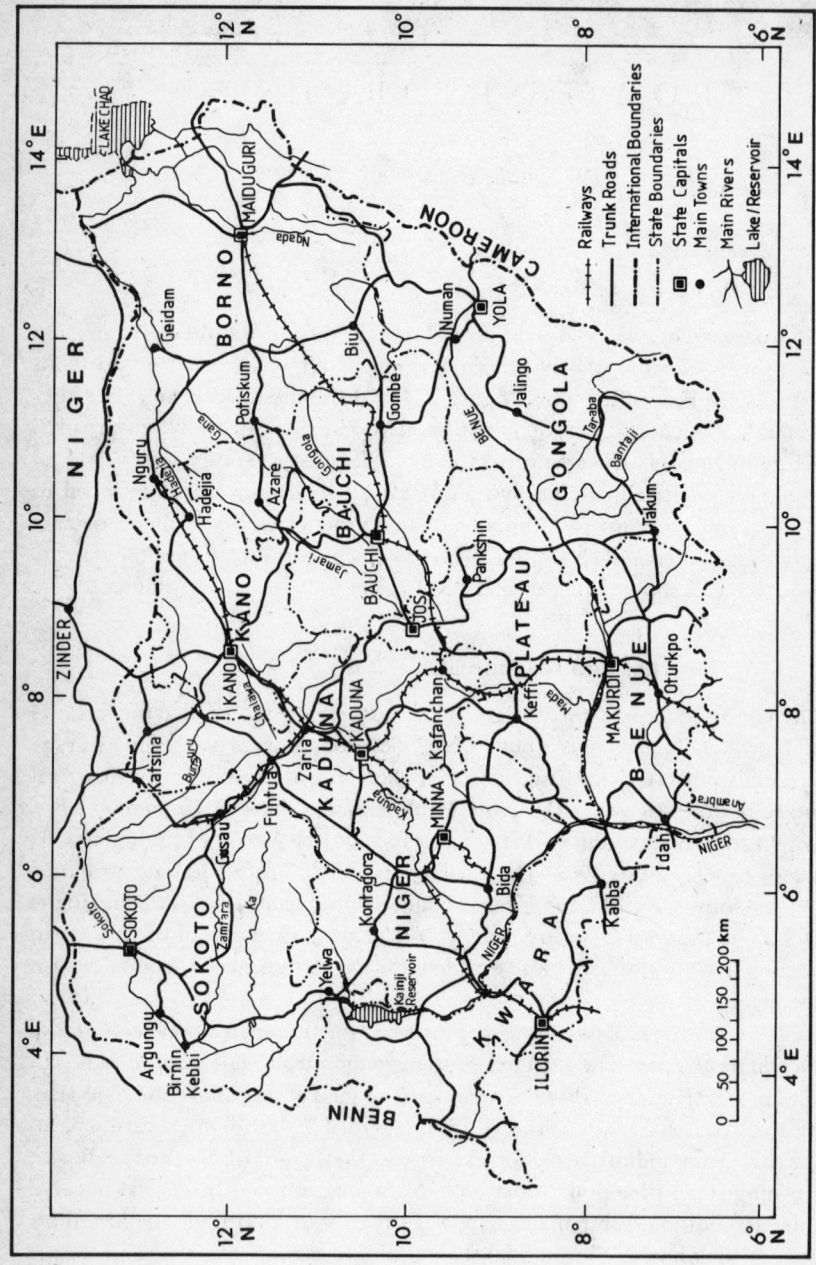

Figure 2.1 Northern Nigeria. (In 1987, after this map was prepared, Kaduna state was split into Kaduna and Katsina state.)

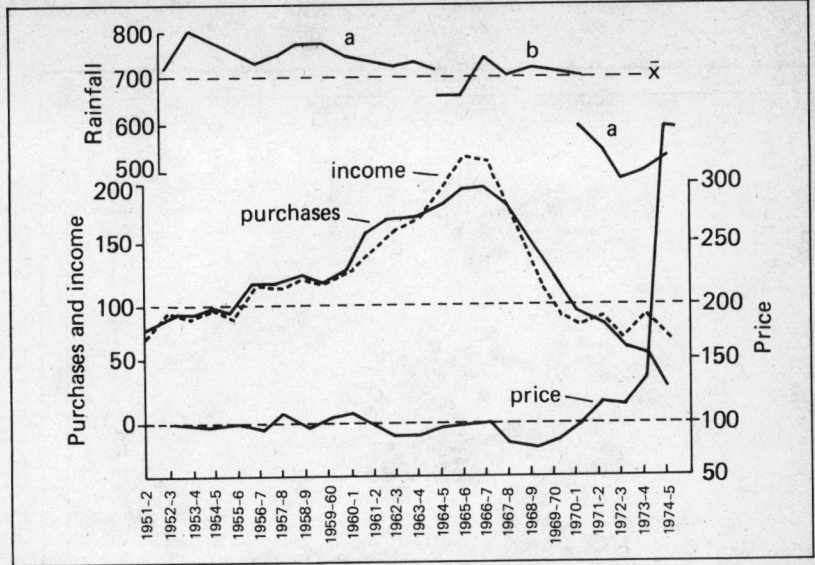

Figure 2.2 The groundnut boom in northern Nigeria. Rainfall is shown in millimetres for selected stations north of Lat. 12°N: (a) Sokoto, Katsina, Kano, Nguru, Potiskum and Maiduguri; (b) Kano and Maiduguri; the three-year running mean for all stations is shown against the overall mean. Indices for the official groundnut price and purchases by the Northern Nigeria Marketing Board and for 'producer income' (purchases × price) are three-year running means based on 1959–60

motivated by the need to stabilise their income from the sale of groundnuts, on which they had become dependent for their cash requirements. An output of one and a half million tons was considered feasible by 1980 (FAO, 1966: 164).

The groundnut (*Arachis hypogaea*) was ideally suited to the well drained, light sandy soils of Nigeria north of Lat. 11°30′N (figure 2.3). These soils, derived from loess deposits and dune sands, are easily cultivated by light hoes; the 90–120-day varieties are well adapted to the length of the rainy season (including ripening on residual soil moisture for 20 days) and, until the seventies, the area was relatively free from diseases and pests (Mortimore, 1968; Kowal and Kassam, 1978: 265–71; Hogendorn, 1978: 40–3; McTainsh, 1984). The groundnut, a nitrogen-fixing legume, was integrated into a mixed cropping system on small family farms[2].

Farmers' incomes from the sale of groundnuts were used to pay taxes, buy consumer goods and food, including, where necessary, making up any deficit in the production of staple grains (e.g. Mortimore and Wilson 1965: 79–89). In addition to farmers' earnings, the groundnut generated profits for the middlemen (licensed buying agents), the marketing boards (which provided

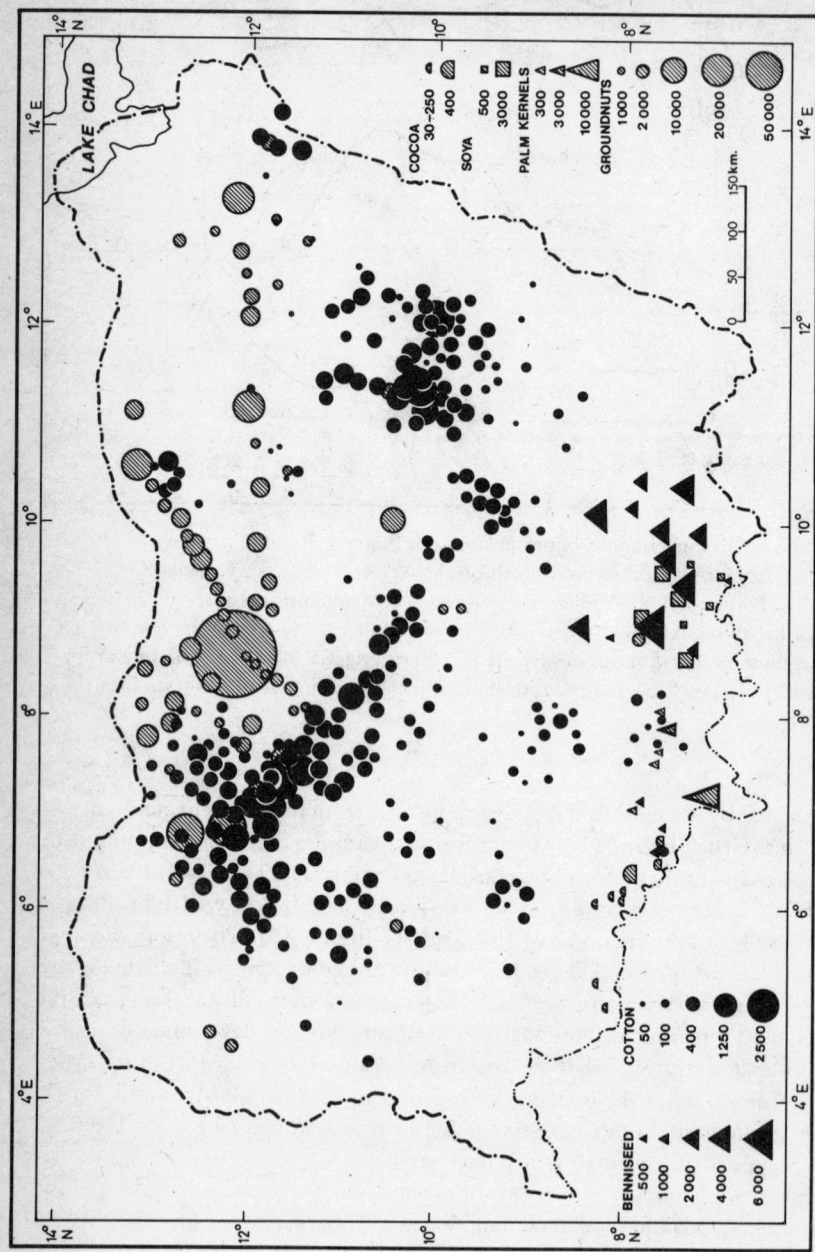

Figure 2.3 Average crop purchases (tons) by the Northern Nigeria Marketing Board, 1963–65

government with revenue for investment in the urban sector: Olatunbosun, 1975: 74–97), the Nigerian Railway Corporation and road transport firms, the crushing mills (about 40 per cent of the crop was crushed locally), and the retail and service sectors in Kano and the smaller buying centres.

Cotton early proved unable to compete with groundnuts on the lighter, drier soils, and had a more southerly distribution, except to the west of Kano (figure 2.3). Cotton production, like groundnuts, defied an unfavourable price trend by increasing from 1952 until 1969. In the mid-sixties about 100,000 tons of seed cotton were purchased each year, collected into ginneries, and the lint divided between export and the Nigerian textile mills, then in process of rapid expansion.

The minor crops handled by the marketing boards (benniseed, soya and ginger) were produced much further south. Mention should be made, however, of air-cured tobacco, which was grown on a quota system for cigarette manufacture in Nigeria, in both rain-fed and riverain areas north of Lat. 11°N. The income from about 4,500 tons of carefully supervised production was very significant locally (Agboola, 1979: 143–7).

Increasing quantities of the staple food crops appeared on the market in response to rising prices, improving transport links between rural markets and the growing cites, and food deficits, which were temporally or spatially variable. The most commercialised food crops were rice and the leguminous cowpea. The first was grown in the flooded valleys of the northern river systems; the second (*Vigna unguiculata*) on rain-fed upland farms where it was intercropped with grain and groundnuts. Both supplied markets throughout the country, especially in the south (Mortimore, 1979). Other crops of commercial importance, grown in the river valleys, were sugar, wheat, cocoyam, sweet potato and vegetables; and, on rain-fed land, cassava, bambarra nut and sorrel.

Two staple grains fed the rural population of the Dry Zone. The first of these, bulrush millet (*Pennisetum typhoides*; H *gero*), grows to a height of 1.5–2 m in as few as 55 days, and produces a head of grain over a metre in length. It can be grown with only 200–50 mm of rainfall in light, sandy soils, and has a remarkable tolerance of drought in the planting period (Kowal and Kassam, 1978: 249–50). On sandy soils it may withstand over 30 days without rain, and withered plants may resuscitate even later. On heavier soils it is more vulnerable. This crop is the major staple throughout the West African semi-arid zone. There is also a late-maturing variety of bulrush millet (H *maiwa*) whose importance is steadily diminishing.

The longer-season crop, known in Nigeria is guinea corn (*Sorghum bicolor*; H *dawa*), becomes dominant southwards, with the longer rainy season and heavier soils. It needs 120–35 days and about 600 mm of rainfall (Kowal and Kassam, 1978: 241); it grows three or four metres tall (or higher) and produces a bunch of grain in the head, which may be compact or dispersed, according to the variety. Guinea corn matures for a month or more after the end of the rains, and its

Table 2.1 *Estimated area and average annual production of some major crops in Northern Nigeria, 1965–67*

	Area ('000 ha)	Production ('000 tons)
Guinea corn (*Sorghum* spp.)	5,119	3,496
Millet (*Pennisetum* spp.)	4,357	2,318
Cowpeas (*Vigna unguiculata*)	3,132	553
Groundnuts (*Arachis hypogaea*)	2,235	1,196
Maize (*Zea mays*)	449	360
Cotton (*Gossypium* spp.)	306	212
Rice (*Oryza sativa*)	122	132

Source: Federal Office of Statistics (1972)

capacity to use residual soil moisture gives it the advantage over millet in certain conditions.

Agricultural statistics in Nigeria are unreliable, but official estimates of the cropped area and production of some major crops for the mid-sixties are given in table 2.1. Earlier estimates are still less reliable, but they suggest that a major increase took place in the first four crops in the previous decade.[3]

The ecological stability of most farming systems was ensured within acceptable limits by manuring (or the occasional use of fertilisers), by rotational bush fallowing, and by migration into less densely populated areas (Mortimore, 1971). Domestic energy requirements – in all rural and many urban areas – were met predominantly from fuelwood cut from living trees in the bush or on farmland. While this undoubtedly posed a threat to the long-term survival of the woodland, the stocks of economic or 'farm' trees were maintained at stable levels.[4] Intensive off-take of fuelwood was not incompatible with the harvesting of a wide range of products (fruit, edible leaves, animal browse, bark, pods, various medicines), both for the market and the domestic economy, in the farmed parklands of Northern Nigera (Mortimore, 1972; Pullan, 1974). Other sources of rural energy were crop residues and charcoal.

The livestock holdings of northern Nigeria were probably not in excess of the carrying capacity of the range, but they were unequally distributed (with the exception of pigs) in favour of the tsetse-free, but less productive, northern pastures (De Leeuw, 1976). Data for 1965 are given in table 2.2, but, lest the possibilities both for error and for change should be underestimated, the table also gives some figures for 1975. It is likely that the earlier figures underestimated goats, sheep and pigs.

The northern herds supplied the greater part of the nation's meat, and the cattle herds of the nomadic, semi-nomadic and sedentary graziers were managed in such a way that most males were sold when fully grown, and the females retained for breeding and milk production. Only a few cattle remained

Table 2.2 *Estimated numbers of livestock in Northern Nigeria, 1965 and 1975 (thousands)*

	1965[a]	1975[b]
Cattle	10,256	8,243[c]
Goats	14,317	23,222
Sheep	4,448	7,691
Donkeys	2,085	720
Horses	431	258
Pigs	181	898
Camels	17	18
Poultry	41,716	83,254

Sources: [a]FAO (1966); [b]FAO (1975); [c]figure for 1971 (Commonwealth Development Corporation, 1971)

in the same location throughout the year, owing to seasonal fodder or water shortages. Transhumance on a generally (but not exclusively) north–south axis was characteristic of most cattle herds and of many sheep and goat flocks. Migratory circuits were influenced negatively by tsetse infestation, and positively by access to grazing, water and market outlets for milk products. Many circuits were subject to a southward shift as formerly unhealthy pastures were opened up by tsetse control projects; later this drift was to be reinforced by the effect of drought north of Lat. 12°N. Animal husbandry, even when practised by nomads, was fully integrated into the market system; and complex patterns of interaction had developed between nomadic and sedentary populations on the one hand, and between animal and crop husbandry on the other (Fricke, 1979; Van Raay, 1975; Mortimore, 1978a).

A changing environment

The agricultural and pastoral systems were (and still are) exposed to a number of variables whose movements were beyond the control of the farmers and herders themselves. The six following are among the most important: prices, rainfall, population, land supply, technology, and government intervention. They were to assume greater significance in the droughts that followed.

Prices for tropical primary products were determined ultimately by the world market, which meant the level of demand in the industrial economies. Attempts were made to cushion primary producers against short- or medium-term price fluctuations. The Nigerian commodity marketing boards, one of whose objectives was price stabilisation, had, however, abandoned the role of trustee for the farmer and assumed that of the tax collector by the mid-1950s

(Helleiner, 1966: 152–84). Trading surpluses were not passed back to the farmer, but spent on development projects in other sectors.[5] It has been argued that groundnut farmers were responsive to price incentives, and in particular to their price expectations for the forthcoming buying season (Abalu, 1975); and in the fifties, a price differential of S£5 per ton had brought about a complete conversion from standard to special-grade nuts in only five years (Helleiner, 1966:66). On the other hand, unavoidable expenses – such as marriages, clothing, house repairs and food – were no respecters of groundnut prices. Producer prices stagnated after 1951, but the trend in production continued upward until the second half of the decade (figure 2.2; Agboola, 1979: 127). Livestock breeders were more fortunate: meat prices were governed by the internal market, which was growing rapidly, and there was a strong incentive to increase herd size and off-take (Ferguson, 1967).

Rainfall, the second variable, will be discussed in greater detail in chapter 6. While no trend was apparent in total annual rainfall before 1970, the variability between years was high, and so was variability within the season. Agricultural droughts of regional extent had occurred in 1903, 1913, and 1926, and more localised ones in the 1940s and early 1950s, but the record for the 1960s was good, with several years above the 60-year mean. Droughts introduced a major element of risk into the husbandry of crops and animals. The response of farmers, lacking any predictive capability, was to minimise risk in cropping practices, by mixing crops with different types of drought tolerance, by accumulating a surplus in good years, by investing profits in livestock or other saleable assets, and by having alternative opportunities available for obtaining income or assisance. Livestock breeders could adapt their grazing circuits, or sell more animals. Conversely, the good rains of the sixties had encouraged the extension of cultivation into marginal areas (more so across the border in Niger), and the enlargement of herds.

Population numbers are not independent of the production system in a strict sense, but they cannot be manipulated in the short term. At the family level, the need for labour and for enough male children to guarantee the long-term security of the family were dominant influences on reproductive behaviour, but at the level of the village or region increasing numbers generated increasing competition for resources, especially land. And the distribution of the population was markedly uneven (figure 2.4). Reliable estimates of vital rates in the north of Nigeria are few. Studies in Hausaland in 1974–78 (Bradley *et al.*, 1982a, b) indicated crude birth rates of 55 per thousand, crude death rates of 26 per thousand (but with infant mortality rates of 170 per thousand live births), and a rate of natural increase of 2.9 per cent.[6] A high gross fertility rate of 221 per thousand suggests a response to high infant mortality, low expectation of life (39 years at birth), and low levels of preventive and curative health services. No decline in fertility was then conceivable (or is, indeed, today) without prior progress in these areas.

The fourth variable was land supply, which in the sixties had begun to

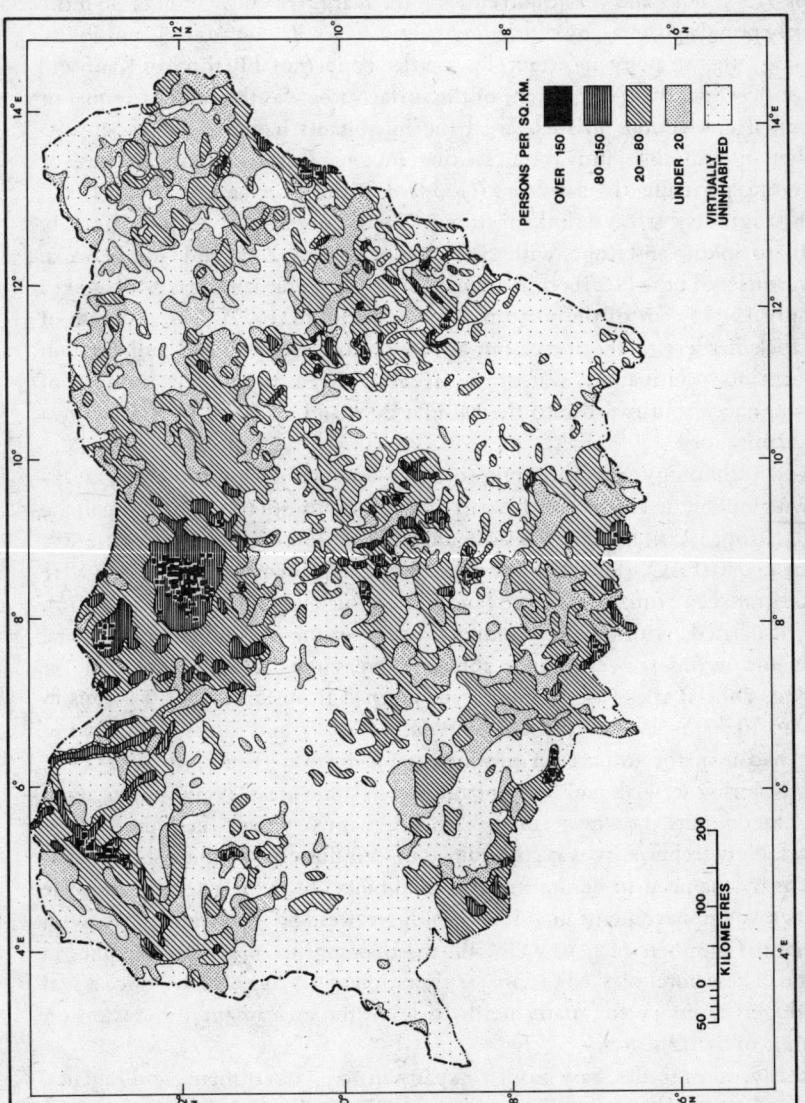

Figure 2.4 Density of the rural population in northern Nigeria based on R. Mansell Prothero and D.H. Birch (1958), *Northern Region, Nigeria, 1:1,000,000. Population Density*, London: Directorate of Overseas Surveys

diminish at an uneven rate, both spatially and temporally. The enormous increase in groundnut production, and also in cotton, had been bought at the price of bringing additional land into cultivation, and by substituting for other crops. Additional land was found either at the margin of cultivation or, in more densely populated areas, by shortening fallow cycles. There was little possibility of using either strategy in certain close-settled zones (notably those of Kano and Sokoto), where over 80 per cent of the surface area was then under annual or perennial cultivation; in such areas the inhabitants had to choose between further intensifying cultivation, seeking income from off-farm sources, or migrating permanently elsewhere (Goddard, Mortimore and Norman, 1975). Such migratory streams linked areas of land scarcity and surplus, notably northern Sokoto and Kano, with the Southern Sokoto, Zaria and Gombe areas. Nevertheless, in the Northern Region as a whole, arable land was estimated to be only 13 per cent of the total in 1965 (Mortimore, 1977). The response of livestock herders to the increasing appropriation of woodland and riverain pastures for cultivation was then, as always, to shift their circuits of transhumance southward into the Middle Belt, and eastward into Adamawa and Cameroon.

New technology was beginning to affect land use in significant ways. Ox-drawn ploughs, first introduced as part of a programme of mixed farming in the 1920s (King, 1939), had made a slow start, but by 1965 the number in use was about 45,000 (FAO, 1966: 202; Laurent, 1968) and growing rapidly. They were concentrated in commercially advanced areas such as Gombe (Tiffen, 1975), and associated with larger holdings and wealthier, innovative farmers. Inorganic fertilisers were being applied increasingly to crops destined for market. Official sales (partly subsidised) jumped from less than 1,000 tons in 1950 to 16,700 in 1965 (Northern Nigeria, 1966: 79), and this is considered to have had a major impact on groundnut production (Wells, 1974: 282–3). Tractor hiring services had been introduced in some areas. New seed varieties and other inputs, however, made rather less impact than their proponents hoped. New technology was constrained in its diffusion by capital shortages as well as by the need to demonstrate its credibility. In the livestock sector, the sixties saw improvements in animal health services and in water supply, with substantial numbers of new wells and the opening up of artesian supplies in Borno. These boreholes led to a major alteration of grazing circuits, since it was no longer necessary for many herds to leave the area in the dry season on account of water scarcity.

Finally, the sixties saw some expansion of government and agency intervention in the rural sector (Forrest, 1977). Following the failure of the Niger Agricultural Project at Mokwa (1947–54: Baldwin, 1957), the regional Ministry of Agriculture was reluctant to get involved in direct production (except for small irrigation schemes), and instead favoured capital-extensive programmes such as 'mixed farming' and fertiliser distribution. A major programme of co-operative development took place, but instead of furthering local participation and initiatives it tended to be hierarchical and uniform (King, 1975, 1977). Pump irrigation schemes were set up on the Hadejia-Yobe, Kaduna, Niger and several smaller rivers; and dams on the Rima. These were expected to be the forerunners of larger-scale developments in the future. In the livestock sector, there were a number of attempts to 'rationalise' or improve upon existing grazing systems, including rotational grazing schemes in some forest reserves, and cattle ranches. None had unqualified success. Conversely, the activities of the Forestry Department shifted noticeably from reserving woodland in extensive areas (which was unpopular with farmers, increasingly hard pressed for land) to establishing plantations of quick-growing exotics (Fishwick, 1970).

These variable elements of the environment should be set against the relatively unchanging factor endowment of soils, surface and groundwater resources. Superimposed on the pattern of the major soil groups a crucial distinction for land management was between the soils of the upland (H *tudu*), generally found over 80 per cent or more of the surface, and the lowland or *fadama* soils, where groundwater was available within the rooting zone of some crops and grasses throughout the year, and where perennial surface water, or shallow groundwater, could be used for irrigation in the dry season (Turner, 1985). Thus the *fadama* soils, usually finer in texture and richer in nutrients, could be used for farming or grazing throughout the year. The *tudu* soils could support only rain-fed farming, and grazing was restricted, in the dry season, to crop residues, dry grass of the last rainy season, or browse.

The vegetation communities were unstable and discontinuous, as farming, grazing, cutting and burning acted upon the variable pattern of soils. Already the broad ecological zones (Southern and Northern Guinea, Sudan, and Sahel savannas) were becoming harder to distinguish from the patchwork of farmed parkland, newly cleared fields, fallows, reserved woodland and various types of secondary woodland that reflected both ecological and land-use competition.

Core and periphery

Social access to the use of land was a pivot of Northern Nigeria's agrarian economy. Such access was (and is still) governed by the land law, by administrative practice based on the interpretation of this law, and by custom.[7] Land law recognised two forms of tenure: statutory and customary. The first, then confined mostly to urban areas, was subject to a Certificate of Occupancy issued by the regional (now state) government, usually for a period of 99 years (renewable), on payment of a nominal rent, in recognition of the governor as trustee of all land on behalf of the people. The law also recognised customary tenure, by which was meant Shari'a Law in Muslim areas and a variety of customary systems elsewhere. Such title was not registered, and depended on the testimony of village heads and their deputies; it could be defended in local courts, but only statutory tenure could be challenged or protected in the higher courts. The costs of acquiring statutory title deterred rural smallholders from attempting to convert their customary titles, mostly acquired through inheritance (although sale, renting, loaning, pledging and gift were all practised). The implications for equity of maintaining two systems, the one associated predominantly with urban entrepreneurial capitalism and the other with agrarian family-based livelihood, were not adequately foreseen. The oil boom had not yet hit the land market.

Furthermore, the law gave the regional government powers of acquisition over land needed 'in the public interest',[8] on payment of compensation for lost crops, improvements and disturbance; these payments were below the market value of land, where it was changing hands under customary tenure. This practice was supported by theory and by assumption: the theory being that all landholders enjoyed the usufruct, ultimate ownership resting in the state;[9] and the assumption being that persons displaced would be able to find free land elsewhere. Both theory and assumption were debatable, but some years had still to pass before government acquisition would be met by, first, passive resistance (on the Kano River Project in 1974), and then by violence and death, on the Bakolori Project in 1978.

Neither statutory nor customary title was available to the users of grazing land. Such land was common, and might be used by farmers to expand their holdings, or by fuelwood cutters, as well as by sedentary, semi-nomadic or nomadic pastoralists. In some areas, the regular visits of nomadic groups, perhaps remembered from a time before the farmers settled, were given recognition by village or district heads as conferring rights of grazing; but these might not be defensible in the local courts, nor against other graziers. The most effective way for graziers to establish rights was to set up compounds and farms of their own, and maintain some livestock in the area throughout the year. *De facto* recognition of such rights was given, for example, in north-eastern Kano State, where disputes were settled by the district head administratively, not in the courts. Even these rights were not exclusive.

Mention should also be made of the forest reserves which, having been gazetted in conservation areas (such as watersheds and steep slopes) and in uninhabited woodland, were unfenced and had to be protected by forest guards. Grazing was usually allowed, but cultivation, hunting, burning and cutting – including for browse – were forbidden. Friction between guards and graziers, and pressure for de-gazetting in areas where farmland was scarce, were not uncommon.

Thus several sources of tension in land matters were becoming apparent in the 1960s, and the signs pointed towards the progressive marginalisation of many rural land users. Unprotected by any system of formal registration, they faced the new demands of the seventies when, in some areas, the state appeared to abandon its colonial aspiration to the role of protector, and to resemble more closely a predator.

The administrative structure of Northern Nigeria, like the land tenure system, was grounded in the fundamental duality of the colonial theory of indirect rule. Only urban areas – and they but in part – were administered directly by the regional government. Nearly all rural dwellers found themselves under the 'native authorities', whose historical roots were deep. In the absence of effective elected bodies, political influence tended to aggregate upwards through a hierarchy of administrative units that closely reflected the pattern of settlement.

The settled population was divided into village areas, under the authority of village heads, who were appointed by the district head, received small salaries, but had no staff. Such an area might consist of a large compact village, itself further divided into wards, under ward heads; or of several small compact villages, each an administrative ward; or of an area of dispersed compounds, subdivided into wards. Village areas were grouped into districts. The district head was appointed by the native authority, and had a small office and staff, a vehicle and funds. Districts were grouped into native authorities which might vary in size from the vast emirate of Kano (with 24 districts and three and a half million inhabitants) to tiny Misau, itself smaller than one of Kano's districts. The size of an administrative headquarters tended to reflect its tributary area and population. The provincial and, finally, regional governments maintained the full apparatus of a professional administration in the provincial headquarters and in Kaduna, the regional capital.

The pastoral nomadic clan group, through its elders, recognised and was in turn recognised by the district head of a given area; lower down, the elders would negotiate with village or ward heads over such matters as the joint use of wells. Disputes between farmers and graziers (for example, over crop damage caused by livestock) were taken, if necessary, to the Shari'a court. There, however, pastoral interests were rarely well represented, even where pastoral groups had become sedentary. Non-Muslim minorities might also recognise, through their elders, the priority of Muslim village and district heads. Significantly, such groups (notably the Maguzawa) lived in dispersed settlements.

Such a hierarchical system had important implications. The further from the centre (in terms of distance, settlement pattern, ethnicity or religion), the less visible was the presence of government. Settlements associated with administrative headquarters prospered from the employment and investment provided by the public purse. Little employment or investment occurred anywhere else, notwithstanding the fact that native authority revenues were based on community tax paid by all adult males, and also included cattle tax paid on each animal. Thus a pronounced urban bias was manifest in the political economy.

The market system reinforced this urban bias: trade made fortunes in Nigeria, rather than land or productive investment. The location and inauguration of a rural periodic market was, of course, a political decision – of great import for the village whose candidature was successful. The level of economic activity was also related to the market's accessibility for urban-based traders, who appropriated profits from bulking and bulk-breaking activities, as well as from retail trade. Agencies for purchasing produce on behalf of the marketing boards were issued on the basis of patronage, linking the marketing and political hierarchies, and channelling multiplier effects towards the city. Small producers were powerless in the face of cheating. And behind the city lay the embrace of the world economy.

The peripheral status of the communities described in this study, in relation to the political geography of Northern Nigeria, compounded the environmental marginality of their farming and livestock-raising enterprises.

FAMINE IN HAUSALAND

By the end of 1972, a vast area of West Africa was in the grip of the worst drought that had been experienced for six decades (Derrick, 1977). Rainfall had begun to fall below normal after 1968, but in 1972 the deficit was sufficient to cause crop failures of 50–90 per cent, and widespread mortality among livestock (Jiya, 1974; Mukhtar, 1974). Estimates of the population affected by drought, both pastoralists and farmers, range up to 22 million. At the same time a catastrophic harvest failure occurred in Ethiopia, with reports of 50–100,000 deaths. By the time that the drought was over, the governments of Niger and Ethiopia had fallen.

Impact of drought

Danbatta district of Kano State (figure 2.5) contained over 150,000 people, organised into 50 administrative village areas, situated on a plain of hummocky dune sands having – till 1970 – an annual rainfall of about 800 mm. The scale of the harvest failure in 1972 was not generally appreciated outside the district until early in the following year. In July 1973, when the next rainy season should have begun, a survey was made of the knowledge and ideas of the village heads (Mortimore, 1973).

The rains of 1972 had begun late and ended early. Guinea corn was affected worse than millet, but some farmers harvested little of either: from less than half down to as little as 10 per cent of their expected yields. Most households had consumed all of their subsistence grain months before the survey. As for livestock, the village heads estimated that there had been losses by death of 4,400 cattle, 6,400 sheep, 5,000 goats, 500 donkeys and most of the fowls – for what such estimates are worth. Mortality in some herds was reported to be 30–50 per cent, and hardly any young had been bred during the year. Cattle tax was being collected at the time, accentuating the distress.

Almost everywhere, groundwater levels had fallen, leaving dry wells and necessitating repeated deepening. Prices of foodstuffs had risen by 400 per cent, and smaller markets were sometimes without grain. Only by selling their goods or animals could most people afford to buy. Yet it was said that even the richer

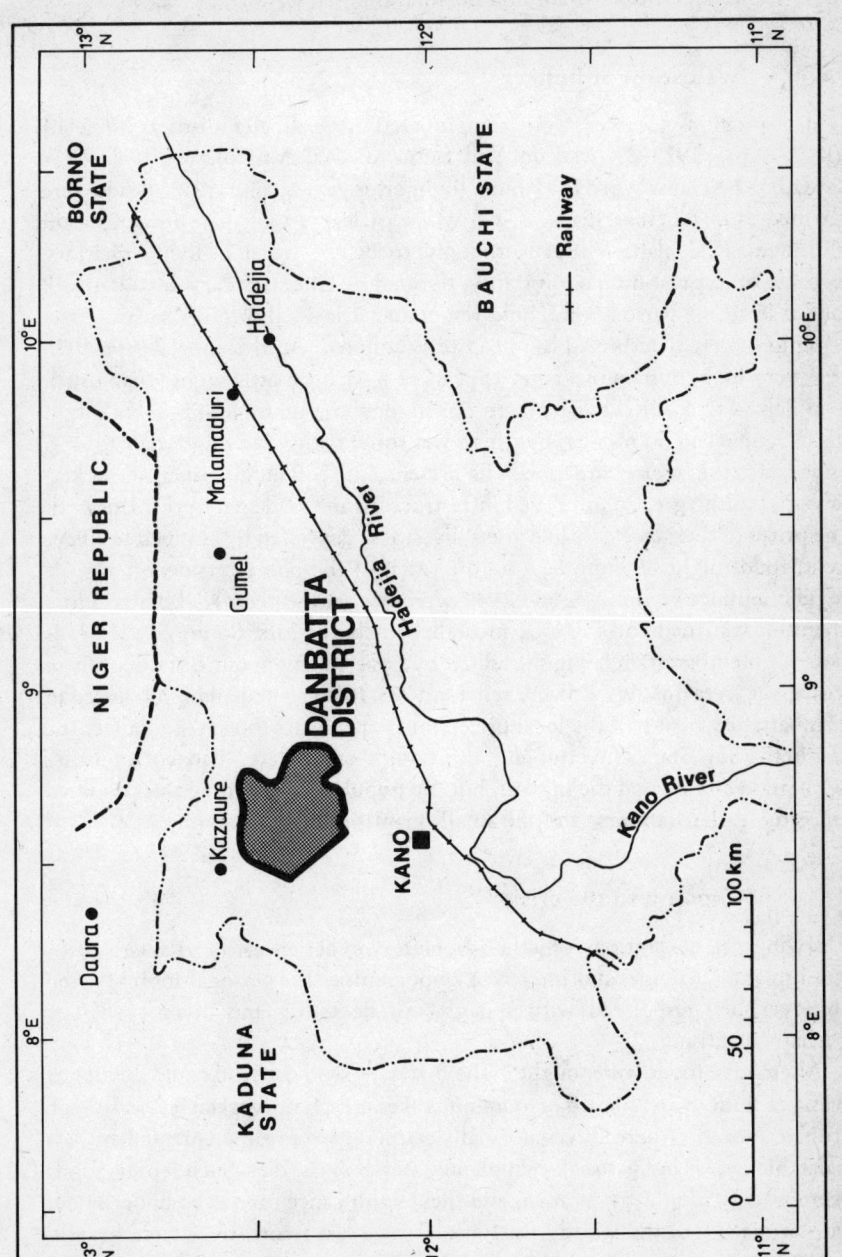

Figure 2.5 Danbatta District in Kano State

farmers were in difficulties. No deaths or epidemics were reported, but people were leaner and weaker than in a normal 'hungry season'.

Evaluation of hunger

Three previous famines were remembered in Danbatta district: 1913–14 (K'ak'alaba), 1926–27 (Mai or 'Yan Buhu, or Mai Amaro), and 1942 ('Yar Gusau). The Hausa word for famine (hunger) is *yunwa*. The village heads were divided as to whether 1972–73 was worse or less serious than previous food shortages. K'ak'alaba is usually thought to be the worst in living memory because of its great human mortality, testified by contemporary documents,[10] but at least one person who could remember it insisted that 1973 was worse. Opinions were determined by the criteria adopted. Animal mortality in 1973 was very high; human mortality appears to have been little higher than usual. People said that in K'ak'alaba there was money, but no food to buy with it; in 1973 people had no money, but there was some food. One might debate as to which was the 'worse' situation. The presence of food in the principal markets was due to the greatly improved infrastructure and trade networks. But with the prices of their chief saleable assets, livestock, depressed by as much as 90 per cent, food might be quite beyond the reach of the poorer farmers.

The famines of 1926–27 and 1942 were less serious than K'ak'alaba. Little mention was made of livestock mortality in those years. Some village heads could remember grain being subsidised by government in one or both of these years, yet were not aware of any relief in 1973. In fact, rice had been released in Danbatta town, at half the local price, but in quantities too small to affect the market. Kano State Government, along with other states, had been putting both grain and rice on the market, but the population had more than doubled since the earlier famines, and the small quantities made little impact.[11]

Response to the crisis

Drawing on generations of ethno-botanical experience, social relations of kinship and clientage, and improved opportunities for personal mobility, the impoverished responded with a degree of flexibility and diversity that is perhaps surprising.

Alternative foods were sought in the market – by those who could buy them (cassava flour, wheat bread, groundnut cake and cheap vegetable foods) – or from the bush, where leaves normally gathered were supplemented by less palatable foods, and gradually supplanted starch in the diet. Such famine foods were usually sought by women, and their significance cannot be understated. For animals, too, the woody vegetation was a last resort; trees were heavily lopped and even the tough, fibrous leaves of the dum palm (*Hyphaene thebaica*) were pressed into use, while the price of fodder in the market escalated out of reach, except for the cattle trade and for rich persons' horses. The wealthier were expected to help the poor, either by hiring their labour or by gifts and loans; but such help was said to be in decline. Indebtedness increased. When kinship links had no more to offer, and the animals had died or been sold, a farmer might have to sell or pledge his property or land. Movement to the city for the dry season might yield income, but must needs cease when the planting season came: and the new millet would not be ready for three or four hungry months.

The famine intensifies

In 1973, the first planting on 6 June was lost and a second was attempted on 30 June. The rains continued irregular both in time and space. By July 12, extensive areas in northern Kano had not had a planting rain nor showed new natural growth, except for on the trees. Where planting was achieved, germination rates were poor and wilting extensive. Since the rains of 1972 had ended, in some places, before August was out, there was good cause for pessimism. Even if the rains continued into September, the longer-season crops (guinea corn, late millet, groundnuts and cowpeas) were unlikely to yield well.

Table 2.3 *1973 and 1974 compared in Roni and Danbatta Districts*

	Percentage of village heads reporting ($n = 67$)		
	More in 1973	More in 1974	Other answers
Food and health			
1 Hunger	15	83	2
2 Sickness caused by hunger	19	27	54
3 Food available in market	72	25	3
Decreasing responses			
4 Selling of labour	67	33	0
5 Borrowing of money	65	35	0
6 Selling of manure	60	38	2
7 Selling of personal property	68	28	4
Increasing responses			
8 Selling of animals	21	79	0
9 Selling of wood, mats, etc.	16	82	2
10 Dry-season migration	3	84	13
11 Selling of land	35	60	5
12 Begging	13	72	16
Unchanged			
13 Obtaining gifts	52	48	0

In Danbatta and neighbouring Roni district, the comparison of information given in July 1973 and June 1974 by 67 village heads gives some indication of the intensifying impact of drought after the rains did, indeed, fail again in 1973 (table 2.3).

The judgements summarised in this table suggest a number of conclusions. Firstly, the food situation continued to deteriorate, but cumulative hunger did not provoke a major increase in sickness, on which most respondents gave answers indicating no trend (51 per cent reported no unusual sickness even in the second year). Secondly, a group of responses declined in frequency, although by no means did they disappear. These had one characteristic in common: they were used within the locality and their decline was a consequence of the decline in money supply. Thirdly, a group of responses increased in frequency. The first three of these gave access to the market system outside the region, whose buoyancy derived ultimately from oil. Land selling included pledging (or temporary alienation for cash) to the richer farmers. Finally, assistance provided through social networks did not change conclusively. A picture, if indistinct, emerges of rural communities, notwithstanding impressive resilience, being nudged towards an increasing dependence on the market system.

Danbatta District had been a major groundnut producer. But groundnuts went the way of the subsistence crops in 1973. However, if export production was stopped, the failure of subsistence production was more fundamental. Danbatta District was north Nigeria in microcosm. This short account provides a vignette for the analysis which follows in chapters 3, 4 and 5.

POSTSCRIPT ON THE GROUNDNUT

The groundnut, paradoxically, escaped lightly in 1972. Owing to the usual delays in shipment, a large quantity of this crop (which exceeded half a million tons) still awaited evacuation at Kano, and elsewhere, after the end of the second calamitous season of 1973. A Central Bank report (1973), widely quoted in October, published the utterly unrealistic estimate of a crop of 600,000 tons in 1973; a month later, the Federal Commissioner for Trade hoped that although production was down to 250,000 tons, the country would obtain more foreign exchange owing to the higher world price. That increase was not passed on to the farmers. In the event, the Board purchased a mere 44,000 tons. In December, however, the Marketing Board came under pressure to raise the producer price because, owing to the high prices offered in local markets, virtually none of the decimated 1973 crop had been offered for sale to its licensed buying agents. The crushing mills, whose capacity had recently increased, were dependent on the Board for their quota of the nuts (the proportion of groundnuts exported as oil or cake had risen from two-fifths to more than half between 1969 and 1972: Ogundana, 1978: 340). Yet groundnuts are acceptable and nutritious food; furthermore the cake could have been used

Postscript on the groundnut

to save *Nigerian* livestock. But it was not until well into 1974 that the export of groundnuts, oil and cake finally ceased (Central Bank, 1974). Meanwhile, the mocking presence of the pyramids provided a poignant demonstration of the export orientation of the 'colonial' type of economy.[12]

As the seventies progressed, the groundnut economy in the north of Nigeria was hit by a combination of aridity – the average rainfall in the next decade was two-thirds of the long-term mean; rosette disease (Yayock, 1977) – which throve on the drier conditions and benefited from new year-round breeding habitats for the aphid fly in the new irrigation projects; and the failure of official producer prices to keep up with the escalating local markets. Even seed became scarce, and the mills began to distribute imported oil. In short, Nigerians were eating all their groundnuts, and these were being grown further south. The collapse of the groundnut economy in the Dry Zone became one facet of the general decline of the agricultural sector during the oil boom years.

DROUGHT IN THE 1970s

Keep them alive in famine

Psalm 33:19

The story of the droughts of 1972 and 1973 in Nigeria has been effectively told by Van Apeldoorn (1978a, 1981), and by others for West Africa.[1] There were signs of trouble as early as 1969, but these were ignored – perhaps because of their patchy distribution – until a full-scale harvest failure occurred in 1972. Even this disaster was uneven in its impact. But by the end of that year, there was no doubt that the Dry Zone of West Africa was facing its worst food crisis for more than a generation. My purpose in this chapter is not to reproduce what has been done, on a more general level, elsewhere, but to examine some extensive data collected amongst drought-affected communities within the lifespan of the disaster (which may be considered to have ended, for the time being, with the rains of 1974).

THE DROUGHTS OF 1972 AND 1973

The droughts that occurred in 1972 and 1973 had major significance on all four of the dimensions – meteorological, hydrological, agricultural and ecological – defined in chapter 1. Expectations of rainfall, although not stated quantitatively in terms of the 'normal', nevertheless guide decisions made by indigenous land users. In the semi-arid zone, expectations are defined in terms of the growing possibilities for the major crops, and for pasture and browse. Beyond the northern limit of rain-fed agriculture (in the arid bioclimatic zone), only the second dimension is relevant. From the agricultural standpoint, rainfall must be considered in relation to, firstly, the rate of evaporation, secondly, crop water requirements, and thirdly, soil water storage (Kowal and Adeoye, 1973).

At its simplest, the spatial dimensions of drought may be described as a southward displacement of the isohyets (figure 3.1). In Kano State, the isohyets for 1972 moved 90–120 km south-westwards from their normal positions and, in 1973, a further 50–70 km. More than half of the entire state was, in effect, transferred to a bioclimatic zone that was incapable of supporting the incumbent life systems of crop and animal husbandry.[2] As this transgression of the isohyets marched across the population density surface (see figure 2.4), increasing numbers became exposed to the direct or indirect effects of drought: at least 10 million Nigerians in 1973, by conservative estimates (Van

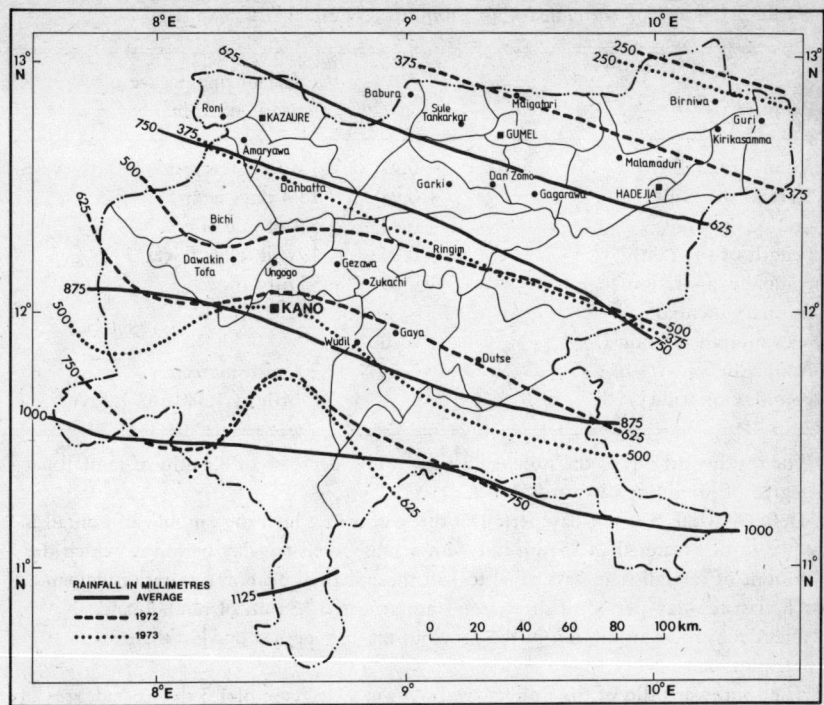

Figure 3.1 Displacement of rainfall isohyets in 1972 and 1973, Kano State

Apeldoorn, 1978a: vol. 1, 88–9), compared with 12 million in the rest of West Africa.

The direction of this displacement was southward. West African rainfall is significantly correlated with latitude (Cochème and Franquin, 1967; Kowal and Adeoye, 1973; Kowal and Kassam, 1978: 70–9), but there is also a southward dip in the isohyets as they extend eastwards across Nigeria. This is because the source of moisture is the Atlantic Ocean whose coastline runs from west to east, with a southward trend in the east of Nigeria. The seasonal movement of the rain-bearing winds, following the Inter-Tropical Convergence from south to north and back again, is not impeded by major mountain barriers. According to Kowal and Kassam (1978: 70–93), using data from over 50 rainfall stations throughout West Africa, latitude correlates significantly with all the major parameters of rainfall (table 3.1).

In 1972 these relationships were broadly maintained, but the slopes of the regression lines altered (Kowal and Adeoye, 1973). In higher latitudes, the rainfall deficit became more pronounced, and the intersection of the regression lines suggested no rain north of Lat. 15°40′, compared with the normal maximum advance of the rains to 18°30′. The start of the rains was retarded

Table 3.1 *Latitude correlated with rainfall in West Africa*

Parameter	r	Effect of one degree of latitude northwards
Annual rainfall (P)	−0.91	131 mm less rain[a]
Start of the rains[b]	+0.94	13.4 days later
End of the rains[c]	−0.84	5.7 days earlier
Length of the rains	−0.94	19 rain days fewer[d]
Annual evaporation (E_o)	+0.69	95 mm more
Annual potential evapotranspiration (E_t)	+0.70	
Annual deficit ($P - E_t$) (index of aridity)		200 mm increase (north of 1,380 mm isohyet)

[a] The southward dip of the isohyets is shown in a decrease of 8.6 mm of rainfall per degree of longitude eastwards.
[b] Defined as the first ten-day period in the season in which the amount of rainfall is equal to or greater than 25 mm but with a subsequent ten-day period in which the amount of rainfall is at least equal to half the potential evapotranspiration demand.
[c] The last ten-day period of the season with at least 12.5 mm of rainfall, and potential evapotranspiration in the previous ten-day period not less than precipitation.
[d] The southward dip of the isohyets is shown in a decrease of 1.3 days per degree of longitude eastwards.
Source: After Kowal and Kassam, 1978

progressively northwards, so that the rainy season was shortened by about six days more than normal for every degree north of Lat. 10°N.

In 1973 the situation worsened. Figure 3.2A shows the regression lines for rainfall in 1973 and for an average year, based on the records of 29 stations situated between Lat. 9° and 13°N (Kowal and Kassam, 1973). In this year, according to the intersection of the regression lines, the advance of the rains had terminated by 14°N.

The northern limit of millet cultivation, and thus of rain-fed arable farming (approximately equivalent to the northern limit of the semi-arid zone: see chapter 1) may be defined theoretically in terms of its total moisture requirements and growing period, making due allowance for soil water storage. According to Kowal and Kassam (1978: 109–10), this boundary approximates to the average annual rainfall isohyet of 348 mm, but everywhere north of the 588 mm isohyet there is a risk of crop failure through insufficient rainfall more than once in ten years. In terms of the length of the rainy period, short-season cultivars requiring at least 55 days of rainfall can meet this requirement (with a risk of failure not more than once in ten years) where the average length of the rainy period is 69 days. Based on the regressions for the

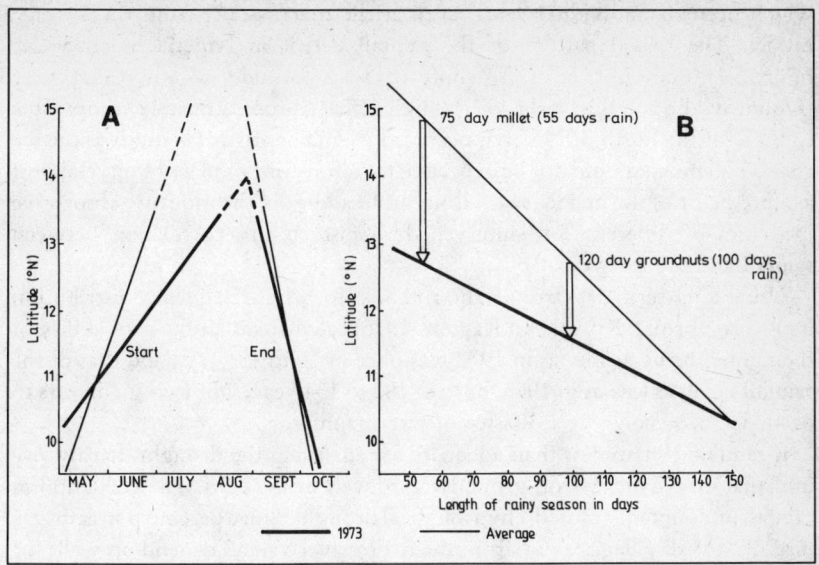

Figure 3.2 (A) Regression lines for the start and end of the rains in 1973 and in an average year; (B) length of the rainy season and the southward shift of the northern limit of millet and groundnut production in 1973 (after Kowal and Adeoye, 1973; Kowal and Kassam, 1973, 1978: 109–10).

Nigerian area, this occurs at Lat. 14°40′ (Kowal and Adeoye, 1973). Actually, successful millet cultivation took place during the sixties even further north, at Lat. 15°N in central Niger.

In 1973, the minimum rainy period of 55 days required by short-season millet was met at about 12°40′ (figure 3.2B). It was as if the northern boundary of arable farming had moved southwards by a distance of over 300 km or two and a half degrees of latitude. Similarly, the effective northern limit of groundnut cultivation moved from about 12°45′ to 11°30′ (Kowal and Kassam, 1973), and total losses were predicted north of 12°20′. Van Apeldoorn (1978a: vol. 1, 57–78) went further to calculate the length of the growing season in terms of the water budget at 16 stations distributed through northern Nigeria. Two growing seasons were defined:[3] (a) a shorter season assuming 100 mm of stored soil water at the end of the rains – enough for the maturing of shallow-rooted crops such as guinea corn, groundnuts and cowpeas; and (b) a longer season assuming 200 mm of stored soil water – enough for deeper-rooting, late-maturing crops such as cotton. In 1973, the reduction in the length of the growing season for shallow-rooting crops ranged from 20 to 80 days, north of Lat. 10°N, and for deep-rooting crops from 30 to 90 days.

Such gross patterns hide local variations, which the coarse-grained distribution of rainfall stations usually misses,[4] and also the irregular departures which occur in individual years, even at the macro-scale, from the average model. The spatial pattern of the rainfall deficit in Nigeria in 1973 (see figure 6.1) does not, in fact, conform with latitude in any exact way (Oguntoyinbo and Richards, 1977). High deficits, proportionately comparable to those of the north of Nigeria, occurred in small areas as far south as the sea coast. Too much should not be expected, therefore, of a model of isohyetal shift as a predictor of the incidence of drought. In addition, the moisture absorptive capacities of different soils intervened to disturb any correlation between rainfall and the growth of plants.

Using long-term data from Kano and Sokoto, whose frequency distribution is close to normal, Kowal and Kassam (1975) calculated that the probability of the rainfall being as low as in 1972 was once in 10 to 13 years, and that of the rainfall being as low as in 1973, once in 100 to 134 years. For two such years to occur in succession was a disaster of rare magnitude.

If rainfall data are less than adequate for analysing the drought, hardly any information is available on groundwater levels in 1972–74. The reduction in rainfall undoubtedly caused a hydrological drought. Since perennial streams are rare, almost all villages away from the major river valleys depend on wells for much of the year for domestic use, and for livestock. In northern Kano state, in the first six months of 1974, it was reported in 11 per cent of village areas that virtually all wells had ceased yielding; in 25 per cent, that about half had ceased; and in 50 per cent, a third. Virtually all yielding wells were below their normal levels at the end of a dry season. Conditions were most serious in the north-western district of Kazaure, where government road tankers were being used extensively to supply water to villages.[5] So far as is known, however, water scarcity did not cause any villages to be abandoned, or significant permanent population movements.

631 VILLAGES

The Danbatta survey was extended to four other districts in the early rainy season of 1973, when it was already apparent that a second year of drought was a possibility. When these fears were realised, a new survey was carried out in 20 districts, including 631 village areas, between February and July 1974.[6] Three of the previous years' districts were revisited, to observe the intensification of drought conditions.

The survey had the modest objectives of measuring, where possible, the social and spatial impact of drought and its dependent variables, and identifying the principal forms of response to hardship. In the virtual absence of earlier studies, a reconnaissance objective could best be achieved by standard survey methods. Lacking much idea of variance to guide in sampling or stratification, a comprehensive cover, interviewing every village head, was attempted in all districts in Kano State north of Lat. 11°30′ (the approximate southern limit of the disaster area). In the event, time ran out; 20 were covered (figure 3.3).

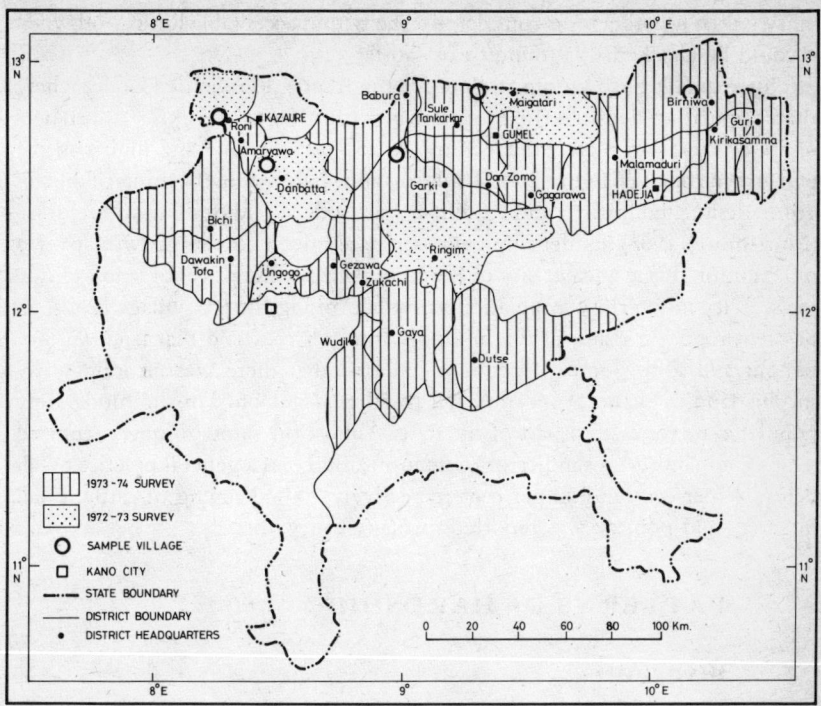

Figure 3.3 Survey areas

Limitations of time and cost also prevented any attempt to sample more extensively in other parts of Nigeria.[7]

The choice of village heads as interview subjects calls for some comment. These officials have sometimes been portrayed as rapacious and unreliable sources of information. They were, however, the official channels of information and authority between district heads and villagers; a major initiative on the part of government normally called for a meeting of them all under the tree outside the district head's house. Since the primary objective of the survey – which was defined in terms of social rather than academic priorities – was to set out the situation quickly in a report to the State Government, the use of normal political structures seemed desirable. Secondly, the hierarchical structure of authority, whose approval and co-operation is a necessary condition of such research, could be mobilised rapidly in the 20 districts at whose headquarters the village heads were ordered to report. This strategy was cost-effective; selective samples at the family level in the villages would have cost much more in time and travel. Thirdly, the subject matter of the survey was so grave, and the village heads were living so close to the situation – although not required to report on individuals – that their reliability (taken together) was considered to be within acceptable limits, and when it could be checked was found to be satisfactory.

This area falls both within and outside the Kano Close-Settled Zone, where rural population densities rise progressively from 150 persons/km² at distances of 50–80 km from metropolitan Kano to 250 persons/km² and higher in the peri-urban zone (see figure 2.4). While this zone falls within the limits of the soil group designated 'brown and reddish-brown soils of arid or semi-arid regions' (Mortimore, 1968) its development is due as much to the growth, over a millennium, of the ancient city of Kano. In the zone the prices of land indicate the scarcity induced by population growth. Among the 631 village heads, 53 per cent reported rising prices, and only four per cent said that land was not bought and sold. Some 74 per cent reported that there was no longer any surplus land, and the proportion fell to 50 per cent only in the moderately populated north-eastern part of the state. Almost the same numbers reported the use of manure – an indicator of intensification – as a general practice on all fields (74 per cent and 52 per cent respectively). The keeping of cattle, small livestock and poultry was reported virtually everywhere.

PATTERNS OF HARDSHIP

Crop losses

Short-season millet and long-season guinea corn provide the farmer with a mixed strategy against unpredictable rainfall conditions. An early start and strong ending to the rains favour guinea corn, which can withstand either drought or waterlogging during the rainy period. Millet makes the best of a late start and early finish to the rains, and can survive long intervals between falls in the planting period. Such intervals (up to *seven* weeks, as in 1974 in some areas) can be critical for the timeliness of planting, effective germination, and rooting of crops; an early ending to the rains may inhibit development of the grain or pods; while late heavy rain can impede ripening. Recognising that either or both of groundnuts and cowpeas may fail in any year, farmers may spread risks between early millet – resowing, if necessary, several times – and late-maturing guinea corn – which may fail completely, but brings a bonus if the rains end well. Beyond the northern limit of guinea corn, all depends on millet.

Crop yields, for the farmer, are the prime indicators of drought intensity. But no one normally quantifies subsistence output, and it is notoriously difficult to obtain reliable retrospective estimates. Village heads were asked instead to estimate average yields in 1973 as a fraction of those normally expected (which probably err on the optimistic side of the mean).[8] It was understood that 1973 was a worse year than 1972.

The answers given for five major and two minor crops are shown in table 3.2. The superior performance of early millet stands out. Only 16.5 per cent reported nearly complete failure of this crop, notwithstanding the fact that

Patterns of hardship

1 A farmer recovers the bird-damaged remnants of his millet crop, near Babura, north Kano State (October 1973). In the background are devastated groundnut and grain fields invaded by secondary regrowth (*Guiera senegalensis*). In the distance, the Harmattan haze has already begun to obscure the outlines of scattered farm trees

Table 3.2 *Reported crop yields in 1973 as a fraction of 'normal'*

	Percentage of village heads reporting ($n = 631$)	
	one quarter or less	One tenth or less
Major crops		
Guinea corn	95.5	31.9
Early millet	81.1	16.5
Late millet	92.3	30.5
Groundnuts	97.4	37.3
Cowpeas	98.1	40.0
Average for major crops	92.9	31.2
Minor crops		
Maize	100.0	70.2
Rice	92.2	68.1

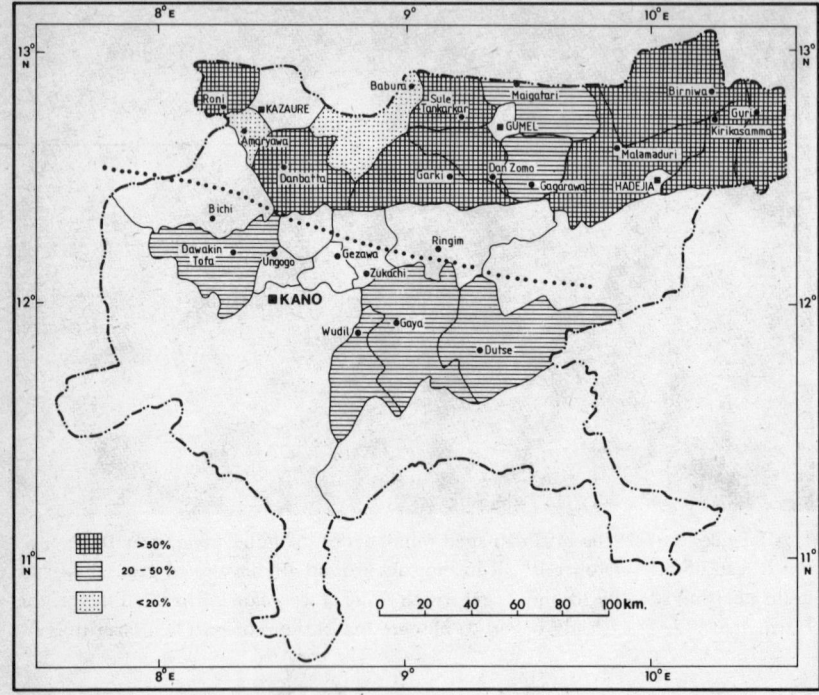

Figure 3.4 Effects of drought on crop yields, 1973 (17 districts): villages reporting 10 per cent of normal or less

some areas in the north of the state received little more than 250 mm of rainfall, badly distributed. The table also shows that maize is ill adapted to such conditions and did even worse than rice, whose cultivation is largely confined to wetter sites.

A clear contrast emerges between southern and northern districts (figure 3.4). More than half the villages reported one tenth or less of normal yields for the major crops, in nine of 13 districts surveyed in the north of the state. The proportion was lower further south. But the irregularities that occurred in the pattern were consistent with the capricious distribution of the rainfall. Working with estimates provided by the administration, Van Apeldoorn (1978a: vol. 1, 84–5) confirms this general picture; even in the south of the state, losses may have exceeded 50 per cent.

Livestock losses

It is impossible to speak with any exactness of the numbers of livestock in Nigeria, owing to tax avoidance (where cattle were concerned), herd division and mobility (sometimes extending across several administrative boundaries),

2 Emaciated cattle attempting to graze regenerating dum (*Hyphaene thebaica*) in barren farmland near Babura, north Kano State (July 1973). These animals had survived the dry season of 1972–73, when mortality was generally very high, but the rains (and the expected growth of grass) were more than a month late and about to fail a second time. Belonging to a sedentary (farming) Fulani family, they were too weak to walk south in search of pasture

divided ownership among members of the same family, some resistance to counting livestock and ineffective record keeping at the lower levels of administration.[9] Numbers for cattle were best known, owing to their importance in the north–south cattle and meat trade, and it is possible that by the late 1960s the national herd was declining in numbers owing to excessive off-take for the market, including a significant proportion of pregnant females (Federal Ministry of Agriculture, 1974: 198–210). Adverse terms of trade and rising food prices forced livestock owners to sell. The effect of the drought years of 1972 and 1973 was to send up the recorded sale and slaughter of cattle in the north of Nigeria by more than 25 per cent to more than 1,219,000 in 1973 and 1,296,000 in 1974 (Van Apeldoorn, 1978a: vol. 1, 119). These figures included cattle imported from Niger and Chad; nonetheless their full significance lies in the fact that they were deducted from a diminishing herd, and sold for prices that had completely collapsed.[10]

This market perspective heightens the impact of abnormal mortality attributable directly to drought. Such mortality was higher in the first dry season (1972–3) than in the second (1973–74). Since both crop residues and rangeland must, on average, have been better in the first year, this anomaly can be explained only in demographic terms. Mortality, sales and unrecorded 'emergency' slaughter, together with the southward relocation of nomadic

Table 3.3 *Reported livestock mortality, 1973*

	Average per village area	Total
Cattle	173	105,876
Sheep and goats	282	168,918
Donkeys	65	38,350
Horses	33	4,422

herds, were sufficient in the first season to reduce the numbers to within, or near to, carrying capacity in the second. The available evidence indicates that such a Malthusian event did indeed occur, attended by much hardship.

Table 3.3 gives aggregate estimates of losses through death in the first year. Only round figures were given, and these were even more impressionistic than the estimates of crop losses. They include deaths among both nomadic and sedentary herds and flocks, and may include some emergency slaughtering also.

Mortality in the sample areas represented 20–25 per cent of the cattle population of the state, officially estimated at about 500,000, even though 18 rural districts and metropolitan Kano were excluded from the survey. Some of the excluded districts had large cattle populations, so even making due allowance for the inclusion of some imported cattle in the mortality figures, it can be confidently stated that Kano lost at least a fifth of its herd.

Sheep and goats, which are easier to estimate when combined, were lost in larger numbers still. The loss of donkeys, which were kept exclusively for transport, together with many camels (whose mortality was not investigated), radically changed the economics of rural transport and eventually paved a way for increased motor penetration of the countryside. Horses, always kept for personal use by wealthier members of the community, were probably the best remembered: they competed with humans for grain, and many starved slowly and conspicuously. Fowls suffered very heavy mortality from outbreaks of disease, practically disappearing from many villages.

These estimates are not excessive when compared with the results of four local studies conducted in admittedly hard-hit areas near the northern border of Nigeria: Illela (north of Sokoto), Sandamu, and Rijiyar-Tsamiya (near Daura) and Mungurun (north of Hadejia). The sizes of both nomadic and sedentary herds and flocks of cattle, sheep and goats fell between 1971 and 1974 by values ranging from 56 to 82 per cent (estimates made in 1976 by Aliyu, Ndaks, Ahmed and Daudu; reported in Van Apeldoorn, 1978a: vol. 1, 114–18). Some holdings were eliminated, and sedentary livestock owners suffered as badly as nomadic. Van Apeldoorn concludes that, at the local scale, the drought may have reduced the livestock population in the hardest-hit area of Nigeria to one-third or less of its pre-drought size; but at the macro-scale, the information available led him to conclude that it is impossible to pin down the losses to the Nigerian livestock economy as a whole with any accuracy (Van Apeldoorn, 1978a: vol. 1, 113).

Prices

Price data in northern Nigeria should be handled with circumspection. Both the government's crop and weather reports and survey data are subject to errors related to units of measurement, conversions, currencies, and reliability. The scarcity of cash (which will be referred to again below) restrained the effective demand for foodstuffs. The supply was constrained by the ability of the trading system to deliver foodstuffs from the south (where shortages and inflation were also rampant) to remote northern markets.

Table 3.4 assembles some of the data obtained in this survey and, for comparison with conditions before 1972, prices of staple foodstuffs reported by Hill (1972: 124–40) in Batagarawa, a village in northern Katsina, for the period 1968–70. The picture is incomplete, but it is clear that prices rose to a peak in the early rainy season of 1973, fell slightly in the ensuing harvest season, and then rose again as the failure of the harvest made itself felt. There were occasions when no grain at all was offered for sale in rural markets, particularly in the dry season of 1972–73. The fact was inescapable that very few could pay the prices being asked. Traders responded to the national shortage of grain by moving cassava flour into rural markets at a relatively competitive price, and its property of expanding in the cooking pot made it a widely used substitute among people for whom an illusion of adequacy was better than no food at all. The only cheap food available was *kuka* or baobab leaves, which sprout in May and are dried and sold throughout the year as a soup (*miya*) ingredient. They could be obtained at ₦6/ton in the early rainy season of 1974.

In figure 3.5 price data from the government's Crop and Weather Reports, as simplified by Van Apeldoorn, have been averaged for five major markets in northern Kano State (Kano, Danbatta, Kazaure, Gumel and Hadejia).[12] This procedure reveals a clear and broadly consistent pattern for the prices of guinea corn and millet for the three years 1972, 1973 and 1974. Data from the present survey have been added: for Danbatta District in July 1973 and for northern Kano in October–December 1973 and March–June 1974. For the first two the correspondence is satisfactory; for the last, there is a divergence, also shown in the last column of table 3.4.[13] Price inflation started in September even before the end of the failed harvest of 1972, and rose to an unprecedented peak in July and August 1973. There was an uncertain fall in the ensuing harvest season to levels still far above ₦ 100/ton, followed by a second rapid rise to an even earlier peak in March and April 1974, and a price plateau until better rains fell in July and August. Finally prices fell sharply, reaching – briefly – levels only a little higher than those obtaining early in 1972.[14]

Table 3.4 *Price of some major foodstuffs, 1968–70 and 1973–74* (Approximate in ₦/ton)

	Katsina, 1968–70		Danbatta District July 1973	Kano (north) Oct.–Dec. 1973	Kano (north) Mar.–Jun. 1974	CWR[a] 5 markets, Mar.–Jun. 1974
	low	high				
Guinea corn	20	80	163–175	128	181	155
Early millet	22	85	150–160	129	181	139
Groundnuts, shelled[b]	44 (58)	113	(70)	147 (94)	220 (94)	
Beans	44	170	246–314	259	344	
Rice	n.a.		299–336[c]	316	429	
Cassava flour	n.a.		n.a.	n.a.	160	
Locust bean flour	80	188	n.a.	n.a.	360	

[a] Crop and Weather Reports
[b] Marketing board prices for shelled groundnuts in parentheses
[c] Small quantities of government relief rice were released at ₦ 150/ton.

Sources: Hill (1972: 124–40); Van Apeldoorn (1978a: vol. 1, 144–6); survey data.

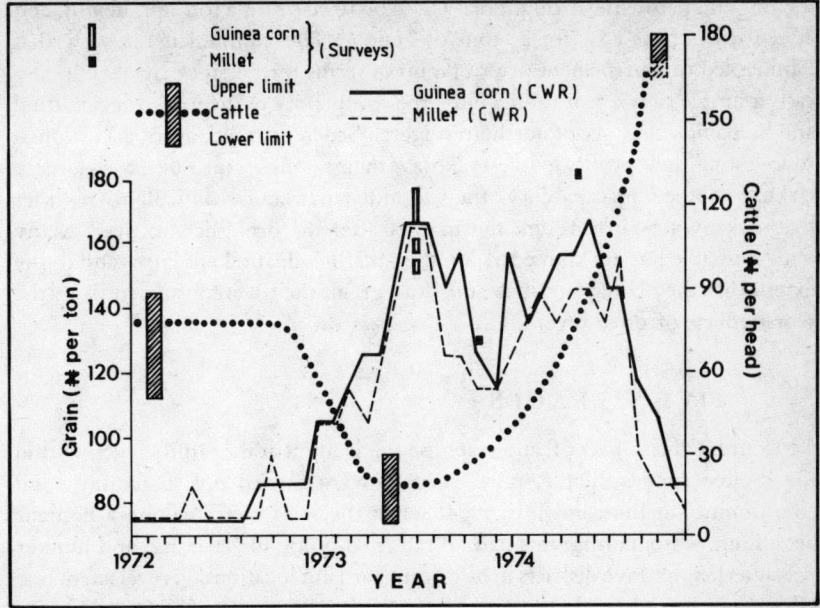

Figure 3.5 Grain and cattle prices, 1972–74. Sources: Crop and Weather Reports (Van Apeldoorn, 1978a); field data

There were widespread reports of an acute shortage of money in the rural sector. Groundnuts, until 1971 the major revenue earner, failed north of Kano in 1972, and everywhere in 1973. With inflowing funds thus diminished, the level of trading activity in markets was visibly reduced. The sale of assets, especially animals, at depressed prices generated little income, and traders quickly exported money realised from market grain sales. To the embarrassment of village heads, who also served as revenue collectors, tax defaulting and moonlight flits were commonplace.

Information on livestock prices was not collected systematically. The collapse of cattle prices between January and July 1973 was experienced so widely that it hardly requires verification (figure 3.5). In January, prices up to ₦40 per ton were still being paid for healthy cows at markets in northern Hadejia and Gumel. By July, the price had fallen to ₦2–6, the lower limit being set by the price a sick cow could fetch if slaughtered for meat in the villages. By November 1973, prices had recovered to ₦20—25 in northern Gumel, and the supply of residues from the ruined crops of 1973, together with the reduced animal population, ensured that they did not collapse again in 1974.

The fall in cattle prices that occurred in 1973 (and those of other livestock are believed to have behaved similarly), represented a loss of 90 per cent in nominal value to their owners. The terms of trade moved from roughly one cow for a ton of grain before the drought of 1972 to 6–10 cows for a ton at its height, and afterwards to one cow for $2\frac{1}{4}$ tons of grain.[15] This contradicts the view that famines led to a permanent deterioration of terms for livestock owners. In any case, a simple two-sector model hides the complexity of the mixed agricultural and breeding economy of northern Nigeria. Sedentary cultivators, accustomed to investing their profits in livestock, saw their savings wiped out, and, largely lacking a breeding capability, they found it extremely difficult to re-enter livestock ownership through the market after the drought was over. Many semi-nomadic livestock breeders, on the other hand, also kept farms and to the extent that they could produce and store grain, they were insulated from the worst effects of price reversals.

FIVE VILLAGES

To examine the impact of, and response to, drought at the family level, within the limited time available, rapid surveys were carried out in remote rural communities in June and July 1974, when the rains were making a hesitant beginning, seasonal migrants had all returned home for farming, and hunger was at its height. Five districts in northern, hard-hit locations were selected, one village area in each, and one hamlet of each. Small samples of household heads were selected from tax lists provided by the hamlet heads, making a total of 125. The locations are shown in figure 3.3; they were distributed widely in order to compensate for the effects of local variations in rainfall and other factors. Since the size of the hamlets varied considerably, the samples ranged from 75 per cent to less than 5 per cent of the population.[16]

Questions of definition plague micro-studies in northern Nigeria. The co-residential unit, which occupies a fenced or walled assemblage of huts (H *gida*) having a single entrance, and under the leadership of a household head (H *maigida*), may be neither a producing nor a consuming unit. Farming practice recognises both joint and individual arrangements, and combinations of both; off-farm earnings may be pocketed by individuals, and women have independent sources of income. Although the small family (H *iyali*) is usually defined as those persons dependent on a man for their daily food, people may live or farm together and eat separately. While co-residential houses contain mainly kin, unrelated persons may also live there. For the purposes of this enquiry, the family is understood to mean a man and his immediate dependants (*iyali*), except in relation to certain questions where the *gida* was the more appropriate unit of enquiry.[17] The *iyali* comprises the smallest cell in the matrix of obligations and decisions that determine individual action in response to opportunities both in the village and elsewhere.

The family heads assigned themselves to several ethnic groups: Hausa (35 per cent), Fulani (27 per cent), Manga (23 per cent) and others (15 per cent). Two subsamples consisted predominantly of Fulani, one of Hausa, one of Manga, and one was divided between Hausa and 'Arewawa' (migrants from the north).

All five had ethnic minorities. The age range (from 18 to 72 years, with a mean of 38) encompassed all the stages of the family developmental cycle, from young adulthood to old age. The great majority had lived all their lives in their villages. Only five per cent had received any primary education, but 60 per cent had received some koranic education, often away from home. The mean size of household was 7.5 persons, with male–female and adult–child ratios both of 1:1.04.

IMPACT

Crops and animals

In a good year, the first planting succeeds with full germination, but when rainfall at the beginning of the season is erratic or partial, or complete failure occurs, or young plants wither, additional plantings are necessary to make up the desired stand density. In 1973, the modal number of millet plantings was three, and 39 per cent of farmers planted four or more times. The number increased from west to east, consistent with the southward slope of the isohyets. Together with the late dates of planting, this experience was a portent of disaster. It is well known that at such times the poorer farmers are heavily dependent on the market or on clientage networks for supplies of seed.[18]

The impact of drought on the production of some main crops is estimated in table 3.5. Informants were asked first to state the output from those farms under their own control in 'former' times (H *da*, before the onset of drought conditions). As a representation of expectations, rather than actual yields, this provided a basis for evaluating the perceived impact of drought, which was the starting point for adaptive strategies.[19] The reported yields for 1972 and 1973, from which the means were calculated, were subject to wide variation, could not be verified, and were among the more sensitive items of information collected.[20] Nevertheless, the picture is clear. The yields of 1972 were less than one-third those of good years, and those of 1973 less than one-tenth. Early millet proved the hardiest crop but, had it been grown exclusively, losses would have been only marginally less serious. The main market crop, groundnuts, was written off and so were cowpeas, also usually sold. For every crop, 1973 was a worse year than 1972, and the worst since 1913.

Table 3.6 shows the losses of livestock owned by members of sedentary farming families during the two years preceding the survey in June–July 1974. Most of the losses occurred between January and July 1973 when the shortage of feed was acute. The figures are subject to error (as shown by the row 'not accounted for'). Sales exceeded deaths: this was because animals were sold, if possible, when death from starvation seemed imminent. When the survey took place, the crisis was over for the surviving livestock, whose condition, however, was reported still to be poor. Foliage, grass and even crop residues from the previous year were still available.

Table 3.5 *Effects of drought on the production of some main crops*
(Mean for 125 family heads in five villages)

	Unit	Former yield	Yield in 1972	Yield in 1973	Percentage of former yield 1972	1973
Early millet	bundle or basket[a]	58.5	23.4	6.9	40	12
Guinea corn	bundle or basket	47.6	13.2	2.1	28	4
Late millet	bundle or basket	14.0	4.7	0.5	34	4
Maize	100 kg sack	1.9	0.4	0	21	0
Groundnuts (shelled)	100 kg sack	9.5	3.5	0.7	37	7
Average					29	7

[a]The heads, cut in the field, are normally stored in the granary and threshed when required. A *sanfo* (basket) or bundle (H *dami*) may weigh 40 kg, equivalent to 25 kg of threshed grain. Both units are variable (cf. Smith, 1955: 237–40; Hays, 1975; Hill, 1972: 222).

Table 3.6 *The livestock account, 1972–74*
(125 families in five villages)

		Cattle	Sheep and goats	Donkeys	Horses	Camels	Fowls
1	Livestock, June–July 1972	350	837	93	24	17	701
2	Mean per family	2.8	6.7	0.7	0.2	0.15	5.6
3	Deaths	30	44	14	9	8	144
4	Sales	132	320	47	10	8	268
5	Not accounted for	14	114	−3	−3	1	30
6	Total losses	176	478	58	16	17	442
7	Livestock, June–July 1974	174	359	35	8	0	259
8	Mean per family	1.4	2.9	0.3	0.05	0	2.1
9	Percentage reduction	50	67	62	67	100	63

Table 3.7 *Meals taken on the day before interview*
(Percent; 125 families eating together in five villages)

Food	Description	Morning	Midday	Evening
Grain- or cassava-based				
Tuwo[a]	millet or guinea-corn porage	27	6	28
Gari	cassava flour, with or without milk, leaves, cakes	23	14	25
Fura or kunu	millet flour in milk or water	16	15	7
Total having some starch content		66	35	60
Vegetable foods only: leaves, pods, pulp		21	12	34
Nothing		12	52	6
		99	99	100

[a]Usually eaten with *miya*, a seasoned vegetable relish.

Food

Retrospective questions on family grain reserves indicated that those who had carried forward a reserve from before the drought consumed it during the dry season of 1972–73, and the failure of the harvest of 1973 affected almost everyone. Although the content of one's granaries is normally a confidential matter, 53 per cent admitted them to be empty before the harvest of 1972 and 64 per cent in 1973. Thus at a conservative estimate, more than half the families had virtually no reserves between December 1972 and September 1974.[21] At the time of the interviews, only 4 per cent claimed to be eating grain from their own reserves; 59 per cent were dependent on the market and 36 per cent were not eating it at all, or only very occasionally.

Normally three meals are taken each day, the principal one being in the evening. Although the noon meal may be light, it is common to eat grain at all three. Table 3.7 shows two possible responses to food scarcity: foregoing meals altogether (more than half found they could do without a midday meal); and substituting inferior alternatives for grain (*gari*, or *kunu*, or just leaves). A standard for comparison is provided by Dayi village (50 km west of Kano) where a one-day survey was carried out in the wet season of 1976 (Longhurst, 1984: 24). Among 47 families, 74 per cent ate *tuwo* in the morning and 91 per cent in the evening; 85 per cent ate *fura* at midday; only four of 141 meals were missed; only five contained no grain; and only one consisted of *gari*. The third response – simply reducing the amount provided at each meal – could not be evaluated without measurements. But some indication of the inadequacy of intake is given by the number of families reporting hungry children: 58 per cent of those eating twice a day, and 35 per cent of those eating three times.

About a third of the families interviewed were from the better-off section of the community, enjoying starch-based meals twice a day, but only a third of them ate *tuwo* on both occasions. The substratum of poor families – a quarter to one half of the sample – ate only vegetable food twice or thrice each day, or missed two out of three meals. Overall, a picture of severe hardship is conveyed, but not one of starvation (although this possibility cannot be excluded). Exceptional sickness was not reported, but many complained of stomach pains (the significance of plant medicines at this season is returned to below).[22]

The geographical distribution of scarcity, measured by three indicators (table 3.8), shows that there was a gradation of hardship from west to east (in addition to that commonly observed from north to south), reflecting the pattern of the isohyets.

Savings

An attempt was made to measure the impact of drought on savings by comparing the distribution of ownership of livestock – the commonest form of investment – early in 1972 and at the time of the survey. Livestock ownership was already unequal in terms of location and degree of specialisation. Two

Table 3.8 *Food supplies by village, from west to east*

Indicator	Village				
	1 West	2	3 ⟵⟶	4	5 East
1 Number of families reporting empty granaries by December both years[a]	7	8	14	15	20
2 Numbers of families missing one or two meals on the day before interview[a]	6	18	12	18	22
3 Number of families reporting hungry children[b]	10	7	16	16	15

[a] $n = 25$
[b] $n = 25$ minus childless couples

Table 3.9 *Number of families not owning livestock*

	Early 1972 ($n = 125$)	May–July 1974 ($n = 125$)	Percentage increase
Cattle	55	95	73
Sheep and goats	26	64	146
Fowls	40	67	67
Donkeys	76	100	32
Horses	105	119	13
Camels	114	125	10

subsamples (in the western districts of Roni and Danbatta) owned 71 per cent of the cattle and 58 per cent of the sheep and goats; these villages were of predominantly Fulani ethnicity. Their flocks and herds also tended to be larger. On the other hand, only one in five families had no small livestock in 1972, and one in two had no cattle. The quartile of largest owners had half of the cattle, sheep and goats.

The losses made no difference to the proportion of livestock belonging to the largest owners, but they eliminated many families from ownership of animals (table 3.9): an increase in non-owners of small livestock of 146 per cent being particularly relevant. Insofar as livestock may be regarded as a repository of savings, drought brought about a significant concentration in the social distribution of rural capital.

Some further illumination is provided by table 3.10. A comparison between the two predominantly Fulani villages and the remaining three shows that

Table 3.10 *Losses of cattle, sheep and goats*

		Fulani villages		Other villages	
		Cattle	Sheep and goats	Cattle	Sheep and goats
1	Number of livestock in 1972	250	484	100	355
2	Number of livestock in 1974	168	290	6	47
3	Mean size of herd or flock in 1972	6.9	9.9	2.9	7.4
4	Mean size of herd or flock in 1974	6.5	7.1	1.5	2.6
5	Reduction in the number of stock (%)	33	40	94	87
6	Reduction in the number of owners (%)	39	14	55	42

livestock ownership took a greater beating in the non-Fulani villages. This does not reflect any difference in the degree of commitment to agriculture but clearly does support the view that sedentary Fulani have a stronger commitment to livestock.[23] It also suggests that for sedentary as well as nomadic owners, to maximise the size of the herd or flock is the best insurance against loss. This is conventionally explained in terms of unequal access to common pasture, so scarce in the neighbourhood of some of these villages that cattle were commonly contracted out to transhumant herders for most of the year. What is particularly relevant, however, to the circumstances of small sedentary owners, is the superior position of the owner of a breeding herd. This is the only alternative to the market for those wanting to recover their livestock interest after a time of loss. Finally, the table shows that the animals lost were not transferred to richer members of the community, but died or were sold outside the locality. Thus savings were further concentrated, but not redistributed, in the community: the better off survived with more than they had before – relatively, but not absolutely; however, their potential for recovery was stronger.

RESPONSE

Adapting to drought

The shortness of the growing season (from three to five months) severely limits the possibilities for adapting farming practice during the lifetime of a one-year drought. The crop mixture may be altered in successive replantings; manure inputs may be reduced; an attempt may be made after the harvest to find land for dry-season irrigated cultivation. Since this drought lasted for two years, farming practice in the second year (1973) and in the ensuing normal year (1974, when the rains appeared to be failing again in May and June) may indicate the adaptive strategies available. The informants were asked how their practice differed from usual (less, the same, or more) in relation to the main operations of planting, manuring and weeding, and to the selection of crops.[24] Their answers presented an ambivalent, even contradictory pattern, owing to the intervention of a multitude of unchecked variables, such as sickness, seed stocks, labour supply and personal preferences. The same questions, however, were directed to village heads in Roni and Danbatta districts, and from the results, the approximate dimensions of a consensus can be gleaned.

A rational response to reduced yields would be to extend the area planted and increase plant spacing. A trend in this direction was reported by the village heads; but the shortage of seed, exacerbated by several replantings, obscured any such trend in the five villages. Even in normal times, poor farmers buy, borrow or are given seed, but at such a time as this, dependence on such sources increased dramatically.[25] Capital for land clearance (to pay extra labour) was exceptionally scarce.

A second area for response was to reduce fertiliser inputs. The village heads reported a trend towards using less manure in 1973, and this trend strengthened in 1974. In the five villages, the livestock population had been decimated (including the donkeys needed for transporting manure from the village animal pens to the fields). The application of manure under dry conditions raises soil temperatures; the effects of chemical fertiliser are believed to be even worse. On the other hand, according to the village heads, time spent on weeding increased: the fight against weeds is more critical when the crop is weak.

A third response is to alter crop mixtures. Among the five major crops, early millet strengthened its dominance; a noticeable shift took place away from guinea corn, and late millet declined even more. The strongest shifts, however, were away from groundnuts and cowpeas. These trends can be traced to two causes. Changes of emphasis among the grain crops were within the farmers' normal range of options, and clearly influenced by the crops' differential tolerance of drought. But the decline of groundnuts and cowpeas was equally due to a shortage of seed, which is normally bought, rather than reserved, for these commercialised crops.

In a farming system one of whose major objectives is to withstand drought (including the possibilities of moisture stress at different times in the growing cycle, and of bird, grasshopper and other pest attack), the scope for spontaneous adaptation is limited. Changing the proportions of the major grains was open to everybody, but to sacrifice diversity might increase the risk of loss. To extend the planted area required capital, at high risk. To reduce manure inputs was a sound response to dry conditions although it augured ill for soil nutrient status in the longer term. So did the decline of the legumes. The limited changes observed were responses to the inevitable, rather than experimental. The system was resilient, but it failed in 1972 and in 1973. Nevertheless, it bounced back, battered but still buoyant, in 1974.

Adapting to poverty

Finding alternative opportunities

A well-established strategy for augmenting income in Hausaland, other parts of northern Nigeria and adjacent Niger, is *k'wadago*, or working for wages on a daily basis on the farms of wealthier people, or on other activities such as handling earth for builders (Hill, 1972: 105–23). Times of general hardship tend to broaden recruitment, but also restrict the availability of wage money. Much *k'wadago* involves movement out of the village, therefore. In the five villages, 68 per cent of the respondents personally resorted to it in both years of drought.

Some 33 per cent cut and sold firewood in both years. This is obtained from trees in the bush and from lopping farm trees; it is sold in the village, in the nearest market, along intercity highways, or to dealers from urban areas.[26] Transport (usually by donkey or camel) is a prerequisite if any distance is involved; this restricts participation, since accessible woodland is already heavily cut over. Rights to cut may be conserved by the village head, and are subject to local authority by-laws: but at such a time as this, restrictions were more honoured in the breach, and the ensuing free-for-all further debilitated an already impoverished woodland.

Another product freely available in the bush is the foliage of the dum palm (*Hyphaene thebaica*), which regenerates spontaneously in the form of slowly growing shoots from underground rhizomes, easily reached. This material (H *kaba*) is used for making mats, ropes and baskets, tasks which are time-consuming but easy to learn. The small number of regular *kaba* workers is augmented in times of economic hardship, and in the five villages, 55 per cent of the families were resorting to it. Undertaken both in the village and on migration, this work meets an urban demand which showed remarkable elasticity in 1973–74, when traders would visit rural periodic markets to assemble truckloads for urban markets where the effects of the oil boom were stronger than those of the drought. By working virtually from dawn to dusk, a man might earn enough to buy grain for one family meal.

Further alternative sources of income were the full range of normal secondary occupations, such as trade, barbing, weaving and donkey hire. The demand for them was seriously depressed, and they remained confined to specialists.[27]

Liquidating assets

Losses of livestock by sales and by death have already been described, and followed a pattern observed before in time of food scarcity (Grove, 1952: 20–1).

As a strategy for economic survival, selling animals diminishes in value as the distress widens socially; it is of little use if everyone does it at once and the prices collapse. The loss of assets this way in 1972–73 was unquantifiable and disastrous.

The sale of items of personal property, such as clothing and other goods, was reported to be taking place in nearly all of the 631 villages, but among the family heads was admitted by only 20 per cent (in one of the years) and by 9 per cent (in both years). Poverty and remoteness, however, affected both the supply and demand sides of the pawnbroker economy.

Compulsion to sell productive resources – manure, transport animals and land – had more serious implications for impoverishment in the longer term. Manure is considered a productive resource, rather than a marketable by-product, because all farms, in normal times, need more than they get. Actually, both the demand for and the supply of manure were reduced; nevertheless, 19 per cent of respondents sold it in both years. Fewer than 10 per cent admitted selling land. These villages, again because of their poverty and remoteness, lacked buyers: and free land was still available in some. But the village heads reported land sales (which included pledging) in no less than 80 per cent of their villages. Land and other sales were significantly correlated with out-migration and indebtedness.[28] No relationship, however, was found with proximity to markets, whose presence might have been expected to facilitate such transfers if outsiders were involved. The resident better-off, however few, appear to have still possessed the resources needed (cf. Van Apeldoorn, 1981: 61).

Mobilising social networks

Society incorporates a number of systems whereby the redistribution of wealth takes place from the relatively well-off (H *masu dan hali*) to the common people (H *talakawa*). Such redistribution might take place via the kinship and clientage systems, or on the basis of Islamic injunctions to assit the needy, or as alms given to koranic scholars (H *almajirai*) of all ages. In precolonial Hausaland, it has been argued, such redistributive functions, presided over by the urban titled class (*masu sarauta*), comprised a 'moral economy' that has subsequently been seriously eroded by the impact of the market (Watts, 1983a: 104ff.); a part of an idealised system of reciprocity and redistribution that was irrevocably altered by the monetisation of rural production (Raynaut, 1977). Ninety-five per cent of the village heads said that the *masu dan hali* (in which class they probably included themselves) had insufficient food for their own families and were not able to assist the *talakawa* with gifts of food or money. In this respect the times compared unfavourably with the past. Nevertheless, 27 per cent of the family heads in the five villages had received gifts of food, and 10 per cent gifts of money, in both years; and a further 12 and 9 per cent in one year only.

The related strategy of personal borrowing was used by 29 per cent in both years, and by a further 19 per cent in one year. Little is known about indebtedness owing to the secrecy surrounding the forbidden practice of usury; survey reports from Hausaland are hard to reconcile (Williams, 1981). About half the village heads reported that it was more common than usual; but while the need for loans must have increased hugely, the scarcity of cash and lack of security acted as constraints.

Begging in public (H *bara*) was previously unknown in the five villages, except for the socially approved practice of almsgiving to the *almajirai*; it is quite striking that one in ten family heads admitted to it during 1973 and 1974. However, the assistance was obtained in nearby villages or small towns rather than in the major cities (where increased numbers and the extended provenance of beggars were noticeable).

Not surprisingly, associations between these strategies and related variables tended to be weak. Of those who admitted to begging, a majority reported their granaries empty by December in both years, did *k'wadago*, received gifts of food, and said their children were hungry. Those who borrowed money were slightly better off; they too did *k'wadago*; they had property to sell, they were as likely as not to sell firewood or *taki*(manure), they might receive gifts of food, but rarely of money; and they did not beg. A third group needed neither loans nor *k'wadago*. Thus were the tendencies to inequality consolidated in time of stress.

Adapting to hunger

The collapse of subsistence production is shown vividly in table 3.11. For nearly everybody, the household store was empty. Those who ate, bought their food

Table 3.11 *Carbohydrate and animal protein foods in use, May–July 1974*
(Percentage of 125 families in five villages)

Food		Description	Users	Obtained from	
				Market	Store
1	*Tuwo*[a]	grain porage	63	59	4
2	*Gari*	cassava	61	61	0
3	*Fura*	millet flour in milk or water	10	10	0
4	*Kulikuli*	groundnut cake	4	4	0
5	*Nono*	soured milk	2	2	0
6	*Madara*	fresh milk			
7	*Nama*	meat, any type	0	0	0

[a] Usually eaten with *miya*, a seasoned vegetable relish.

Table 3.12 *Plant foods used in famine, northern Kano State, 1972-74*

No. Scientific name[a]	Hausa[b]	Kanuri[c]	Parts used[d]	Remarks[e] Families (%) (n = 125)	Villages (%) (n = 631)	Other authorities
A Most common species[f]						
1 *Leptadenia hastata* (Pers.) Dec'ne	ya'diya	njera	leaves	60	64	
2 *Ficus* spp., including:	ce'diya		fruit, some leaves, aerial roots (E)	35	53	(E) (P)
F. abutilifolia (Miq.) Miq.	yandi					
F. capensis Thunb.	uwar yara					
F. platyphylla Del.	gamji	ngabara				
3 *Cassia obtusifolia* (Linn.) Irwin and Barneby (Syn. *C. tora*)	tafasa	tafasa	seedlings	44	19	(U)
4 *Sclerocarya birrea* (A. Rich.) Hochst.	daniya, danya	kuma	fruit, leaves (?), kernels	35	20	(P)
5 '*Loranthus*' spp. (parasite)	kauci	burangu, borongu (D)	leaves	30	41	(B) under *Tapinanthus globiferus*
6 *Boscia senegalensis* (Pers.) Lam. ex Poir.	hanza	tabila	berries	14	—	(B) (P)
7 *Urena lobata* Linn. (wild type)	yakuwa	karasu	leaves, calyces, flowers	10	—	
8 *Moringa oleifera* Lam.	zogale	alinga	leaves, roots, young pods, seed oil (E)	9	6	(E) (P)
9 *Parkia biglobosa* (Jacq.) Benth. (Syn. *P. clappertoniana*, *P. filicoidea*)	'dorawa	runo (D)	pod pulp, seeds, flowers (E)	7	—	(U) (E) (P)
10 *Adansonia digitata* Linn.	kuka	kuka	leaves, seeds, pod pulp, green fruit, young root, bark (E)	no data		(E) (P)

B Other Identified species[g]

11	*Acacia albida* Del.	*gawo*	?	pods and leaves fed to animals;★	(P)
12	*Acacia ataxacantha* DC	*kwandariya, kwandari*	?	identification uncertain (D)	
13	*Ampelocissus africanus* (Lour.) Merr. (Syn. *A. Grantii*)	*rogon daji* (D), *inabi* (K)	fruit	root has medicinal uses (D); *rogo* is cultivated cassava[h]	
14	*Balanites aegyptiaca* (Linn.) Del.	*aduwa*	fruit, seed oil, leaves, flowers and resin (E)		(E) (P)
15	*Boscia salicifolia* Oliv.	*zure*	berries, leaves (D)	may be confused;[i] but★	
16	*Capparis sepiaria* Linn., var. *fischeri* (Pax) De Wolf; *C. tomentosa* Lam.	*janibaibai* (D), *k'abdodo, haujari* (K)	fruit, leaves (D)	poisonous?[j] may be confused	(P)
17	*Cassia sieberiana* DC.	*marga*	*kiskatigrai*		
18	*Ceiba pentandra* (Linn.) Gaertn.	*rimi*	*kuci* (K)	?	sweet extract from stems	(P)
19	*Celtis integrifolia* Lam.	*zuwo*	*nguzo* (D)	leaves (?) seeds (D)	★	(P)
20	*Cleome gynandra* Linn. (Syn. *Gynandropsis gynandra*)	*gasaya*	*knasi*	leaves, fruit	eaten by Hausa but not normally by Manga in Dagaceri	(U)
21	*Combretum collinum* Fres., ssp. *geitonophyllum* (Diels) Okafor (Syn. *C. verticillatum*)	*tarauniya, taramniya* (K)	*katakkara* (D), *katankara*	seedlings, leaves	★	
22	*Commiphora africana* (A. Rich.) Engl.	*dashi*	*kabi*	?	★	
23	*Crateva religiosa* Forst. f. (Syn. *C. Adansonii*)	*ingidido, ungudidi* (K), *gudai*		leaves, fruit (D)	not known in Dagaceri	(U)
24	*Desmodium velutinum* (Willd.) DC. (Syn. *D. lasiocarpum*)	*'danka'dafi, damgere* (K)		leaves (D)	horse fodder (D)★; identification uncertain[k]	
25	*Detarium senegalense* J.F. Gmel.	*taura*	*gatabo* (D)	fruit, kernels (D)		(U) (P)
26	*Euphorbia balsamifera* Ait.	*aliyara*		shoots (D), leaves	★	

(Contd.)

Table 3.12 (Contd.)

No.	Scientific name[a]	Hausa[b]	Kanuri[c]	Parts used[d]	Remarks[e] Families (%) (n = 125)	Villages (%) (n = 631)	Other authorities
27	*Ficus ingens* (Miq.) Miq., var. *ingens* var. *tomentosa* Hutch. (Syn. *F. kawuri*)	*shirinya*	*kazu, ja-ja* (D)	fruit	*		(P)
	Ficus polita Vahl.	*k'awari, k'awuri* (K)		?	*		(P)
28		*durumi*	*rita* (D)	fruit; young leaves (E)	*		(E) (P)
29	*Guiera senegalensis* J.F. Gmel.	*sabara*	*kaseshi* (D)	leaves?	*		(E)
30	*Hippocratea africana* (Willd.) Loes.	*gwa'dayi*					
31	*Hyphaene thebaica* (Linn.) Mart.	*goriba*	*karjim*	fruit rind	used extensively in arid zone		(B) (P)
32	*Lannea acida* A. Rich.	*faru*		young leaves, fruit, gum (E)			(U) (E)
33	*Maerua angolensis* DC.	*ciciwa*		leaves (D)			
34	*Maerua crassifolia* Forsk.	*jiga*		leaves (D)			(B)
35	*Mangifera indica* Linn.	*mangwaro*		fruit; gum (E)			(E) (P)
36	*Mitragyna inermis* (Willd.) O. Ktze	*giyayya*	*kawi*	?			
37	*Nymphaea lotus* Linn.	*bado*	*dambi* (D)	rhizome, seed (D)	water lily, rather rare in N Kano		
38	*Piliostigma thoninngii* (Schum.) Milne-Redh. (Syn. *Bauhinia reticulata*)	*kargo*	*kaghril, kalur* (D)	leaves, pods, sometimes stems, bark (E)			(E) (P)
39	*Physalis angulata* Linn.	*matsarmama, faraduwa, -rus* (K)		fruit (D)			
40	*Sesbania sesban* (Linn.) Merr.	*alambu, zamarke* (K), *checheko*		flower, fruit (D)	*zamarke*: arrow, shafts[1]		

41	*Senna occidentalis* (Linn.) Link	*rairai, rai dore, majamfari*		young leaves, unripe pods, sometimes roots, seeds (E)	* (E)
42	*Tamarindus indica* Linn.	*tsamiya*	*tamzu, tamsugu* (D)	pods; leaves, fruit, seeds, flowers (E)	(E) (P)
43	*Terminalia macroptera* Guill. and Perr.	*kandare*		?	leaves have medicinal uses (D)
44	*Vetivaria Nigritana* (Benth.) Stapf.	*jema*		roots (D)	roots used to purify water; plant also used [m]
45	*Vitex doniana* Sweet	*'dinya*	*ngaribi*	leaves, fruit (E)	(U) (E) (P)
46	*Ximenia americana* Linn.	*tsada*		fruit, leaves, seed oil (E)	(U) (E)
47	*Ziziphus spina-christi* (Linn.) Desf.	*kurna*		fruit	not known in Dagaceri (E) (P)

C No record of use

48	*Amorphophallus dracontiodes* (Eng.) N.E.Br.	*gwazar giwa*		tuber	root or rhizome is *kinciyar* (Hausa); poisonous; requires two days' boiling (U)
49	*Boerhaavia repens* Linn.	*babba juji*		seed, leaves (D)	(B)
50	*Brachiaria lata* (Schumach.)	*gariji*		seed	(B)
51	*Butyrospermum paradoxum* ssp. *parkii* (G.Don) Hepper	*k'adanya*		fruit, flowers, seed oil (E)	rare in N Kano (U) (E) (P)
52	*Cenchrus biflorus* Roxb.	*k'arangiya*	*ngibbi* (D)	seed	[n]
53	*Colocynthus vulgaris* Schrad. (Syn. *Cirullus colocynthus* (Linn.) Schrad., the bitter melon)	*Kwartowa*		fruit	bitter gourd, requires long boiling (D) (B)
54	*Dioscorea dumetorum* (Kunth) Pax	*rogon biri*		tuber (D)	poisonous, requires prolonged washing (D); famine staple in the Sudan (Corkill, 1949)[h] (E) (P)

(*Contd.*)

Table 3.12 (Contd.)

No.	Scientific name[a]	Hausa[b]	Kanuri[c]	Parts used[d]	Remarks[e] Families (%) (n = 125)	Villages (%) (n = 631)	Other authorities
55	*Dioscorea praehensilis* Benth.	*magoraza*		tuber (D)		a plant called *bazara* was in use in 1974	
56	*Echinochloa colona* (Linn.) Link	(several)		seed		(D) has no record of a Hausa name	(B)
57	*Eragrostis* spp.			seed			(B)
58	*Gardenia erubescens* Stapf. and Hutch.	*gau'de*	*kingerr* (D)	fruit (D)	*		(U)
59	*Oryza barthii* A. Chev.	*shinkafar gyado*		seed			(B)
60	*Panicum laetum* Kunth.	*baya*		seed	o		(B)
61	*Panicum turgidum* Forsk.			seed		nor reported in N Nigeria (D)	(B)
62	*Raphionacme brownii* Sc. Elliott	*rujiya*	*katakirri* (D)	tuber (D)			(U)
63	*Saba florida* (Benth.) Bullock	*ciwo*		fruit (D)			(U)
64	*Solenostemon rotundifolius* (Poir) J.K. Morton	*tumuku, tamaka*		tuber (D)	p		(U)
65	*Stylochiton lancifolius* Kotschy and Peyr	*gwandai*	*ngurra* (D)	leaves, rhizome (D)	poisonous; requires repeated boiling of young leaves or washing of rhizome; root or rhizome is *kinciyar* (Hausa)		(U)
66	*Trianthema portulacastrum* Linn.	*rogon yata*		leaves (D)	h		
67	*Trianthema pentandra* Linn.	*gadon maciji*		?	poisonous[q]		
68	*Ziziphus mauritiana* Lam.	*magarya*		fruit	common further north		(B)

[a] After Dalziel (1937), Hutchinson and Dalziel (1954–68), Keay, Onochie and Stanfield (1964), Stanfield (1970). The help of H.M. Burkill (Royal Botanic Gardens, Kew) and Brain Harris (Department of Biological Sciences, Ahmadu Bello University) is gratefully acknowledged.
[b] After Bargery (1934), Dalziel (1937), Keay, Onochie and Stanfield (1964). The advice of Mark Duffill is gratefully acknowledged. Where alternatives are listed, the name cited by our informants in the Kano survey of 1974 is indicated by (K).
[c] Where not attributed to Dalziel (1937) the Manga variant is given as used in Dagaceri. These names have not been checked with a standard orthography.
[d] Where information was not obtained from field sources, the authority is given. Much additional information on uses is provided by Dalziel (1937), now under revision (Burkill, 1985).

[a] Figures are percentage frequencies of mention by informants who were asked what they (their families) or people in general (in their villages) were eating. Plants not considered edible by Manga informants in Dagaceri are shown with an asterisk.

[b] Some plants in list A were identified in the Herbarium of the Department of Biological Sciences, Ahmadu Bello University, Zaria.

[g] Other famine foods mentioned occasionally included: *dangwara maza* (Bargery, 1934: 'a common edible weed'); *dusa* (bran, normally fed to animals); *gawari* (?*gawayi*, charcoal); *jagindi* (groundnut gleanings); *katuri* (Bargery, 1934: *ka turara*, a 'dumpling made only in times of scarcity'); *tafasa ruwa* (a small unidentified plant which grows in wet places); *tururuwa* (Bargery, 1934: a variety of black ants which collect corn; a reference to collecting grain from ant silos, also reported in 1913–14 by an eye witness interviewed near Danbatta in June 1974); *zangare* (stripped head of guinea corn); *zogalen zomo* (probably *daniya*, *Sclerocarya birrea*).

[h] *Rogo* normally means cultivated cassava *Manihot utilissima*. In the nineteenth century, this was an inferior food widely used by the poor as a substitute for grain, especially in times of scarcity (Duffill, 1986). In the twentieth century, its cultivation became widespread and it was often sold. Cassava flour (*kwaki*) was imported from further south, especially in times of shortage (see table 3.13). There are several unrelated wild plants whose rhizomes are more or less edible, known as 'bush cassava' (food no. 13), 'monkeys' cassava' (no. 54) and 'boys' cassava' (no. 66). In the famine of 1972–74 *rogo* was frequently nominated as an alternative food, but it has to be assumed that these references were to cultivated cassava, unless identified by an alternative name (as food no. 13).

[i] This and other *Boscia* spp. (food no. 6) may be confused.

[j] Some authorities say fruit and leaves are poisonous; some edible.

[k] A specimen of *damgere* (a fodder plant) was identified as *Zornia latifolia* Sm. by the Herbarium of Ahmadu Bello University. '*Danka'dafi* may describe any plant with sticky burrs.

[l] The same name may be used for other plants. Dalziel (1937) gives *checheko* for *Sesbania pachycarpa* DC, but specimens of *tchachako* (*shashako*), a fodder plant, were identified as *Sesbania sesban* by the Herbarium of Ahmadu Bello University.

[m] Plant used to mark field boundaries near Katsina (Hill, 1972: 235) and for thatching (Bargery, 1934).

[n] *Cenchrus biflorus* seeds were widely used in the Sahel. *Karengia* (Arabic *khaskanit*) was eaten from Borno to Timbuktu in the time of Barth and Nachtigal (Lewicki, 1974: 44), and Hastings (1925: 111) recorded its use in the famine of 1913–14; this report was confirmed by an eye witness interviewed near Danbatta in June 1974. Not recorded in Bargery (1934).

[o] The name *fonio* (Fulfulde for cultivated *Digitaria exilis*) may be extended to this plant (EMASAR II, 1977: 14).

[p] *Solenostemon rotundifolius* (*Coleus dysentericus* in Dalziel, 1937) is cultivated in some areas, but the informants are believed to have been referring to wild *tumukun biri*.

[q] *Gadon maciji* (snakes' bed) was reported in use in 1913–14 by an eye witness interviewed near Danbatta in June 1974. The name may be applied to other plants of rubbish heaps (such as *Boerhaavia repens*; Dalziel, 1937).

*Not considered edible by Manga informants in Dagaceri.

Sources:
(B) Listed as used in the Sahel (Bernus, 1980b)
(D) Dalziel (1937)
(E) Used as food (additional to medicinal) in Wudil district of Kano (Etkin and Ross, 1982, 1983)
(K) Name cited by informants in Kano survey of 1974.
(P) Listed as a common tree on West African farmed parklands (Pullan, 1974)
(U) Nineteenth-century famine foods mentioned by Al-Hajj 'Umar (Mischlich, 1943; Duffill, 1986)

in the market, and the grain substitute, cassava, was as common as the staple *tuwo*. As table 3.7 suggests, a significant proportion could not afford to eat either, and were dependent on vegetable foods. Animal protein had almost vanished.

At such a time, resort is made to certain edible plants (table 3.12), whose strategic significance in rural food supply in West Africa is not limited to times of drought or to lowland Muslim cultures (cf. Seignobos, 1979). It is the work of women and children to collect most of these foods. The list of species used may give some indication of the severity of the food shortage, for it appears that some famine foods, used in the past, did not have to be used this time. There is no record, for example, of the use of the tuber of the poisonous *Amorphophallous dracontiodes*, which can be rendered edible only by repeated boiling. But the digging out of termitaria for the tiny stores of collected grain found therein, so vividly described by Hastings (1925: 111) in the famine of 1913–14, occurred in Kano and was photographed in Borno in 1973 (*New Nigerian*, 19 December 1973).[29] In table 3.12, the plant foods cited frequently are given in list A; in list B, additional plants nominated by the village or family heads which have been identified; and in list C, a number of foods from other sources, whose use was not recorded in 1972–74. The table is certainly incomplete. List A under-records the true frequencies of many plant foods. For example, *Cassia obtusifolia* seedlings spring up in large numbers in the early rains, and are easily gathered; with *Leptadenia lancifolia*, it was almost universally used in its season. Most of these foods consist of foliage (H *ganye*) which is boiled and mixed, if possible, with a little grain or groundnut cake, to make it go further.

As Duffill (1986) emphasises, the majority of the leaves, roots and fruits that feature in such a list are used extensively in better times to supplement the staples, and not only by the poor. But they are brought into different and more intensive use in time of shortage. However, the significance of this variety of edible plants goes further. Along with many other plants having no known food uses, they form part of a rich herbal pharmacopoeia (Dalziel, 1937; Etkin, 1981; Etkin and Ross, 1982, 1983). Indeed, a clear distinction between food and medicine cannot be made. The time of flowering and leafing of many trees and shrubs is also significant, anticipating or coinciding with the early rains, at a season when the greatest hunger is experienced, the greatest physical demands are made on farm labour, and there is the greatest exposure to diseases (such as malaria) that are related to conditions of high humidity.

Thus, when subsistence production fails, dependence is transferred first to the market and then, as cash resources are depleted and the season advances, on to the farm trees and the bush. The importance of trees in sustaining animals as well as humans in times of drought may be overlooked when the weather is normal. Such trees should never be destroyed, either by woodcutting or to make way for mechanised farming, if this source of emergency food is to be conserved.

EVALUATION

By now it is clear that a disaster, whose immediate cause was meteorological, generated patterns of impact and response ranging from ecological to social and economic. These patterns elucidate the relationship between famine and drought – and point to some of the other factors involved. In the folklore of drought-related disasters in Hausaland, the meteorological drought (H *fari*) is remembered less than the ensuing famine or hunger (H *yunwa*). This is clear from the names given to such events (table 3.13): names which often incorporate a certain wry humour, and which may vary considerably from place to place.[30] As a chronological guide, however, the memories of the family heads proved to be unreliable, less so with regard to the major famines that are remembered well (K'ak 'alaba, Mai Buhu and 'Yar Gusau) than to minor, even quite recent events.

The method used for dating the food shortages after 1942 was to plot the frequency distribution of each one recalled, in years before the present survey (1974), accepting the mid-point as its approximate time of occurrence. Less common names were eliminated. The dates of the earlier famines are well known. The later food shortages cannot be checked from the meteorological records, because they were sometimes limited in geographical extent, and may not have been due to drought. The pattern is a chart of the collective recollection of hardship, rather than a proper chronology. The impression of an increasing frequency of food shortages is probably due to improving accuracy of recall, and more respondents in the younger age groups. Nevertheless the occurrence of at least two shortages in the fifties and sixties which were sufficiently widespread to affect five villages distributed over 200 km is a necessary corrective to the idea of advancing agricultural prosperity during these decades (chapter 2). A further interesting characteristic is the geographical specificity of many of the names, though not necessarily the occurrence, of the lesser shortages. By contrast, the major famines are known widely by the same name, although at the time of the survey there was still not a complete consensus about the name of the contemporary famine – Kakaduma; nor on the date of its commencement.

Both the village and the family heads were asked to compare the present famine with its predecessors, although it was understood that while most could remember 'Yar Gusau, information about Mai buhu and K'ak'alaba was often secondhand. Forty-five per cent of the village heads, but no less than 93 per cent of the family heads, judged the present famine to be worse than all three; the greater pessimism of the family heads was perhaps to be expected in view of their northerly distribution. With the exception of five districts, the spatial pattern of pessimism (figure 3.6) agreed with that of reported crop failures (figure 3.4). Over 60 per cent of the village heads who reported less than 10 per cent of normal expected yields of all five major crops (guinea corn, early and late millet, groundnuts and cowpeas) gave pessimistic evaluations. Fewer than

Table 3.13 *Some names given to famines in northern Kano*

	Villages arranged from west to east				
Year (s)	1	2	3	4	5
1914	K'ak'alaba[a]	K'ak'alaba Sude mu gaisa[b]	K'ak'alaba	K'ak'alaba	K'ak'alaba
1927	Mai buhu[c]	Mai buhu	Mai buhu	Mai buhu	Mai buhu
1942	'Yar Gusau[d]	'Yar Gusau	'Yar Gusau Macuwari	Macuwari	Macuwari
1947–49			Kwajaja[e]	Mai zarara[f]	
1954–57	'Yar tiya[g] 'Yan Dikko[g]	Mai dankwano[h] Kwajaja	Mai zarara Mai amaro		Kwaki babba[i]
1964–65	Kwajaja Uwar sani[j]				Kwaki karama[i]
1967–68	Gudogu	Macuwari	Mai budu[k]	Mai budu Dan Kyari[l]	
1971	Mai agogo Jalolo, falolo[m] Macuwari				Kakaduma
1972–74	Kakaduma[n]	Kakaduma	Kakaduma Kakaduba[o]	Kakaduma Kakaduba Kaguduma[p] Albazu, A bazama[q]	Kakaduma Kakaduba

Accepted translations of these terms are difficult to achieve. They are based on words which may have more than one meaning, or whose rendering has been altered by popular repetition, or whose meanings are geographically localised or historically obsolete. Most are hausa words but some may be Kanuri. The suggestions given in the notes below are based on Bargery (1934); on notes kindly provided by Yusufu Kankiya of the BBC (who has investigated famine names in the Katsina area) and Graham Furniss of the School of Oriental and African Studies, University of London; Hill (1972: 231); Baier (1980: 98); and informants in the field.

[a] Baier suggests *K'ak'ala'ba*: 'how (does one) hide (from famine)? Or, 'how to walk stealthily?' as for a nefarious purpose (Bargery), avoiding other people. Kankiya and Furniss suggest a construction based on an alternative meaning of *k'ak'a*, 'in great quantity': the 'great constriction', possibly onomatopoeic. Other names used for this famine include *Gyallare* (meaning not known); *Doguwar yunwa* ('the long hunger'); and the descriptive epithet *ka ba sa a ba ka domin dawa*, or 'you give a bull and you are recompensed with a bundle of guinea corn'. Hill records the name *Malali* near katsina.

[b] 'Eat everything before greeting' (*sude*: completely finish off remains of food in a vessel by wiping round with a finger). Rules of hospitality require that a visitor should be offered food, but the image conveys the preciousness of every crumb.

[c] Or *Mai dan buhu*, 'the little sack famine': a reference to the use of small sacks used in relief work at the time. Also known near Katsina as *Kwana* or *Kona* (Hill).

[d] 'The one of Gusau'. It has been suggested that there was a migration towards the area of Gusau which was less affected by famine.

[e] According to Kankiya and Furniss, *kwajaja* means 'dried up' or 'desiccated'; according to Baier, *gwajaja* is a word conveying helplessness.

[f] 'A person dressed in rags'; an alternative meaning offered by Kankiya's informants is 'pulled out' or 'stretched'.

[g] Or *Tiya tiya*, which is a small grain measure in the Katsina dialect, suggesting scarcity. Furniss suggests 'thimbleful time'. *'Yan Dikko* refers the famine to the time of Sarki Dikko.

[h] *Kwano*: 'headpan', or 'money' (rare); a reference to searching for manual work in the towns or the tin mines.

[i] 'Great' and 'little' cassava flour (*kwaki*), a reference to the use of this grain substitute, imported from the south. Also *Mai garin kwaki*.

[j] According to Kankiya, also used to denote the famines of 1927, 1942 and 1949; according to Hill, that of 1954.

[k] *Budu*: a bundle of thatching grass or fodder. The name may refer to its scarcity at the time or to its abundance relative to food.

[l] Possibly 'the famine of Kyari's time'.

[m] *Zololo*: 'like a ghost or skeleton' according to Kankiya and Furniss. The same name is used for the 1967 famine in the Katsina area.

[n] *Duma*: a gourd, either the small bottle gourd or the larger kind used for drumming. Kankiya takes this series of words to be based on *k'ak'a*, 'great quantity', as in *K'ak'alaba* (1914), and not *kaka*, 'harvest'. Alternative names for the 1972–74 famine in the Katsina area are *Yar balange*: 'a time of extreme scarcity when others cannot help' (also used for the 1942 famine near Katsina: Hill); *Takojilo*: 'skinny'; and *Jagindi*: 'dragging along 'as a beggar on the streets.

[o] *Duba*: 'look for'.

[p] Possibly based on *gudu*: 'flee' or 'run away'.

[q] *Baza*: 'spread out' or 'vanish'; *bazama*: 'bolt' or 'run away'.

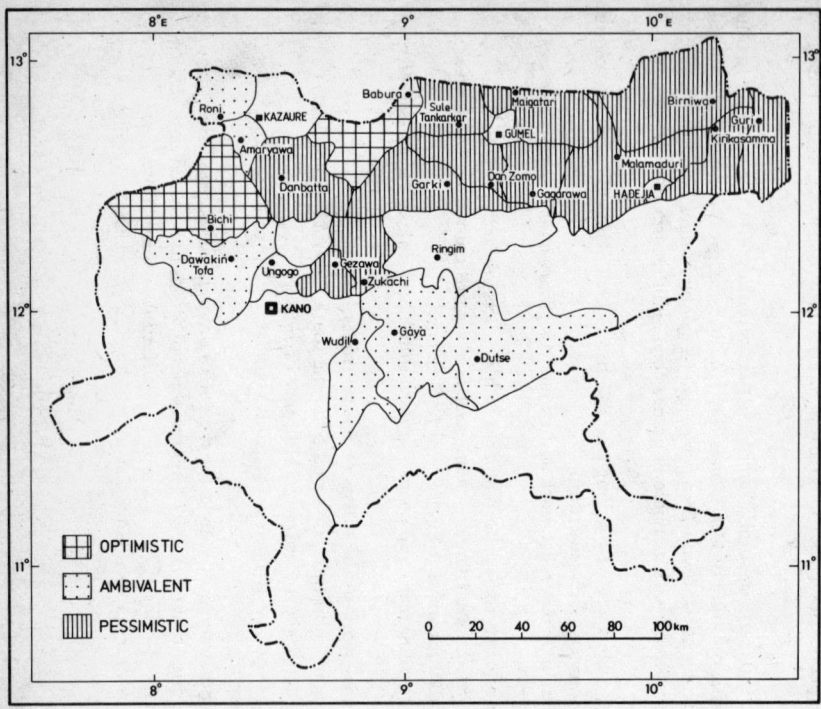

Figure 3.6 Evaluations of the 1972–74 famine, Kakaduma

30 per cent thought it less severe than its predecessors.[31] Table 3.14 shows the reasons given in support of these evaluations. Their answers, much simplified, show that the perspective of the village heads, who had a broader view of the economic factors and were probably better informed about the earlier famines, differed from that of family heads in the five villages, whose outlook reflected the priority of subsistence. Few were content with statements about the rainfall. Cause and effect were interwoven in evaluative judgements. But the importance attached by village heads to improvements in transport and food supply since the earlier famines throws proper emphasis on the operation of the infrastructure and market system. In 1914, victims of the drought set off on foot southwards, and many died on the road.

Among the pessimistic village heads, some drew attention to the change that had occurred in the man–land ratio, implying that a larger population had adversely affected food production per capita:

There were less farmers in the past than now. In the past, a measure of millet cost only $1\frac{1}{2}$–2 pence, but now up to 4s 6d. The population of the past was not as much as now, so food was more,

Table 3.14 *Reasons given for evaluations of the famine (per cent)*

	Village heads (n = 631)	Family heads (n = 125)
Pessimistic evaluations (worse than earlier famines)		
1 Inflation (high prices, food or money scarcity)	55	50
2 Rainfall deficiency	9	12
3 Low yields (affecting all crops: for two or more years)	7	24
4 Out-migration ('running away')	2	
5 Animal mortality	2	
6 Resort to eating leaves	1	7
7 Wide geographical extent	1	
8 Others	6	
Optimistic evaluations (better than earlier famines)		
9 Transport improvements	17	—
10 More food, more money	14	—
11 Government assistance	4	7
12 Lack of human mortality	2	—
13 Others	1	—

Note: Some informants gave more than one answer.

or to related increase in population movement:

Harvest was not as bad [in the earlier droughts] as now. People were not running away from their villages into towns, but settling in their villages. Now people are running away from their place to another place,

or to the breakdown of the 'moral economy'.

In the time of K'ak'alaba there was food. If you had no food, a friend or relative could give you food. But in this case you do things on your own'

and one or two even referred to the advancing Sahara hypothesis, which had been publicised in the media.[32]

Government relief was not a major factor influencing judgements. The government of Kano State was carrying out a major distribution from April 1974 onwards in certain areas. About 46 per cent of the families had received small amounts – less than 50 kg – in early distributions in 1973–74. From April 1974, a distribution of grain intended for seed took place in four arbitrarily defined areas: the emirates of Hadejia, Gumel and Kazaure and the district of Babura, at rates of about 17 tax payers (assumed to represent 102 persons) per bag in Hadejia to 1½ bags per family in Kazaure, usually on production of tax receipts. A second distribution was planned for the growing season (Kano State Drought Relief Committee, 1974).[33] But if a somewhat random occurrence, relief was welcomed, and 94 per cent of the family heads expressed the view that in time of famine the government's responsibility was to give food or help. Perhaps this was an ominous harbinger of dependency, but in 1972–74 no one was under any illusions about his responsibility to see to his own family's needs, or face starvation.

The famine was thus interpreted in the context of socio-economic change. But to restrict folk-perception to this dimension would be to ignore the significance for individual action of the religious and moral framework in an almost entirely Muslim society.[34] As the effects of the droughts of 1972 and 1973 bit deeper, prayers for rain were, of course, offered in accordance with the practice of the Prophet, especially in 1973 and in May–June 1974 (when it appeared that they were failing again). Additionally, a link was soon made between the lack of rain and the increase of social evils, in particular prostitution. In 1973, 'unmarried women' were driven out of various towns by the authorities, amidst some press publicity.[35] In the following year the press reflected widespread criticism not merely of the sins of the people, but of their rulers (Hinds, 1978), whose moral leadership was becoming ineffective. Conflicting explanatory modes found expression:

> although the scientists suppose that the Sahara has entered Nigeria, the learned of the Islamic religion suppose that it is due to some evil habits which the people of this land are exhibiting. It requires censure from those whose responsibility it is; this means the traditional chiefs...
>
> (M.G. Ibrahim, translated and quoted in Hinds, 1978)

Such a conflict was less obvious in the answers given by the village and family heads to the question: 'Why did God allow this calamity?' (H *bala'i*: table 3.15). An analysis of the glosses offered by some of the more responsive informants is illuminating. There were two interpretations of the theme of sin: some said that

Table 3.15 *Reasons given for the calamity (per cent)*

Simplified category	Village heads ($n = 631$)	Family heads ($n = 125$)
It is the will of God	58	66
Lack of rain	23	24
Social evils, sins	16	10
Others	3	0
Total	100	100

Evaluation

the calamity was a punishment, others that it was a warning sign to society to mend its ways. The list of social evils was extensive. For example:

> Men have no pity for their brothers. Women refuse to get married. Young boys are not listening to their parents' consent. Chiefs refuse to pass justice. Bribery and corruption are sophisticated. Religious instructors are only after money. Even a bad or good thing, provided you give them money, they will just carry it out for you, even murder. All these, they are enough [for God] to punish us for our sins.

and:

> Rich people are not helping the poor ones nowadays... they spend their money extravagantly on prostitutes and they are disobeying God's laws as well... God is punishing them for their sins.

Amplifications of the concept of the immutable will of God included the idea that the time (H *zamani*) was ripe, and even that the disaster had been prophesied:

> It is time, and the trends of modern fashion in this world, that bring all these things, for it was written like this; many important malams had been telling people that there will be a time when this type of thing will happen, since long ago,

which some linked to the expected return of the Mahdi; and there were reports that some had set off to Bima Hill.[36]

On the other hand, a few village heads referred to scientific explanations, even in opposition to the moral thesis: 'God has not allowed the drought. It is time and our nearness to the desert', and in another allusion to desertification, progressive deterioration of yields is linked to the will of God: 'Since ten years ago, farm production has been declining until it came to the period of the drought. So it's God's will', while the hand of man was at least partly to blame, in deforestation: 'No forests. If there were many trees, it would attract rain. There is a forest reserve near the village, where more rain is received than in the village.'[37]

A passive attitude reflects, perhaps, the need for an explanation consistent with the helplessness of rural communities faced by forces beyond their control, as much as adherence to any dogma. It certainly does not imply inactivity, where the possibility of a recurrence is concerned. Notwithstanding a devout hope that no such thing would happen[38] – 94 per cent of the family heads stated firmly that they did not expect another drought – only 23 per cent said they would not prepare for it, while 70 per cent intended to produce and store more food in future. Here, on the margins of the good earth, the omnipotence of God excluded neither self-reliance nor social criticism.

4

THIRTEEN YEARS IN THE LIFE OF A VILLAGE

...the jar of flour did not give out *I Kings 17:16*

ONE VILLAGE

The Manga village of Dagaceri Karaguwa (figure 4.1) is approached across the hummocky surface of a former dunefield, now stable under secondary woodland and cultivation. Similar villages, ranging in size from less than a hundred to more than a thousand inhabitants, occur at intervals of 4–10 km, westwards towards Hausaland and eastwards into the homelands of the Kanuri (of whom the Manga are a subgroup). With the exception of the prominent baobab, full-grown trees are few, and widely spaced. But the site of the village, atop its dune, is first recognisable from its wooded appearance: protected trees, and the perennially green foliage of the neem.

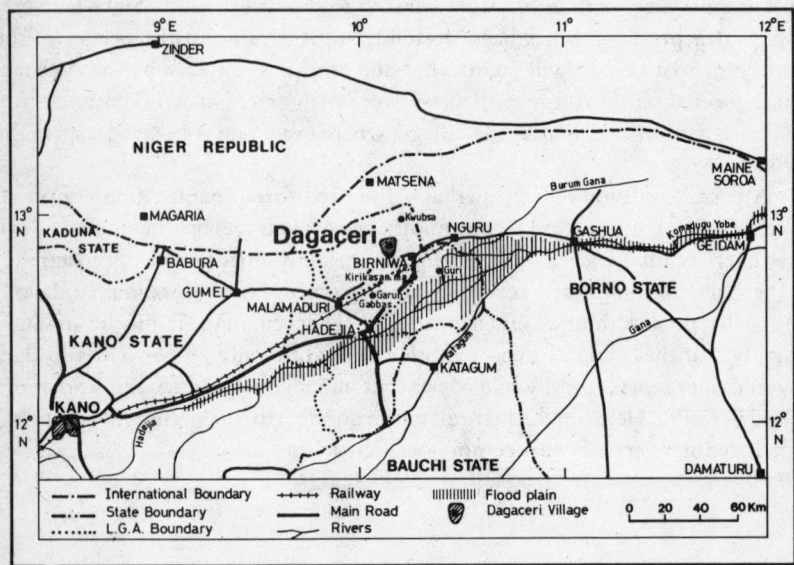

Figure 4.1 The location of Dagaceri

One village 83

At the village perimeter, the visitor passes the all-important clusters of granaries, elevated above the ground, built of timber, thatch and straw. He then passes over the middens and into one of the sandy streets that form a rectangular grid based on the *dendal* (figure 4.2). This is an extended square, oriented from SSW to NNE (towards Mecca), and is found in nearly all Kanuri settlements. At the eastern end of the *dendal* is the house of the *Dagace* or village head, and the principal mosque. Built of mud brick and roofed with metal sheets in 1982, this building (with two others constructed later) replaced an open praying place, where the menfolk assembled several times a day. Apart from a handful of mud walled and roofed (*soro*) houses, which have appeared in recent years, the only other permanent buildings are those of the primary school, built in 1974, and the headmaster's house, both at the western or strangers' end of the village. Other houses are round huts, roofed with thatch, supported by timber frames, and usually walled with mud. They are arranged in roughly rectangular compounds, entered through an entrance hut, a door or merely a gap.

At the outset of this longitudinal study, in 1975, such compounds covered 8.3 ha, and accommodated some 154 households and, in addition, 14 females living alone. The modal household numbered five, and consisted of a man, one

3 Street scene in Dagaceri (October 1986). Overshadowed by neem trees, the fenced compounds contain either round thatched huts or rectangular mud-brick (*soro*) houses with metal doors and projecting roof drains, indicating selective achievement of a modest prosperity notwithstanding recurrent environmental or economic uncertainty

84   *Village*

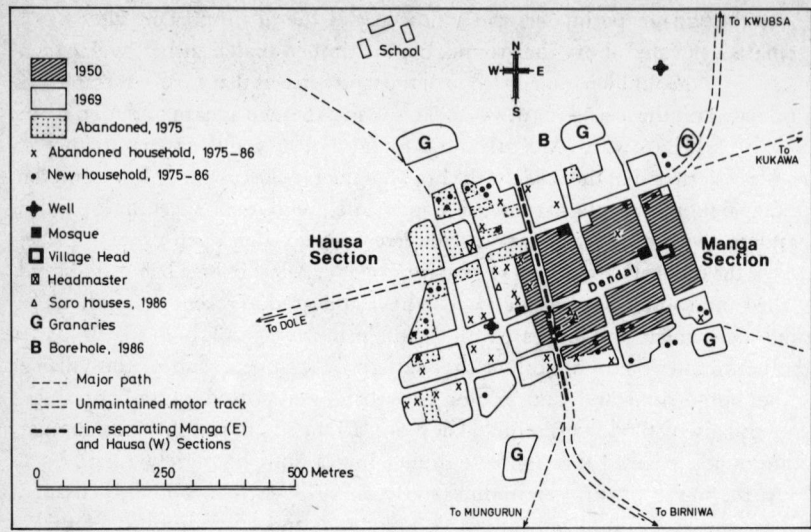

Figure 4.2 Plan of Dagaceri

wife, some young children and one other adult (a grown child, an aged parent, or another relative).[1] Divorce, remarriage and rearrangement of residential units are common among the Kanuri (Cohen, 1967: 42–6). The total population was estimated to be 738 in 1975; the ratio of females to males was 1:2, and children under the age of 15 made up 37 per cent of the total. There were 17 houses that had been unoccupied for more than a year.

Although Dagaceri is a Manga village, half its population are Hausa migrants from Damagaram in Niger who have settled in the village during the past four decades. Forming a community of koranic scholars (H *almajirai*), their numbers were swelled in 1984 by refugees from the drought farther north. Their head is delegated considerable authority by the Dagace, who has allocated farms and building plots to the newcomers over the years. Their houses are found at the western end of the village, and are not visibly different from those of the Manga.

Shade trees (most commonly the neem, *Azadirachta indica*) grow in most of the compounds, along with survivors of the natural flora such as figs (*Ficus* spp.) and the baobab (*Adansonia digitata*); under these, the life of the village is carried on, both inside and outside the houses.

There are a few part-time craftsmen: two tailors, a leather worker, two blacksmiths; a butcher, several drummers and retailers. The *dendal* accommodates the camels or donkeys of itinerant traders during the day, and its sandy expanse provides a stage for children's dances at night. Income-earning activities, such as mat-making, are carried on in shady places; and in the hot-

One village   85

season nights, the men sleep or relax outside on mats. Little wife-seclusion is practised by the Manga; the Hausa are marginally stricter.

Beyond the village and its system of fields, the grazing lands are occupied by settled or semi-settled Ful'be, whose rights are based on prior occupation and upheld by the local government authority (figure 4.3). The hamlet or *ruga* – a cluster of domical straw huts, tethering hooks and occupational debris – accommodates from four to twenty families. The mean size of family is

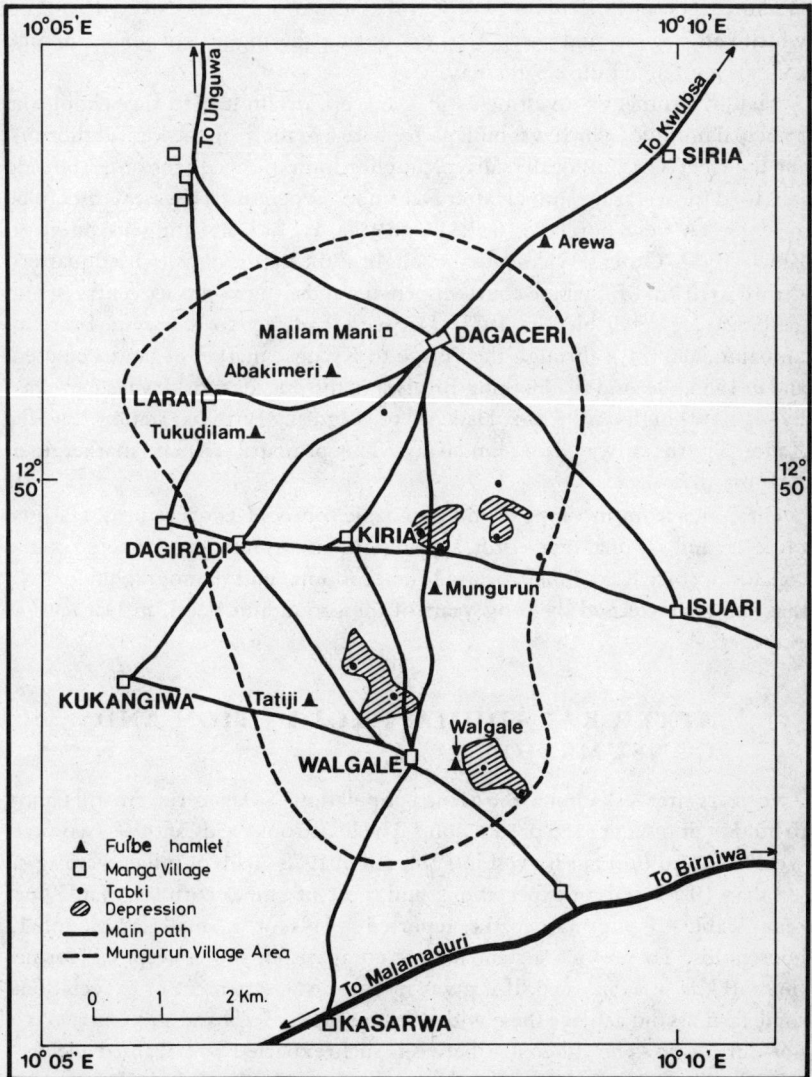

Figure 4.3 The settlement pattern in the vicinity of Dagaceri

eight persons, with rather more adults than children.[2] The huts are smaller than those constructed by farmers, sometimes barely large enough to hold a bed, and less durable. There are no fences surrounding the huts, and no planted trees. Doorways face west in accordance with custom. The *ruga* is the base for transhumance, and for farming operations during the rains. The herds are moved in the dry season to grazings in the Hadejia river valley, 20–50 km south. Abandoned for several months of each year, these settlements appear deceptively temporary, but all five have occupied the same sites for thirty years or more. The *ruga* occupies two or three hectares of scrub-covered land on which cattle, goats and sheep are kept during the night, and young or sick animals are tended during the day.

The government's investments in Dagaceri are limited to the school, the principal mosque (which was built partly with a grant from the local authority), and the wells. One, inside the village, supplies domestic needs; the other, outside it, is used for watering animals after December, when surface water ceases to be available. A new borehole, drilled in 1984, was still awaiting its pump in January 1986. Other services must be sought at the local authority headquarters, Birniwa (13 km SE), where court, dispensary and an agro-service centre (set up in 1982) are available. In 1975, Dagaceri had no road access; later, an unmaintained track through the village to Kwubsa market began to be used; and in 1983, the sand road linking Birniwa to the outside world was superseded by a new highway from Hadejia to Nguru. Birniwa station on the Kano–Nguru railway is 17 km away. The principal weekly market is at Kwubsa (8 km).

Gleanings from incomplete and unreliable sources (census returns, tax lists, cattle tax and vaccination records, and air photo interpretation) converge on the conclusion that Kakaduma reversed the economic and demographic growth that had characterised the long years of the groundnut boom in Dagaceri as elsewhere.[3]

AFTER KAKADUMA: PRODUCTION AND CONSUMPTION

Two years after Kakaduma, the Manga population of Dagaceri were still failing to hold their own in food production.[4] The disastrous yields of those two years were followed by a fair harvest in 1974, but in 1975 yields of millet were again less than 10 per cent of expectations, and those of guinea corn less than 30 per cent. Table 4.1 summarises the reported yields of a small sample of 13 households. To provide a standard for comparison, the concept of 'former times' (H *da*) is again used. If it gives an optimistic statement of expectations, some farmers did achieve these with certain crops in 1974 and 1976, so they are not unrealistic. The difference between such expected and realised yields is crucially important in influencing decisions about food supply and alternative sources of income.[5]

Table 4.1 *Average crop yields* (kg unthreshed)[a]

	1974	1975	Expected
Per family (5.5 persons)			
Millet	2,199	334	3,475
Guinea corn	364	514	1,652
Cowpeas	288	194	1,088
Groundnuts	238	67	1,219
Per estimated hectare[b]			
Millet	373	57	590
Guinea corn	62	87	280
Cowpeas	53	35	216
Groundnuts	58	16	245

[a] Grain on the head, cowpeas and groundnuts in shell. Manga farmers use the basket (H *sanfo*), which contains about 40 kg of grain on the head.
[b] Sizes of farms are estimated from the number of man-days required to do the first weeding. Crops are grown in mixtures, usually two or three in unequal proportions.

Estimates using FAO formulae show the daily energy requirements of a rural population near Zaria to be 2,008 kcal per capita, averaged over the year (Simmons, 1976: 10, 72). Studies in seven northern Nigerian villages suggest considerable variation in energy intake: in Dayi, 80 km west of Kano, it was 1,931 kcal per capita, over the year (Longhurst, 1984: 46–8).[6] Cereals provide 77–80 per cent of energy intake and three-quarters of protein. Using the Zaria estimate, and assuming that twice as much millet as guinea corn is eaten, and no other cereals, 432 gm of flour are required.[7] Allowing for loss of weight in threshing and grinding,[8] the daily requirement of threshed grain per capita is 540 gm, or 190 kg/year, or 218 kg/year after adding 11 per cent for gifts and 4 per cent to cover storage losses, as suggested by Hays (1975). If children under 14 years are treated as half adults, the requirement is 261 kg per adult equivalent.

However, farmers' own estimates of their minimum grain requirements for the year were substantially higher: 359 kg per capita, or 428 per adult equivalent. Their estimates gave approximately 1 kg of threshed grain per capita per day, which is in agreement with the opinion widely given in northern Nigeria that one *tiya* (about 2.5 kg) of grain will last a man just over two days (Hill, 1972: 254). Other authorities give lower figures (120–75 kg for the Sahel countries);[9] an FAO report on Nigeria said that cereal consumption in northern

Table 4.2 *Grain yields and requirements*

	1974	1975	Expected
Per capita (kg threshed)			
Millet	247	38	392
Guinea corn	46	65	210
Production as a percentage of requirement			
Theoretical	135	47	276
Farmers' estimates	82	29	168

Nigeria was 'very high' at 186 kg (FAO, 1966: 21); but famine relief agencies in Africa were working with estimates of 176 kg (0.5 kg/day) in 1984–85. The difference between nutritionally based estimates and the producers' own conception of their needs could be explained by wastage, lower grain quality, or the inclusion of grain for purposes other than consumption.

Table 4.2 shows that the expected yields, if achieved, would provide more than double the theoretical requirement, and significantly more even than the farmers' own estimates. A large surplus is indeed necessary in such a marginal environment. For, after achieving roughly adequate output in 1974, the level of production fell far below the subsistence requirements in 1975, on either set of estimates. Most farmers were buying their grain again nine months later.

Economic inequality between households is a fact of life in northern Nigeria (Hill, 1972: 57–83). In Dayi, in 1976, the average proportion of grain needs met by production was found to be 1.20, ranging from 1.30 for the 'rich' families to 1.07 for the 'poor'.

These data suggest that on average the cultivating 'poor' just meet their grain needs, net of payment for production inputs or hired labour, or for debt repayments for which there is no information. The rich more adequately meet their grain needs under similar assumptions.

(Longhurst, 1984: 43)

Dayi enjoys a more favourable environment than Dagaceri, and 1976 was a favourable year. A balance between years of variable output can be attempted by storage. Redistribution may take place between richer and poorer families. But their impact is very limited.

Even the food grown was not all eaten. Pitiful yields of groundnuts and cowpeas forced producers to sell grain. Those selling were as likely to be in deficit as in surplus. Even after the terrible harvest of 1973, local millet had appeared in the markets.

A CHRONICLE OF ADVERSITY

A consecutive account will now be attempted of the buffetings that this small community withstood between 1974 (in the second year of Kakaduma) and 1986. The account is based on information obtained on 16 visits to the village at different times of the year.[10] The perspective owes much to a baseline study by Ismaila Daudu (1976), who lived in Dagaceri during July and August 1975.

The essential background to such an account is the variable pattern of rainfall and prices, the two principal exogenous factors affecting the village's livelihood. Table 4.3 gives a composite index of rainfall at four stations in the area, for the four critical months of June to September, and for the year. June is

Table 4.3 *Indices of rainfall at Kano, Nguru, Zinder and Maine Soroa, 1970–85*

	June	July	August	September	Total
Base[a] (mm of rainfall)					
Kano, 1931–60	119	210	311	137	872
Nguru, 1942–60	57	137	233	100	565
Zinder, 1931–60	55	153	232	71	548
Maine Soroa, 1931–60	32	106	176	62	418
Composite index (four stations; per cent of normal,[a] averaged)					
1970	30	125	98	66	94
1971	16	73	71	118	75
1972	123	41	62[b]	24	63[b]
1973	37	71	34	57	45
1974	30	133	95	92	96
1975	44	88	92	118	88
1976	92	93	40	126	75
1977	72	73	105	61	79
1978	115	173	89	46	107
1979	142	121[b]	62	76	80
1980	106	122	65	44	83
1981	101	103	59	65	73
1982	70	65	90	48	72
1983	58[b]	43	60	54	49
1984	n.a.	91	25	65	51
1985	92[b]	84[b]	67	88[b]	n.a.
1986	81[b]	75[b]	n.a.	76[b]	n.a.
Mean, 1970–85	75	93	70	70	75

[a] The use of the period 1931–60 to define 'normal' is justified in chapter 6.
[b] Index based on three stations only.
n.a.: Records not available.

the main planting season (unless rain comes earlier in May); July and August are the main growing seasons and, in September, late-maturing crops reach their full development. A major deficiency in June means late planting, resowing, and delayed growth; in July and August, wilting and loss of yields in the early millet; and in September, partial or complete loss of guinea corn or cowpeas. However, differences among these four stations in any month were commonly of the order of 100 per cent, and so the index does not necessarily give a reliable guide to conditions experienced at Dagaceri. The indices are shown graphically in figure 4.4.

Prices of the three major crops, millet, guinea corn and cowpeas, during the same period are also shown in figure 4.4. These are obtained from the Crop and Weather Reports issued each month by the Ministry of Agriculture and Natural Resources, Kano state, and represent the average prices for the state.[11] Like the rainfall data, these series are not necessarily a reliable guide to prices in Dagaceri's local markets; the nearest market included in the series is 50 km away.

With rainfall erratic in amount, late in starting, patchy in distribution, or finishing prematurely, successive rainy seasons soon gave the lie to the idea that drought had departed in 1974. Hardly any two seasons were alike. Nearly every dry season, when the contents of the granaries were reviewed, brought a subsistence challenge to a major proportion of households.

The astonishing upward progression of prices – from ₦50 to more than ₦1,000/ton for grain between 1970 and 1984, and from ₦70 to more than ₦2,000 for cowpeas – ought to have favoured the farmers. But crippled by rainfall deficiencies, and faced with the inflating cost of labour – from about ₦0.25 in the sixties to as much as ₦5.00/man-day in the eighties – very few were able to benefit. On the contrary, births, marriages and deaths did not cease to occur; until 1979, community tax had to be paid; and other cash demands arose from a network of social obligations, debts, accidents and personal requirements. Those needing cash urgently had to sell at harvest, even if they had to buy back grain at two or two and a half times the price later in the season; while high seasonal prices might attract imprudent sales, risking subsistence reserves against the next fickle rainy season. Price fluctuations, of course, always assist those in surplus to accumulate while pillaging those in deficit. Seasonal fluctuations were clearly superimposed on the upward trend (figure 4.4). However, the correlation between monthly rainfall and prices during the four months June – September was less impressive than that between total annual rainfall and the price peak in the ensuing dry season.

The tale from year to year:

1974. Following the disasters of 1972 and 1973, the rains of 1974, after a late start, brought a satisfactory millet crop. At last the prices for the three major crops, which had jumped by 200 per cent since 1970, turned downwards. The incidence of circulation dropped dramatically in Dagaceri; and in West Africa

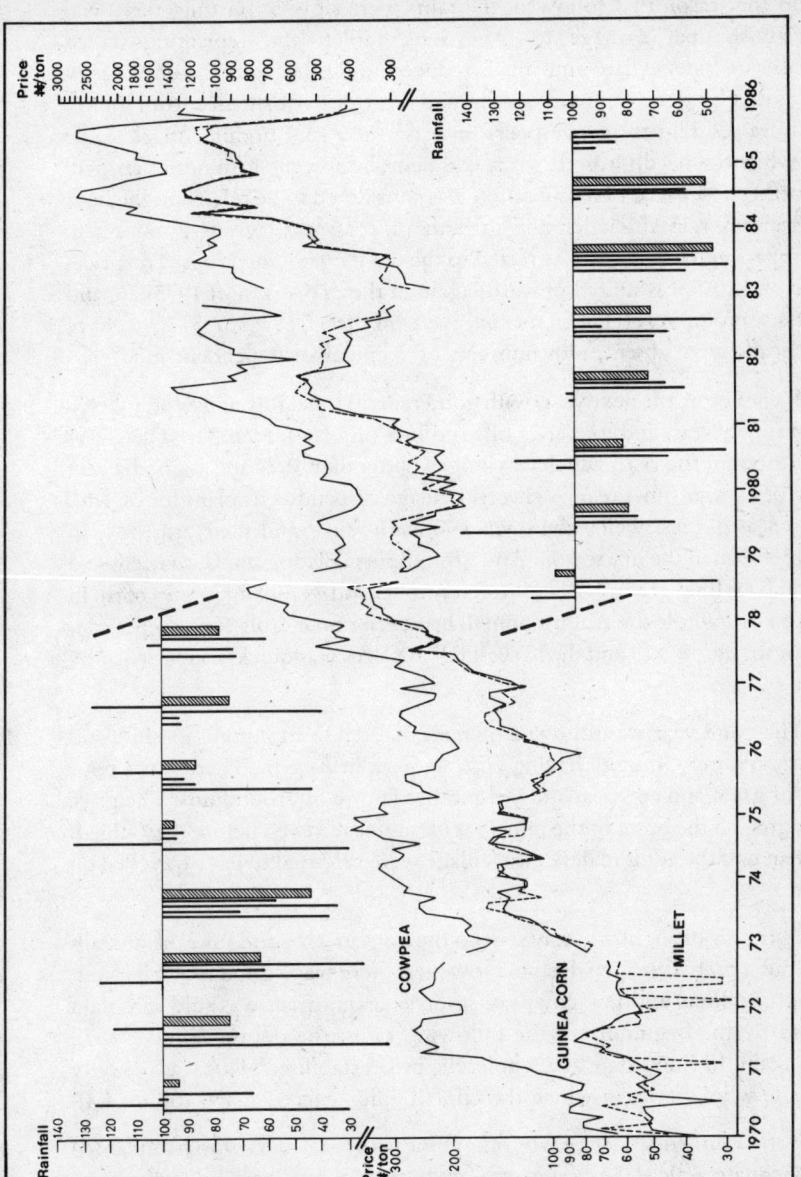

Figure 4.4 Prices and rainfall, Kano State, 1970–85. Prices of cowpeas, guinea corn and millet are average monthly prices for all stations reporting in the state (CWR, Ministry of Agriculture and Natural Resources, Kano state). Rainfall for the months June, July, August and September (black columns) and for the year (hatched columns) is a composite index (base: 1931–60 mean) for the stations Kano Airport, Nguru, Zinder and Maine Soroa (see table 4.3)

as a whole, the level of drought preparedness was relaxed. But there were danger signals in Dagaceri: grasshopper infestation, and poor seed yields.

1975. In the season that followed, the rains were slow in starting, there was heavy grasshopper damage to the young millet, and replantings failed repeatedly in June. A late and much reduced millet harvest was only partly compensated by good September rains and a better performance from guinea corn, more resistant to grasshoppers. Insects[12] and root nodules attacked the beans, while the deadly rosette virus was demolishing the groundnut crop all over northern Nigeria. Pest infestation was considered to be related to the light and uncertain rainfall: when the growth of grass and weeds is inhibited, grasshoppers in particular are attracted to the tender growing crops. The prices of grain and cowpeas surged upwards again in the dry season of 1975–76, and the dry-season movers (H *masu cin rani*) were on their feet again: 37 per cent of family heads were absent, plus numbers of younger or dependent men.

1976. Relief came the next year, with good rains in June, July and September, a good grain harvest, and granaries filled to bursting by December. There was enough food for the year, which was judged better than 1974 and as good, even, as those of former times (*da*). A cheerful village suspended its plans for *cin rani*. Yet cowpeas did less well – there was too *much* rain – and their price rose to over ₦300/ton in the dry season. And groundnuts failed again. Cash needs had to be met by selling grain, at ₦ 120/ton for millet and ₦ 100 for guinea corn. In the region as a whole the August rainfall had been remarkably low (40 per cent of the mean, figure 4.4) and the harvest fall in prices was quickly superseded by another rise.

1977. The good year was followed by rains that fell short in June and July and ended prematurely, though having a strong peak in August. There was a poor harvest of grain and cowpeas and yet another failure of groundnuts. The price fall that greeted the onset of the rains was extinguished even before they ended. More than half the adult males of the village were reported to have travelled on *cin rani*.

1978. This year brought a timely start to the rains in May and June, heavy falls in July, but a premature conclusion. However, there was a good millet harvest, but no groundnuts and few cowpeas; so once again, grain was sold to obtain cash, and by the beginning of the following rains, the poorer families were buying again. In Kano state as a whole, the prices stabilised. This was the only year in the whole series in which the rainfall index exceeded 100 (figure 4.4).

1979. Good rains ended late, favouring guinea corn; farmers who had planted on an adequate scale – the better off – were reported to have adequate food stocks. Yet the local prices of millet and guinea corn climbed to ₦320/ton by the end of the 1979–80 dry season, and cowpeas to ₦600: considerably higher than the state average. These price movements were thought to result from the activities of traders as much as from the rainfall. Fewer travelled in this year.

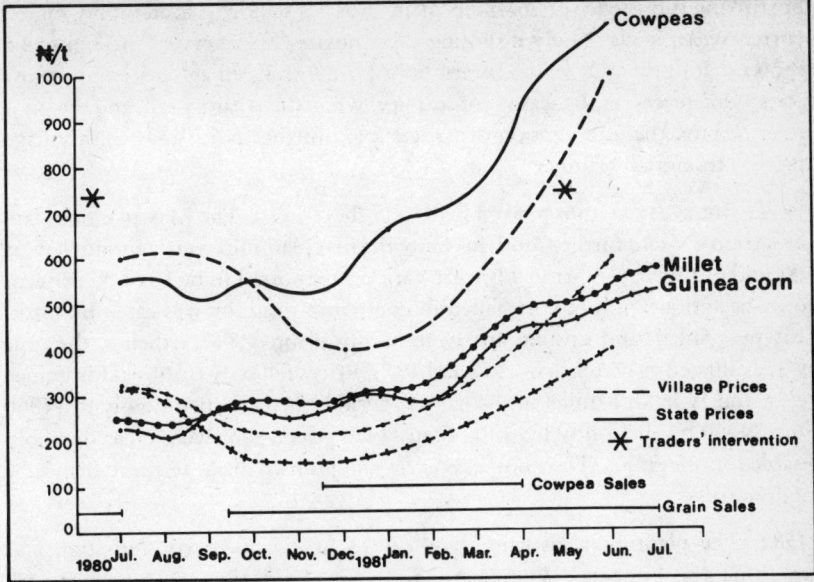

Figure 4.5 Village and state-wide prices, 1980–81

1980. Locally, late rains favoured guinea corn again, which produced 50 per cent more than in 1979, but millet suffered grasshopper damage in the ripening period (August rainfall was deficient) and yielded 50 per cent less: of course much more millet had been planted than guinea corn. The harvest-season prices were ₦160 and ₦220/ton for guinea corn and millet, reflecting this divergent performance (figure 4.5). It was a boom year for cowpeas, which entered the market widely, many farmers selling two to five bags (worth ₦80–200 at the lowest price); the price fell to ₦400/ton, but later climbed steeply. These prices were *lower* than the state average. Even a few bags of groundnuts were sold. So there was no increase in the incidence of *cin rani*. Prices then surged to ₦400 for guinea corn, ₦600 for millet, and ₦1,000 for cowpeas: seasonal increases of 250–300 per cent (figure 4.5). They were said again to be inflated by traders from Niger, where Maine Soroa had received only 70 per cent of its normal rainfall, and only 45 per cent of that expected in August.

Cowpea sales ended in April when stocks were exhausted, but grain sales were continuing in July 1981, when the larger of two village grain dealers said he was taking two tons each week to Kwubsa market, for which he charged commission at ₦5.00/ton, as decreed by the local authority. After a good harvest he expected to carry ten tons, but some of this may have been bought in other villages. Such sales in July may have been due to the off-loading of stocks in expectation of a new crop, together with the unusually high prices; but following the poor millet harvest, some households were already buying or borrowing subsistence grain. If the village was not selling its food supply for the current year, it was surely disposing of its next year's reserve. And figure 4.5 shows that in the glut season, when poor farmers sell, village prices fall below state-wide prices; in the seasons of scarcity, when the wealthy sell, and the poor have to buy, the intervention of traders from further north may push village prices substantially higher.

1981. But the next rains proved fickle and short-lived. The May plantings had to be resown, and further south in Kano, the first plantings were a month late, at the end of June.[13] Millet yields were variously reported to be 10 or 30 per cent of expectations; guinea corn failed to come into head, or was eaten by birds; cowpeas failed, and groundnuts were hardly planted. Nevertheless, the year was evaluated as better than 1972 and 1973. Prices fell only to ₦300 for guinea corn and ₦360 for millet in the harvest period, then climbed slowly to ₦360 and ₦460 by the following July. State-wide prices, too, were unaccountably stable for the grains. The numbers on *cin rani* jumped again to more than fifty individuals.

1982. The planting rains came late, on 28 June, amidst consternation and awaiting grasshoppers: 'We're waiting'; 'we're afraid'. But the August rainfall was fair and well distributed. Rain continued into September; millet yields were much better than in the year before, and those of guinea corn satisfactory. Abalu *et al.* (1983: 27–8) estimated millet yields in the area at 400 kg/ha and solo guinea corn on a few plots as high as 900 kg/ha. Those of cowpeas were patchy, less than 100 kg/ha. The numbers on *cin rani* in the dry season of 1982–83 dropped to twenty. Prices fell sharply in 1982 to ₦200 for guinea corn and ₦260 for millet, the lowest for several years, and the seasonal price rise was modest and late (₦250 and ₦300 in July, 1983): but rapid. This pattern was also true of the state, though the rainfall in Kano, like the regional index, was only 73 per cent of the mean. It was, however, correctly distributed, with a peak in August – opposite to the previous year's distribution.

1983. It was as if the early price rise anticipated the disastrous rains of the following season. The first fall at Dagaceri in late June was followed by 25 days of drought, broken only by a light fall on 20 July. Although some millet survived, yields were less than 10 per cent of expectations overall, and most of the annually cultivated fields near the village yielded nothing. It was the worst year in the series, with the sole exception of 1973. The *masu cin rani* swelled ranks and took to the road, but not in the numbers of that year. The state-wide prices of guinea corn and millet shot up to all-time highs at over ₦1,000/ton in April 1984, and that of cowpeas passed ₦2,000 in June. The fourfold increase in the prices of these crops between 1982 and 1984 was considerably faster than the threefold rise in grain prices that occurred between 1972 and 1974.

1984. In this year, areas of Nigeria south of Lat. 11°30'N had good yields, and foodstuff prices fell by as much as half in some urban markets. Further north, however, there were major failures, following the spectacular disappearance of rain in August, normally the wettest month.

Across the border in eastern Niger, the third successive calamitous season occurred, worse than 1972–74 in some areas, sending refugee families south in considerable numbers.[14] In Diffa department, for example, there was only a month's food supply. The lack of pasture generated an influx of herds into Nigeria, from July 1983 onwards – many of whose owners had never before been there. In Dagaceri, millet yields were up to 100 per cent of expectations on the bush fields, but much lower on manured fields: for many farmers, food stocks for the year were insufficient. Even after the harvest, prices were ₦800 for millet and over ₦1,000 for cowpeas. An alternative source of income for hard-pressed farmers, since there had been a fair growth of grass, was the cutting, raking and bundling of pasture grass for sale to the owners of the starving herds from the north, at prices which rose from ₦2 to ₦7 per bundle between February and June 1985.

1985. Bumper harvest in many parts of Nigeria were due to the excellent distribution of the rainfall. Nevertheless, the composite index for the four stations (and rainfall generally in the Sahel) fell below the long-term mean. At Dagaceri, the year's grain production, however, was little better than that of the previous year.[15] Cowpeas alone did well, and entered the market in quantity. At least twenty men travelled after the harvest on *cin rani*. But the growth of grass was exceptionally vigorous. Since the northern herds had followed the rains back home, the pastures still lay replete and undergrazed in the mid-dry season. Their recovery, in view of the conditions to which they had been reduced in 1984–85, was remarkable.

1986. The Year of the Pests. It would have been reassuring to end this account with fat pastures and falling prices at the end of 1985. But in the next year the rural system was once again at the point of collapse, this time not directly because of either rainfall or price behaviour. During the dry season the jumping rodent *Jaculus jaculus* L. (Tzarza jerboa) arrived unannounced from the north, where it has been commonly observed.[16] By the time the rains arrived it was widespread from Dagaceri in the west to Maine Soroa in the east (figure 4.1). Nocturnal in habit, adept at burrowing in loose sand and apparently addicted to eating grain, this pest (H *maitono*: the unearther) descended on the new plantings, excavating the seed before it germinated, and storing it in underground silos, pursued by aggrieved farmers. The battle of wits continued throughout the planting season; repeated replantings were necessary; and with its long hind legs and kangaroo-like motion, the jerboa was even able to attack low-hanging heads of ripening grain later in the season.

Afterwards came the grasshoppers. Also well-known in the northern Sahel, *Oedaleus senegalensis* has a capability for extensive migration and first appeared in Dagaceri in September.[17] It attacked the unripe grain of millet, leaving the foliage alone (whereas in previous years, grasshopper damage was most serious in the early growing stages of millet, when the leaves were destroyed). What the jerboa left, the grasshoppers took. Both millet and cowpea harvests were disastrous. The local varieties of guinea corn, which were not affected by the grasshoppers, were unable to ripen because the rains finished early.

Having failed to anticipate the risk of grasshopper infestation, government sprang to action and announced food aid and aerial spraying contracts late in October (*The New Nigerian*, 29 October, 1986: 7), far too late to save the millet harvest. The rodent pest had been reported to government agents near Dagaceri long before, without effect. Possibly this was due to the surprise element in the invasion of *Jaculus jaculus*, though an effective regional monitoring system could easily have warned of its approach. Grasshoppers, on the other hand, have a place in the institutional structure of agricultural research. Now, impoverished by declining revenues, government, as before, responded with too little, too late.

Dagaceri faced the subsistence year, 1986–87, with less than three months' food supply. The rainfall, which would have been adequate for millet but for the pests, supported only a growth of pasture. But, as always, there was hope. A new short-season variety of guinea corn, introduced by the State Agricultural and Rural Development Authority, had been tried by a few farmers and the results were impressive – fat heads in a sea of desolation. Next year, it was sure to be popular.

Thus circumscribed by precipitation, prices and pests, the producers and consumers of Dagaceri survived from year to year. Against such exogenous forces, the system could field only a limited armoury of adaptive weapons. Storage and redistribution of food could not compensate for an overall shortage. This shortage was generated, at least in part, by grain-selling. And the search for alternative income-earning opportunities was itself affected by unpredictable events in the political economy.

In 1974, a primary school was opened in the village, two years before the ill-fated programme of Universal Primary Education took off in Kano State (Bray, 1977, 1981). So limited has been the impact of this school, owing to parents withholding their children, inadequate equipment and maintenance, curricular and staffing deficiencies, and examination failures, that fears of a rural exodus based on the experience of earlier primary schools have not materialised.[18]

In February 1976, there was an abortive coup in Lagos that sent seasonal migrants helter-skelter for home: the *masu cin rani* of Dagaceri arrived penniless, some having had to walk the last 40 km. In 1979, elections took place for federal, state and local governments, and soon afterwards, both community tax (₦5 or ₦6/year, payable by male adults) and cattle tax (₦0.75/animal/year) were abolished. Another set of elections in 1983 was followed by a return to military rule in January 1984, accompanied by drastic restrictions on the urban informal sector – and income-earning opportunities for the poor – in state capitals. By 1985, taxation had been reintroduced in neighbouring Borno state, though not in Kano. And during the period under review, the national economy plunged from oil bloom into deepening recession.

In 1982, a new agricultural and rural development authority (partly funded by the World Bank) opened an agro-service centre in Birniwa and began to disseminate improved inputs, fertiliser, tractor hire, advice and credit. 'Integrated rural development' diffused somewhat slowly in the area. Finally, in 1983–84 a new intercity highway linking Hadejia with Nguru brought cheap and rapid transport to Birniwa (13 km).

Such externalities provided the context for the village's continuing struggle to feed itself. For, in the years 1972–86, the search for economic alternatives to agriculture was never completely suspended, and from the admittedly superficial narrative given here,[19] it appears that in only two years (1976 and 1979) did *most* families manage to grow their year's food requirements, still less satisfy their needs for cash, or secure a reserve for the future.

HOUSEHOLD VIABILITY AND CHANGE, 1975–86

A re-survey of residential households in January 1986 makes it possible to evaluate the social impact of the economic uncertainties of the preceding decade (table 4.4, and see figure 4.2). In October 1975 (row 1), there had been 154 residential households having male heads (96 in the Manga section of the village, and 58 in the Hausa). Of these, 101 survived *in situ* (row 2). Nine moved elsewhere in the village (row 3) and 15 were transferred by inheritance or gift to relatives (row 4). The remaining 29 were closed (row 5), but 31 new households were opened in the period (row 7). Since in both communities first marriage normally results in the formation of a new household, the net change in the number of households (row 8) falls short of that which would be expected, given demographic growth on the scale reported elsewhere (Faulkingham and Thorbahn, 1975). On the reasonable assumption that mean household size remained constant, the total population was virtually stable, at 747.

An explanation for demographic stability is offered by out-migration. The number of households closed through departure without further continuity through inheritance or gift (row 5) – six in the Manga section and nine in the Hausa section – exceeded the number of those transferred to relatives (row 4) – seven, all Hausa. It is also noteworthy that the number closed through death without inheritance (row 5) – i.e., the survivors of all eight households disbanded and went to live elsewhere – was equal to the number transferred by inheritance. Table 4.4 shows that the Manga section was more stable demographically than the Hausa section. Losses by out-migration and gains by in-migration were on a higher level in the Hausa section; furthermore, these movements perpetuated the ethnic division of the village, only Manga and Fulani migrants being allocated plots in the Manga section. While some out-migration is consistent with a theory of environmental stress, differences

Table 4.4 *Changes in residential households in Dagaceri, 1975–86*

	Manga section		Hausa section		Total
1 Number of occupied households, Oct. 1975 (with male household heads)[a]		96		58	154
2 No change, 1975–86		72		29	101
3 Moved elsewhere in village		4		5	9
4 *Headship transferred:*					
Died, inherited by wife or son	7		1		
Left, given to close relative	—		7		
Total transferred	7	7	8	8	15
5 *Losses: closed households*					
Died, no inheritance	6		2		
Left, no gift	6		9		
No record	1		5		
Total losses	13	13	16	16	29
6 Total accounted for		96		58	154
7 *Gains: new households*					
Move or marriage	5		5		
In-migration[b]	7		12		
Reclaim of unoccupied site by son	1		1		
Total gains	13	13	18	18	31
8 Net change		0		+2	+2
9 Number of occupied households, Jan. 1986[c]		96		60	156

[a] In addition there were 14 females living alone and 18 houses unoccupied, eight of them since the drought (<5 years).
[b] In the Manga section, three Fulani and four Manga households; in the Hausa section, 12 Hausa households.
[c] In addition there were 11 females living alone.

betwen ethnic communities provide a warning against reliance on such a simplified explanation of demographic stability.

ADAPTIVE RESPONSE: THE FARMING SYSTEM

The farming system of north-eastern Kano state has been designated 'permanent cropping with millet' in a description by a farming systems research team

(Abalu *et al.*, 1983). Its principal characteristics are: a dominance of annual cultivation over fallowing; minimal land preparation before sowing; rapid and often unsuccessful sowings of the principal crops at the onset of the rains, using the long-handled seed hoe (H *sungumi*) for making seed holes, women and children dropping the seed, and covering with their feet; weeding twice with the long-handled hoe (H *ashasha*) in the early stages of the growing season, and perhaps with short-handled hoe (H *fartanya*) later (the *ashasha*, with its forward-working crescentic blade, is effective only when the weeds are at the seedling stage); the dominance of early (70–90 day) millet; the secondary role of guinea corn, either interplanted in the sparse mixtures with millet or grown solo on wetter soils; the complementary role of the legume cowpea, grown in mixtures with the grain; the use of the ox-plough by richer farmers for ridging (but not for weeding); and the small importance of other crops.

How did this system adapt?

Soil management

On the irregular surface of this former dunefield, now the fields of Dagaceri, whose relative relief is less than 15 m and whose slopes are generally less than 3°, the soils are predominantly orange-brown free-draining sands.[20] In the interdunal hollows, where finer material has been transported by surface wash, a darker soil is found with a finer texture. The two are distinguished by their colour as 'white soil' (H *farin kasa*) and 'black soil' (H *bakin kasa*). There is no integrated surface drainage; rainwater either percolates into the soil or runs into the nearest hollow where a seasonal pool (H *tafki*) may persist until half-way through the dry season. The soils are considered to need less rain than the heavier soils of the flood plain of the Hadejia river, where the clays and silts may crack in drought and residual moisture is less available to the crops.

Farming practice distinguishes, as elsewhere, between manured infields (H *karkara*) and bush, or unmanured, outfields (H *gonar jeji*). The first are found close to the village and the second at greater distances. The main sources of the manure are the livestock population of the village, ash heaps and compound sweepings.[21] Supply and transport are normal constraints. A small sample of farms suggested a mean rate of application on the manured fields in the order of 1.8 tons/ha before 1972,[22] with a predictable relationship between distance and frequency of manuring (table 4.5). Manure was less commonly obtained from grazing cattle; however, the Ful'be relied on them exclusively for their own farms.[23]

During Kakaduma, most of the donkeys died. On the basis of experience elsewhere, this must have dealt a critical blow. Headloading is a very inefficient alternative to donkeys, and cannot be attempted over distances of more than half a kilometre. Additionally, the loss of livestock decimated the supply of both household and cattle manure. The number of farmers using the first type fell by a half, and of those using the second by a third, between 1970 and 1974, while those using none increased sevenfold (Daudu, 1976; table 4.6).

Table 4.5 *Manuring and distance from the village of Dagaceri*

Distance zone (km)	Manured fields	Unmanured fields	Fields manured by grazing cattle
0–0.5	10	9	2
0.5–1.75	8	9	3
1.75–3.0	4	14	0

Table 4.6 *The effect of drought on manuring*

	Number of farmers ($n = 100$)			
Type of manure	Before drought	1972	1973	1974
Household	59	54	38	28
Livestock	34	34	33	24
None	7	12	28	48

Source: Daudu (1976: 32–3)

However, a property of organic manure sometimes overlooked is that under conditions of subnormal rainfall and high temperatures it heats the soil to a level which can cause young plants to die. The soil is said to be 'hot' (H *zafi*). Application – which is done stand by stand – may therefore be delayed until the rains seem established.[24] In 1983, the manured fields produced virtually nothing, and in 1984 yields were again reported to be generally better on the *jeji* (unmanured) fields.

The impact of these three negative factors – reduced supply, reduced transport and increased risk of harmful effects – has been to reduce the levels of manuring, and consequently of potential yields and of long-term organic replenishment, in soils already having low fertility. Thus the possibilities of intensification – as a response to population growth, land scarcity and low productivity – are severely constrained. The solution advocated by Abalu *et al.* (1983: 48ff.) – the importation of soil nutrients by means of chemical fertilisers – is subject not only to supply constraints (a rather well known problem in Nigeria), but even to the same harmful effects: there is evidence from all over Kano State that farmers perceive such *takin zamani* (literally, modern manure) as even more dangerous, in drought, than organic manure.

Labour

Adult males not too old to work provide the greater part of family labour, together with boys of about eight years upwards who can help with most field tasks. Married women among the Hausa are not supposed to do field work. However, they do help, especially in planting and harvesting tasks (Norman, 1972: 29; Hill, 1972: 279, 335; Baba, 1975: 94–6). Among the Manga it is quite normal to see women working in the fields.[25] Including women (stated to take part in field work), and boys aged 8–14 years, as each equivalent to half a male adult, the average number of labour units per producing family is only 1.92[26]

Labour bottlenecks may occur in planting and in the first weeding, which begins immediately the newly planted crops show above ground. The urgency of planting derives from the unpredictable arrival of the first rains, the necessity to plant as large an area as possible to counteract the effects of drought if it occurs, the need to re-allocate labour to weeding within a few days, and the decline in millet yields that may result from late planting. Replanting may be necessary if rainfall is poor, necessitating the sowing together of crops which are

4 Manga woman hoeing a groundnut plot at Dagaceri (late July 1978). In the middle distance, cowpeas are growing, and in the far distance, among sparse farm trees, are millet fields. The grassed plot on the left has not been cultivated in the current season

normally staggered. To spread the work, seed may be sown in the soil up to a month before the rains occur (H *binne*). Such a practice risks premature germination, and only those who can afford to lose their seed normally try it; the poor and the cautious await a heavy shower.

Weeding preoccupies the labour force in the ensuing weeks, the first weeding with the *ashasha* being followed immediately by the second weeding (H *maimai*) in August. From daybreak until afternoon, the village is devoid of all but a few women, children and infirm. A small sample of farmers estimated the number of days required for the second weeding: the mean was 27 per labour unit, spread over an average of 32 actual days, which disposed of about a third of the growing season. Since the first weeding may take longer (the young crops are delicate), and both weedings must be completed as soon as possible, a labour shortage may occur. While those falling below the mean could easily complete the work within a month, the remainder were far above it, and depended on the use of hired or shared labour.

Before the drought, the profits of groundnut sales could be spent in hiring labour to clear and cultivate additional land or maintain large holdings in cultivation. Even transhumant Ful'be hired labour. Afterwards, with groundnuts ruined and food production deficient, few could afford to hire, and to take the risk that another harvest failure would destroy their investment. The wage-rate rose from ₦1.70/day in 1970 to ₦3.00/day in 1985. After a bad harvest, income from *cin rani* might be used to pay wages, or labourers might accept payment in clothes.

An increase in the scale of farming is a necessary condition for ensuring a minimal yield of grain under conditions of drought. The adoption of the faster *ashasha* hoe (see below) has facilitated such an increase. But the financial resources with which to risk planting failures, and to hire the labour for clearing and extra weeding, are unequally distributed. About half the farmers in the village hire labour; the rest provide it (Daudu, 1976: 21). The exchange of labour, as usual, serves as a critical indicator of economic well-being, and adaptive capability is inversely related to economic vulnerability.

Land

The average size of farm holding (5.9 ha), estimated for a small sample (table 4.7), is larger in Dagaceri than in areas further south and west, both in terms of absolute size and in the ratio of cultivated land to population.[27] A review of field studies in various parts of Hausaland (Hill, 1972: 236) yielded a range from 1.7 to 6.1 ha, with the great majority falling below 4 ha.

Land is acquired in the first instance by inheritance (H *gado*). Such land is normally cleared arable and, usually, manured. Since inheritance under Shari'a Law is divisible between sons, an inherited holding may be inadequate to support a family. Additional arable may be obtained by gift (H *kyauta*) from a relative, a patron, or one's father before his death (but not usually before the

Table 4.7 *Size of farm holdings in Dagaceri*[a]

Averaging size of holding (ha)	5.9
Average number of dependants	5.3
Average number of labour units[b]	1.9
Area per capita (ha)	1.1
Area per labour unit (ha)	3.1

[a] Calculated from 16 holdings using weeding-time as a proxy for area, with a coefficient of 10.2 labour units/ha, derived from measurements made on five contiguous farms belonging to five separate holdings. These farms measured 1.84, 3.13, 7.10 1.50, 3.95 and 1.70 ha.
[b] One adult male or two women (declared as work force) or two boys aged 8–14 years.

Table 4.8 *Modes of acquisition of farmland*

Mode	Number of fields	Per cent
Allocation	21	35.0
Inheritance	18	30.0
Gift	15	25.0
Re-allocation of arable land	4	6.7
Purchase	2	3.3
Borrowing, pledging	0	0
Total	60	100.0

beneficiary attains the age of 25). Occasionally, arable land may be re-allocated by the Dagace, who is the allocating and registering authority, if land falls vacant through permanent migration. Or it may be bought (H *saye*), borrowed (H *aro*) or pledged (H *jingina*) – all rare.[28] More usually, the seeker of extra land obtains an allocation of abandoned fallow, which carries scrub woodland. Table 4.8 sets out the frequency of modes of acquisition among a sample of 60 fields (H *gonaki*). The frequency of allocation compared with other modes results from several factors: population growth, in particular new settlement by the Hausa *almajirai*; attempts to increase market production; and long-term decline in yields per hectare. Of ten Manga farmers interviewed in 1976, eight had succeeded in adding to their foundation holdings, six of them by breaking woody fallows.

Table 4.9 *Distribution of farm holdings around Dagaceri*

Distance from village (km)	No. of fields	Average size (ha)	Quadrat	No. of fields	Average size (ha)
0–0.5	19	2.32	N	23	2.26
0.5–1.75	15	1.86	E	9	1.83
1.75–3	15	1.64	S	8	1.76
			W	9	1.59
Total	49	$\bar{x}$ 1.97	Total	49	$\bar{x}$ 1.97

Yields are lower, in a dry year, on manured than on unmanured land. A fragmented holding takes account of this difference, as well as the usual ecological risks of micro-variations in soil water, showers, pests and diseases.[29] A sample of 49 fields was found to be distributed evenly in terms of distance from the village (table 4.9), though unevenly in terms of direction; the 'average' holding was distributed among two distance zones and two quadrats, at a mean distance of 1.3 km from the village.

Some 58 migrant Hausa families were allocated land in the thirty years prior to 1975. Fallow fields may be observed and, together with the average size of farm holding (5.9 ha), these facts do not indicate a scarcity of land on the scale known, for example, in the Kano Close-Settled Zone (where farm holdings average 1.4 ha and over 85 per cent of the land is cultivated every year). But this is land of poor quality; manure, labour and capital may be scarce, and constrain the poor in particular from the expansion which they know is an appropriate response to the long-term decline in yields.[30]

Only a few farmers fallow systematically, and then for a mere two or three years between cultivation cycles. Consecutive cultivation was reported for 25, 30 and even 50 years on manured land, and for up to 21 years on unmanured land. But this is far longer than ideal. The acquisition of more land is the necessary condition for taking land out of cultivation. 'Everyone has his own farm and if he leaves it [fallow], there is none he can use.' In addition, the amount of labour required for burning and clearing fallow land is such that, unless communal work groups can be activated, it is considered preferable to take over vacant arable. So the two constraints on fallow practice are a shortage of land and the cost of labour.[31]

An increase in arable land at the expense of woodland, grassland and fallows occurred in an area of 25 km² at Dagaceri between 1950 and 1981 (table 4.10). Roughly a half of this woodland and grassland bears the marks of cultivation in the past, and can therefore be categorised as fallow. The remainder is officially protected grazing land. Amongst the easily identifiable fallow fields, another survey of a larger area (158 km²) around Dagaceri suggested that grass fallows

Table 4.10 *Changes in land use at Dagaceri,*[a] *1950–81 (per cent)*

	1950	1977	1981
Woodland, grassland, fallow	71.3	65.6	60.4
Arable	28.5	34.2	39.4
Others	0.2	0.2	0.2
Total	100.0	100.0	100.0

Source: Air photo interpretation (1:30,000, 1950, Kano State Ministry of Land and Survey; 1:25,000, 1977, Kano State Ministry of Land and Survey; 1:25,000, 1981, Kano State Agricultural and Rural Development Authority)
[a] 25 km² around Dagaceri.

(one or two years only) were increasing in frequency relative to woody fallows (three or more years) between 1950 and 1969 (Mortimore, 1976: 34). This evidence points to the conclusion that the fallow system is breaking down as the area of arable land increases, in concert with population growth, to meet the finite supply of available woodland and grassland.

Cropping system

The wisdom of combining millet (tolerant of drought and fast-maturing) with guinea corn (resistant to grasshoppers, able to use late rain and residual soil moisture) has been noted. The only adaptation possible to this risk-spreading strategy is the abandonment of guinea corn if the season is expected to be too short, as occurs farther north. The second pillar of the cropping system is the mixing of nitrogen-fixing legumes with grains in alternate stands or rows. The principal legumes selected for this role were the cowpea and the groundnut, both of them grown for the market.

The varieties of cowpea normally grown in the savanna take 80–160 days or more to mature, and the harvest may be prolonged into December. They are vulnerable to insect pests, but intercropping with grain reduces this risk (Kowal and Kassam, 1978: 262–3). Rainfall conditions influence the pest population and yields are apt to vary wildly from year to year. Short-season varieties are preferred, and when the yields fell consistently below expectations in the 1970s, a fast-growing variety (called *'dan arba'in* in reference to its reputation for maturing in only 40 days) spread rapidly from Niger and Chad, unremarked by agricultural officials, and arrived in Dagaceri in 1974 (Daudu, 1976: 29). However, it did not replace longer-season varieties, and the wisdom of this was

Table 4.11 *Marketing Board purchases of groundnuts in northern Hadejia, 1963–75*

Season	Tons
1963–64	31,845
1964–65	33,235
1965–66	57,025
1966–67	48,109
1967–68	27,268
1968–69	30,009
1969–70	15,707
1970–71	4,813
1971–72	3,591
1972–73	4,878
1973–74	0
1974–75	177

Source: Ministry of Agriculture and Natural Resources, Hadejia (data for Malamaduri, Birniwa and Garun Gabbas markets)

seen when it failed in 1976. This failure was attributed to too *much* rainfall.[32]

The groundnut has almost ceased to exist as a crop in this area, which in the 1960s produced a tenth of the output of Kano State, itself the producer of half Nigeria's crop. Table 4.1 shows that the collapse began with the season of 1969. The failure of the groundnut to recover its former position was due to both rainfall conditions and the rosette virus, transmitted by an aphid (*Aphis craccivora* Koch.) that can now breed throughout the year on the irrigation schemes. In Dagaceri, yields were zero in 1976, 1977 and 1978; in 1979 none was sown. In 1981 some farmers again tried it, 'with doubting', and their doubts were justified, both then and in 1982. Thereafter nothing more was heard of it. Thus ended the career of a market crop whose early development in Kano had been little short of spectacular (Hogendorn, 1978), and which had sustained the cash economy of the village throughout the sixties.

Before the groundnut boom began, the melon known as *guna* (*Citrullus lanatus*) was grown for its oil-bearing seeds, rich in fat and protein (Dalziel, 1937: 54–5). Its abrupt reappearance, as a market crop, after 1972 (Davies, 1979a) was due to an increasing demand for melon seeds (*egusi*) amongst the urban population of Nguru and other towns. The price of washed seed rose from ₦40 per 75 kg bag in 1976 to a seasonal peak of ₦120 in July 1983. So vigorous was the demand that travelling traders bought *guna* on the farms, and its production was spreading rapidly eastwards into Borno in 1985. This ground-trailing plant is sown on upland fields in September, and uses residual moisture in the sandy soils. Although at risk from livestock (one variety is called *gunar shanu*, cows' melon), it is not harvested until later in the dry season; threshing and washing the seeds is labour-intensive but can be done at the slack time of the agricultural year. However, after the rains of 1983, pest damage reduced yields to vanishing point, and optimism about its prospects seemed premature.[33]

Between 1978 and 1981 some farmers experimented with a variety of *Sesamum indicum* (H *ri'di*) which was said to have been introduced from the Sudan by returning pilgrims in the sixties (Davies, 1979b). It grows late in the season, and is used for making sweet cakes, but its oil content is known. The crop had been officially promoted in 1975–76 in Gumel (80 km to the west), where it sold for ₦500/ton in the market and ₦250/ton to the Marketing Board. But in Dagaceri, no market outlet was known in 1981, so it had disappeared from the farms by 1983. Another minor crop, the Bambarra groundnut (H *gurjiya*), also commanded too low a price to be grown for sale.

Abalu et al. (1983: 29–30) reported the adoption in this area of new varieties of millet—popular owing to their ability to yield well with little rain—and of guinea corn, bearing no resemblance to varieties promoted through formal extension in the recent past. They, too, were thought to have entered the country from the north. Davies identified five early-maturing varieties of guinea corn whose better performance – in years of early terminating rains – had attracted interest, and whose derivation was thought to be in the gene banks of northern Gumel or southern Niger.[34]

In the cropping system, therefore, the experience of one village shows that adaptive response has both drawn on accumulated experience from the past and sought innovative alternatives from elsewhere, in particular to the stricken groundnut. Remote rural communities rarely get full credit for such experimentation even if the strengths of their basic crop mixtures have been conceded in recent years. Neither *'dan arbai'n* nor *guna*, however, have been able to provide an adequate substitute for the groundnut, and the selling of grain has consequently increased. Moreover, the northern provenance of several of the new crop varieties invites a hypothesis of environmental change: in other words, the farming system of the northern Sudan zone is exchanging some of its elements for those of the Sahel, in response to a corresponding shift in rainfall distribution.

New technologies and capital supply

The shortage of farm labour in Dagaceri is not perceived in the village as a consequence of seasonal migration (which dovetails with the requirements of farming), nor to permanent out-migration, but to a long-term decline in yields leading to an inevitable increase in the size of farm holdings. Labour-saving technologies have a recognised importance, but capital is scarce. Three possibilities are available: the *ashasha* hoe, the ox-plough and the tractor.

The long-handled *ashasha* (H) is a weeding tool known in francophone countries as *iler*. A double-edged crescentic blade is pushed forward and backward just below the sandy surface, slicing off the young weeds from their roots. Aside from giving greater comfort than the back-testing short hoes (H *fartanya, magirbi*), it permits considerably more land to be covered in a day. It is well adapted to the light sandy soils, which do not have to be ridged before planting, and to wide stand spacings, but may be supplanted by the short hoes later in the season, when the crop cover thickens and the legumes are spreading among the grain.

The *ashasha* is used widely in Niger, Chad and the Sudan and, significantly, was reported to be on the wane in densely populated areas of Niger in the 1960s, in association with intensifying labour inputs (Raulin, 1964: 22–30). However it has spread rapidly into Nigeria with migrants. Its popularity increased further when most of the plough oxen died or were sold in the early 1970s. As a 'poor man's plough' it had been adopted by all the Manga farmers of Dagaceri by 1980.[35] It spread on to Birniwa, 13 km away, mostly after 1975.

No less than 46 per cent of the farmers of Dagaceri had access to the services of an ox-plough before the drought of 1972, 34 per cent by hire and the remainder by ownership. In 1974, only 4 per cent used a plough. The same experience was reported from neighbouring villages; the cause unambiguous: the loss of the work bulls. This part of Kano state had been in the forefront of a largely uncharted diffusion of the plough. Loans were still available in 1982 from at least one local government authority in the area (Abalu *et al.*, 1983: 41). A deposit of ₦150 was necessary to secure a loan of ₦1,660 for the purchase of a pair of work bulls (₦520), a plough (₦808) and a cart (₦332).

Not until 1979 did plough ownership reappear in Dagaceri. In that year, two farmers bought ploughs, and bulls at ₦200 each; in the following year there were four, in 1983, five and in 1985, seven, when they were reported to be as numerous as before the drought. This 12-year cycle of recovery was not, however, financed from the proceeds of agriculture. One of the first two owners was a tailor, and all obtained money from seasonal migration. A side effect of the return of the plough was the appearance of seven ox-carts by 1985, the beginning of the solution to the transport bottleneck noted earlier.

There were no tractors in the area before 1972. Ownership is probably still beyond the means of anyone in Dagaceri. In 1981 a new hire service at Birniwa, charging a subsidised rate of ₦5/ha, immediately became popular, but from 1983 its inability to satisfy demand, especially at greater distances, provoked scepticism, and to depend on it would have been folly. Elsewhere in Nigeria, the failings of government tractor hire services appear to have led to an increase in private tractor ownership (Kolawole, 1974); in poorer Dagaceri, the ox-plough is seen as the natural substitute.

Having failed to make good the loss of the groundnut – and with

5 The *ashasha* in use at Dagaceri (June 1978). The first rain has fallen and millet has been planted in unprepared ground. Weeding is skilfully accomplished between the rows when the seedlings are barely visible. Beyond the infields, cultivated annually, the village may be seen accumulated capital eliminated during Kakaduma – the farming sector's dependence on intersectoral transfer of capital, and the long recovery period of 12 years in ox-ploughing, are clear evidence of the need for enlarged credit facilities if labour-saving technologies, other than the *ashasha*, are to be extended in the future.

ADAPTIVE RESPONSE: OTHER SYSTEMS

The village herd

The misfortunes of the village's livestock holdings – both a repository of domestic capital and a source of recurrent income – demonstrated the impact of drought in the short term, and also the longer-term difficulty of reconstitution under unfavourable terms of exchange. Among a sample of 25 families, there had once been 39 cattle, owned by 14 individuals; 137 sheep or goats, owned by 15; a number of donkeys, and a camel. All but one cow, and 17 sheep or goats, disappeared after 1972. Only four households still owned animals, other than fowls, in 1974.

The setback had not been reversed by 1985. The price of cattle, having collapsed from ₦60 before the drought to ₦5 early in 1973 in Kwubsa market (and even to zero in remote places), rose to ₦100 in June 1975 and continued to climb thereafter; prices in 1985 ranged from ₦160 for a poor-quality animal to as much as ₦600 for a prize bull. Camels were worth ₦600 as early as 1976. Donkeys fell to ₦10 in 1973, rose to ₦40 in 1976 and to ₦80 by 1985. Small livestock doubled to ₦40 in the same ten years. In terms of grain equivalent, these movements always worked against the farmer, for when terms were favourable to him, grain was scarce.[36] The position was better with small livestock than with cattle. Their advantages – shorter reproduction cycles, lower unit cost, more efficient and adaptable grazing, ease of management by women and children, and greater divisibility as savings – were responsible for their popularity in densely populated areas. The droughts of the seventies underlaid a significant decline in cattle ownership by sedentary farmers, always more important in northern Nigeria than is generally recognised.

The experiences of three families were representative. The first, whose smallholding was supplemented by income from mat-making, had lost three cows and two sheep in Kakaduma; the wife alone had succeeded in acquiring one goat by 1985. A second had lost a camel, a donkey and three sheep. Late in 1984, the head of the family was able to replace the donkey, and was back on the road to earning revenue. The third, although the head was the largest farmer in the village, emerged from Kakaduma with no livestock. In 1985 he had a pair of work bulls, bought the year before, one sheep and a goat. Lacking a breeding capability, and with the terms of trade moving against them, farmers found a return to livestock ownership, via the market, eluding them (as elsewhere in the region: Grégoire, 1983).

Selling fuelwood and fodder

The air photographs suggest a need for caution before accepting the view that woodland has diminished in unit biomass since mid-century; however it has certainly diminished in extent (table 4.10). Fuelwood may be obtained from lopping the branches off farm trees as well as those of more distant woodland. The cutting of whole trees is forbidden by the local authority without licence, but the ban is not effective; and in the forest reserves, illegal woodcutting occurs despite the existence of forest guards.

The trade in fuelwood is limited to the village and local markets, where a donkey load fetched ₦2.50 in February 1985; but along the new highways further away, stacks of firewood appear in response to the growth of intercity traffic, especially following a bad harvest. Its price rises with inflation. The deterioration of surviving woodland is reported, and the pressure will increase if Dagaceri falls within the orbit of the voracious fuelwood dealers of metropolitan Kano.[37]

Dry grass fodder (H *budu*), which is free for the collecting, became commercialised in Dagaceri after 1982. An overall shortage of feed, the influx of herds from farther north, and farmers' own needs for income following poor harvests, combined to create a volatile market in fodder. Grass that had been collected—using locally smithed steel rakes—was not available for grazing; and the dry-season labour surplus of the farming sector could be used to hold the visiting herders to ransom. Prices rose from less than ₦1/bundle at the beginning of the season to an outrageous ₦7/bundle before the rains of 1985. Fallow land was systematically stripped of three-quarters of its dry matter, and herders had to sell animals to buy it, or watch their herds starve.[38] The new trade stimulated demand for blacksmithing and cart transport, and had far-reaching implications for relations between Manga farmers and Ful'be pastoralists.

Mat-making

In times of trouble (H *wahala*) and financial adversity (H *matsuwa*), such as the years of Kakaduma, had it not been for mat- and rope-making 'we would all have gone away from here'. The leaves (H *kaba*) of the regenerating dum palm appear everywhere. Mat-making has been pursued in the village, and in the Hadejia flood plain to the south, for at least three decades. Thereby, the head of a family having a high dependency ratio, a small farm and insufficient resources to hire labour, could spread his own labour throughout the year, and hope to mitigate his poverty until he had sons old enough to work. The number of mat-makers in the village oscillated between half a dozen after a good harvest to 80 per cent of the menfolk in 1973–74.

Gathering the *kaba* is quick, and the skill fairly easy to learn. A man can make two mats in a day; and their price rose from ₦0.25–₦0.35 in 1975 to ₦0.80–₦1.50 (depending on the season) in 1985. Urban traders buy in bulk at

6 Rope- and mat-making gets under way after a bad harvest in Dagaceri (October 1986)

rural periodic markets, directly from producers. But the trend in terms of trade has been adverse. Plastic imitations are flooding the market nationwide. In ten years, the price of grain inflated by 800 per cent, and that of mats by less than half as much. A day's work in February 1985 yielded barely enough to buy one person's grain ration; but the terms became more favourable later in the year. Consequently mat-makers work throughout daylight hours with impressive devotion, especially after a bad harvest.

Transport

The donkey population was decimated in 1973. Not only did this virtually stop manuring and necessitate much headloading of farm produce, fodder and fuelwood: it also introduced a constraint in produce marketing. Before, the village's numerous beasts, along with a few camels, were used to carry groundnuts to the licensed buying agents at Garun Gabbas and Malamaduri (25 and 40 km south-west). During the dry season, some animals were employed in collecting natron (H *kanwa*) from the saline lakes of the Manga Grasslands (a seven- to eight-day return trip) and carrying it to Kano (14–20 days). There were no road links from Dagaceri, and the soft sand does not permit the use of cycles.

A general decline in the use of donkeys, in favour of the ubiquitous motor pick-up, has not reduced the value of the donkey and the camel as sources of income. The commission on a round trip to Kwubsa market is ₦1, and several owners are reluctant to use their animals for carrying manure to their farms for fear of missing such revenue. An increasing demand for fuelwood and fodder from distant locations has set a premium on transport, and seven ox-carts have made their appearance. Donkeys are not bred in the village, and only the male is worked. Both working donkeys and camels must be purchased. Given the high cost of returning to animal ownership, they are obviously associated with capital accumulation.[39]

ADAPTIVE RESPONSE: TWO FAMILIES

The fluctuating fortunes of two families in Dagaceri encapsulate the experiences of the village in microcosm. The first is headed by the largest farmer in the community, and the second is heavily dependent on income from mat-making to supplement agricultural production; the first normally has a food surplus and the second a deficit. Both, however, are assisted by income from *cin rani*, earned by a brother, and a son, of the head.

M is the son of a labour-hiring farmer who died in 1963, leaving his holding to three sons who agreed to work it jointly under the elder's leadership (an arrangement called *gandu* in the literature on Hausaland: Hill, 1972: 38–56, 259–50). One is now married with four young children; one remains unmarried; M and his wife are childless; so family labour is limited to three men, two wives and one older child. Unassisted, they would be able to work only the two inherited farms of 3.5 ha which lie in the manured zone close to the village (1 and 2 on figure 4.6).

However, the brothers wished to sell crops for profit, and obtained four additional farms by gift and purchase in 1965–66, spending ₦25–30 each on two of them, and thereby increasing their holding to 19.1 ha, the largest in the village. At about ₦70/ha, the estimated value of the holding is now ₦1,344. They hired an ox-plough for ₦100 to plough two farms having heavier soil suitable for guinea corn; and they hired labourers for clearing, planting and weeding at ₦1.50/day (around 1970). Thus they expanded their operations; and a good yield in the sixties was 10 tons of millet, 5 tons of guinea corn, and 1 ton of cowpeas (all unthreshed), plus 2 tons of shelled groundnuts from two of the farms. This implies grain yields of 740 kg/ha averaged over the whole, plus yields of intercropped cowpeas of 50 kg/ha. They owned cattle, sheep and goats.

Nevertheless, the droughts of 1972 and 1973 made short work of their prosperity. By 1974 their granaries were empty, they were dependent on the market for grain, all the livestock were dead or sold, M had joined the mat-makers, and his wife the gatherers of *tafasa* and *tabila* berries (see table 3.12). Drought seemed to be the great leveller.

Afterwards, M continued to hire ploughs and labour, sell crops (millet,

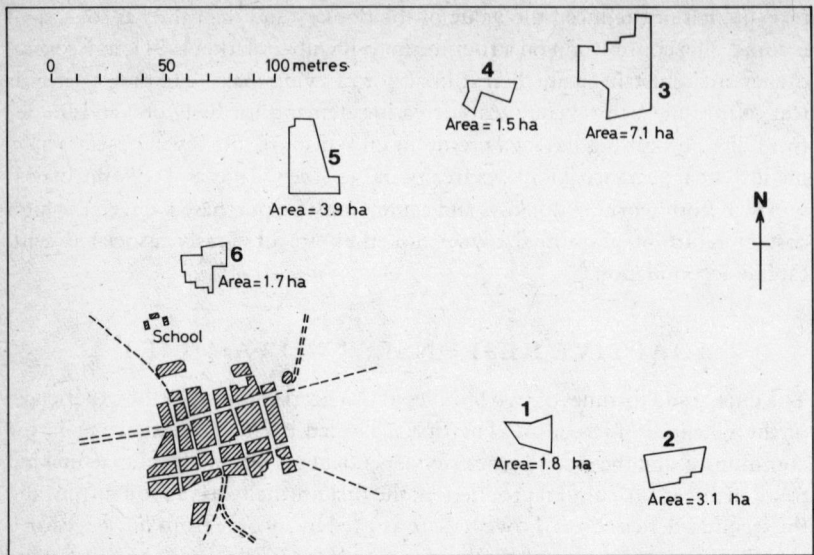

Figure 4.6 M's farms at Dagaceri

guinea corn, cowpeas, a few groundnuts, and *guna*) in the good years, and buy and sell livestock. In 1984 he invested ₦120 in 20 bags of fertiliser. By 1985 he had a sheep, a goat, and a pair of recently acquired work bulls. The rising cost of hired energy had led him to try the new government tractor hire service (which let him down), and then to buy a plough and workbulls, though he was not able to use them in the first year owing to a lack of handling experience. He had seen millet rise in his time from ₦2 to ₦88 a bag in Kwubsa market. It was not price movements which he considered to be a problem, but the uncertainty of yields, which made the hiring of ox-ploughs and labour, and indeed any capital inputs, very risky.

The extent of risk is clear from M's farming operations in 1986, the year of the pests. On his largest farm of 7.1 ha (3 on figure 4.6), he could expect a maximum yield of about 3.75 tons – about 528 kg/ha. At the highest prices of 1984 and 1985 – over ₦1,000/ton – this would be worth ₦3,750; at the price obtaining in October 1986 – ₦500/ton – it was worth ₦1,875. His estimated labour costs for planting, first weeding (110 man-days), second weeding (75 man-days), harvesting and carriage, at ₦4/man/day, were ₦905, allowing a considerable margin even at the lower price. But the rodents ate the seed, and his reserves ran out; he had to purchase more.[40] The grasshoppers demolished his yields, and he harvested only 0.9 ton – worth ₦450. Equivalent losses were made on his other farms.

The ox-plough is not only a more efficient source of energy; it also allows work normally done by hired labour[41] to be transferred to family labour, removing it from the cash account; but its use was restricted to farms 1 and 6 in the manured zone, where guinea corn was planted on ridges and, even there, labour had to be hired for weeding. However, having observed that millet grew well on ridges, M intended to extend the use of his plough to his other fields. Ploughing would probably render the use of the *ashasha* impracticable, increasing the cost of late weeding; only if one weeding could be dispensed with would it save him money.

M intended to stay afloat in 1986–87 by trading in cowpeas. He did not expect to change his plans the following year. His brother still goes on *cin rani*. He does not consider himself to be as prosperous as he was in the groundnut boom before Kakaduma. Nevertheless, he wants to enlarge his holding further, and if so, may decide to fallow two less productive farms. Although still committed to enlarging his operation, with yields reduced, and grain prices falling (in 1986), lasting accumulation may yet linger beyond his grasp – as it did in the 1960s.

At the other end of the spectrum, now in his fifties, A inherited two farms, less than 4 ha. By the late 1960s his dependants numbered six (his mother, elder sister, wife and three young children); he took up mat-making then and has continued in it ever since. His livestock (three cows, two sheep, a donkey and fowls) disappeared in Kakaduma – a cow and a sheep died, and the rest were sold.

Gathering the *Kaba* leaves for six mats takes a couple of hours. A can make two mats between 7 a.m. and 2 p.m., which he accumulates during the year (even in the rains) to sell in bulk to a visiting trader. At prices prevailing in 1985, three mats would purchase one *tiya* (about 2.5 kg) of grain, not more than half what he needed to feed his family for a day. At that time, he was living with a second wife and small son; his son by his earlier marriage, now adult, on *cin rani* for much of the year, and married; and his son's small step-daughter: four adults and two children. During the early seventies, with grain prices lower, the terms of exchange had been more favourable. Grain prices fell throughout 1985, and his position improved. The elder son was successful in goat trading and able to contribute financially.

The rest of the food came from the farm. In 1980 he was allocated a farm left vacant by a migrant; this permitted him to fallow an exhausted farm for three years – considered to be enough – and in 1985 he planned to work all three, about 5 ha in all. For this his son's labour was essential. Lacking working capital, hired labour or ploughing were out of the question. His wife's goat was their only livestock. Yet 1984 had not been a bad year: with over a ton of grain, he had enough to last through to the following harvest at about 3 kg/day, plus his purchases. He considered that his economic position had improved since Kakaduma: for he was now able to feed his family.[42]

A BALANCE SHEET

Under the onslaught of recurring droughts and subsistence crises, Dagaceri has stabilised in terms of demographic growth and net household formation. In-migration from regions worse affected has roughly balanced out-migration, and the indigenous Manga have exerted a tenacious hold on their heritage. Land as well as livelihood shows the marks of the struggle: shortened fallows, diminished fertilisation, and rising scarcity of woodfuel having ominous significance. But the productive systems are resilient. Somehow Dagaceri survived, and during the eighties exported perhaps 10 tons of grain a year (or more) to the market.[43]

If the manuring and fallowing systems appear to be breaking down, access has been gained to the new input packages offered by the Kano Agricultural and Rural Development Authority. Farmers able to produce for the market have lost the groundnut as a revenue earner but gained a fast-maturing variety of cowpea, enjoyed a flirtation with the *guna* melon – perhaps short-lived – and benefited from inflating grain prices. In Kakaduma the village lost the plough but gained the poor man's substitute, the *ashasha*. The reappearance of the plough after a decade is a testimony to the possibilities for accumulating some productive capital, even in these conditions.

The village lost most of its savings but gained a remission from taxation. It lost most of its transport animals but gained access to motor transport. It lost access to cheap food, fuelwood, fodder, labour and land; it gained improved access to the wider economic system, through cheaper travel, and a wider market for its mats, labour or entrepreneurial skills.[44] Self-sufficiency, if it ever existed, has been lost but through the school, and travel, a window has been opened on the world.

Few in Dagaceri would claim unequivocally to be better off than in the heady days of the groundnut boom. On the other hand, no one would deny having improved his situation since the unforgettable days of Kakaduma, notwithstanding rainfall failures and calamitous harvests in the eighties that rivalled those of the seventies. However, economic survival, and even a selective improvement in living standards, have been achieved by calling, increasingly, on resources outside the village system. The key that has unlocked this door is personal mobility.

5

WIDER HORIZONS

Go down and buy grain for us there, that we may live

Genesis 42:2

Spatial mobility is an entrenched characteristic of West African populations, but its importance in the economic strategies of individuals and households is rarely given adequate recognition in sectoral studies. Not all mobility has economic objectives, of course. Prominent exceptions are: the periodic circulation of koranic scholars (H *almajirai*) and of schoolchildren; the pilgrimage to Mecca (*haj*; Birks, 1978); and migration for marriage (e.g. Mortimore and Wilson, 1965: 35–7). The existence of such objectives implies that a strictly economic interpretation of mobility is inadequate. However, an inescapable factor in the system is drought, whether viewed on a seasonal or an annual basis. A sudden intensification of drought, as observed from behaviour in the past, may be expected to produce corresponding changes in the intensity and characteristics of mobility.

MOBILITY IN THE CITY: KANO'S STRANGERS

The population of the old walled city (Birni) of Kano was swelled each dry season by numbers of men and boys who left their villages for a few months in order to pursue koranic study (H *karatu*) or practise secondary occupations (H *sana'a*). Such visitors had a claim, under an Islamic convention of hospitality, to sleep in an entrance hall (H *zaure*) of a city householder (H *maigida*).[1] Short-term mobility (*cin rani*) for *karatu* and *sana'a* is deeply embedded in Hausa tradition, and the destinations chosen are rural as well as urban. But owing to its religious status and commercial development, Kano is the most important objective in the Muslim north of Nigeria.

A survey of 297 Hausa-speaking Muslim strangers (H *bak'i*) was carried out in the Birni and township of Kano in the dry season of 1973–74. Of them, 232 were found inside the walls, and the remaining 65 at various locations in the newer parts of the urban area (figure 5.1). The ethnic and religious affinity linking the city with its historical hinterland ensures its predominance over the suburban communities beyond its wall, where the strangers were fewer, and harder to locate.[2]

Practice of *karatu* was not incompatible with earning income from *sana'a*,

118 *Wider horizons*

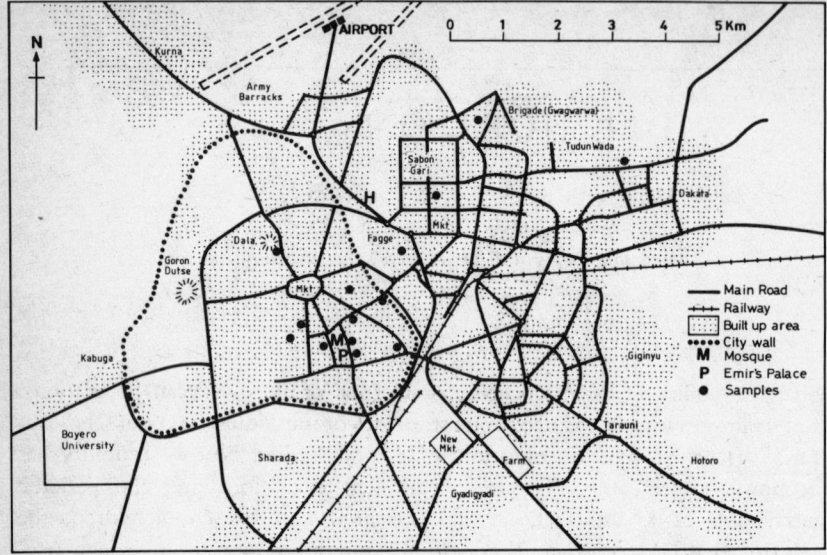

Figure 5.1 Urban Kano, showing sample locations

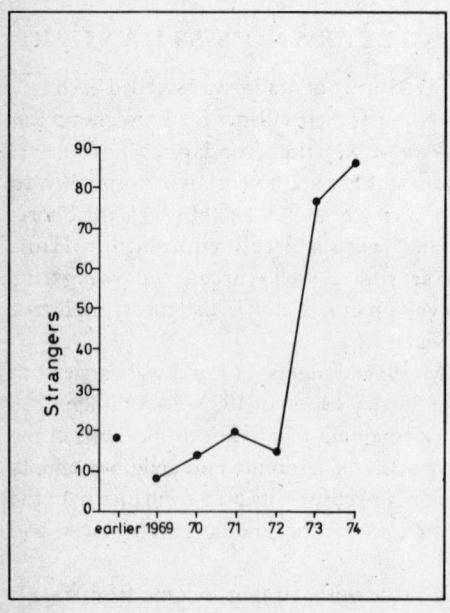

Figure 5.2 First arrival in Kano: strangers interviewed in January–April 1974 ($n = 242$)

Mobility in the city: Kano's strangers

although the more committed scholar might depend only on alms for support. Thus 40 per cent gave *karatu* as the reason for their current visit, and 60 per cent gave *sana'a*; the proportions for previous visits were similar. The system was capable of absorbing a substantial increase in numbers when rural food scarcity became acute. Such an increase occurred in 1973 and 1974 (figure 5.2). The absorptive capacity of the urban informal sector was then approaching a peak, owing to the ramifications of the oil boom in the Nigerian economy.

Origins and distribution

The geographical and ethnic origins of the strangers reflected both the dominance of Kano in Hausaland (within and outside Nigeria) and also the

Table 5.1 *Geographical and ethnic origins of strangers interviewed in Kano urban area in the dry season of 1973–74*

		Location of sleeping quarters			
		Inside the Birni	Outside the Birni	Total	Per cent
Geographical origin					
Niger Republic		27	36	63	22
Kano state:	north of 12°N	68	5	73	25
	south	107	6	113	39
Nigeria:	north-west[a]	13	10	23	8
	north-east[b]	7	4	11	4
	Middle Belt[c]	1	3	4	1
Other places:	Chad, Mecca[d]	4	1	5	1
Total		227	65	292	100
Ethnic origin					
Hausawa		189	24	213	72
Buzaye[e]		26	37	63	21
Others[f]		14	5	19	7
Total		229	66	295	100

[a] Sokoto, Katsina, Daura
[b] Borno, Bauchi, Adamawa
[c] Zaria, Niger state
[d] Including four returning from the *haj*
[e] With Adarawa, 'Faransawa' (from Niger)
[f] Fulani (2), Kanuri (5), and others

predominance of the Hausa in northern Nigeria (table 5.1). Almost two-thirds came from within Kano state, and 72 per cent claimed Hausa ethnicity. After Kano state, the next most important area of origin was neighbouring Niger. The spatial pattern of movement reflected the city's hinterland, rather than latitude (as a proxy for aridity): Kano state, *south* of the latitude of Kano, provided the largest number.[3] The insignificant numbers of Fulani and Kanuri strangers (ethnicities both well represented in the hinterland) derived from the commitments of the farmers to their livestock, and the negative view of Kano held by the Kanuri.

The social geography of urban Kano further underlines the distinction between those who belong and those who do not (figure 5.1). The Birni, although accepting small communities of permanent migrants in the past, is overwhelmingly Hausa in culture. It attracted the great majority of Hausa strangers. The suburbs outside (Fagge, the precolonial caravan terminus; Sabon Gari and Gwagwarwa, areas of southern Nigerian or mixed migrant settlement) were preferred by members of the other ethnic groups. The only important exception was a concentration of Buzaye in Kofar Mata and adjacent wards of the Birni. The Buzaye,[4] although originating from Niger, have enjoyed close commercial and cultural ties with Hausaland for centuries, and many have settled there and adopted Hausa ethnicity.

The situation at home

Scanty information on the situations that the *masu cin rani* had left at home suggests that most of them retained a few assets (figure 5.3). Only 13 per cent claimed to be landless, and less than 3 per cent reported that they had sold land during the last two years. The distribution pattern was characteristic of rural Hausaland, the majority having from one to three small fields. These had been left in the care of parents (57 per cent) and relations (22 per cent). The ownership of animals was also characteristic, and was marked by inequality, and a large proportion of non-owners (between 60 and 70 per cent). This distribution had already borne the impact of drought losses. Those who reported losses by death in the two years 1972–74 were nearly as numerous as the surviving owners of cattle, sheep and goats (table 5.2). None of the 40 Buzaye interviewed said they had any livestock left.[5]

This picture of an impoverished, but not generally destitute, population of strangers is borne out by their family arrangements. Some 18 per cent had no wife or children, and a further 17 had no children. These were typical *masu cin rani* and might be expected in any year. Of those married with children (65 per cent), only a quarter – 16 per cent of all respondents – had brought a wife and some children with them. This was a departure from normal practice, particularly among the Buzaye, and indicated the extreme hardship of the time. The remainder had left them at home, though whether they had sufficient resources to live on is another matter.

Mobility in the city: Kano's strangers

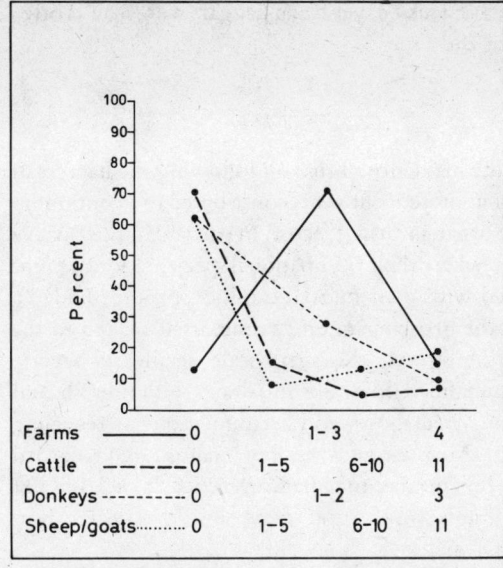

Figure 5.3 The situation left at home by migrants to Kano

Table 5.2 *Livestock ownership, Kano strangers in January–May 1974*

	Reporting deaths			After the drought	
	1–2	3	Total	Owners	Non-owners
Cattle ($n = 241$)	33	24	57	69	172
Donkeys ($n = 244$)	33	3	36	93	151
Sheep and/or goats ($n = 246$)	14	62	76	96	150

Adjustment and employment

Four in ten of the strangers were lodging with kinsmen in Kano, but only two with townsmen. The Buzaye depended to the greatest extent on such ties (six in ten of them were lodging with Buzu who were also kinsmen). Those lodging outside the Birni, not surprisingly, used such ties the least (two or three in ten). Kinship was thus more important than geographical commonality in assisting adjustment to the urban environment. This was not a *maigida* or commercial landlord system (cf. Cohen, 1969). Only three individuals (of 245) said that they were given employment by their host.

Getting income was relatively easy, and 93 per cent said they had found some on the day of their interview. However, distinguishing between *sana'a* (the practice of an occupation), *karatu* (koranic study) – together reported by 63 per cent – and *bara* (begging alms) – acknowledged by 37 per cent – is impossible, since *karatu* often includes receiving alms. The Hausa strangers lodging in the Birni were more ready to call a spade a spade and admit to *bara*. The Buzaye insisted, to a man, that they practised *sana'a*; yet Buzu beggars were seen as often as any in the streets of Kano at the time.

The later seventies

Although there was a drop in the incidence of *cin rani* following the harvest of 1974, the recurrence of drought in subsequent years contributed to a continuing intensification of activity in the urban informal sector. In particular, the Buzaye were conspicuous on the streets, where their occupational specialisation as night guards (Armstrong, 1967) fitted with their martial traditions, and provided a correcting mechanism against the growing rate of crime, itself related to the hardness of the times.[6] On the streets, the small-scale retail and service subsector[7] invaded the traffic lanes, filled sidewalks and drains with rubbish, and covered vacant land with improvised shops, and accumulations of recycling materials such as discarded tyres. An incomplete survey (Gano, 1983) recorded over 10,000 street hawkers; the full number may have exceeded 25,000. It is not known how many were migrants from rural areas, but the system was controlled by urban traders (Khan, 1984).

The threat to urban security that was posed by the growing street population of Kano was hideously fulfilled in the insurrection that occurred in December 1980 under the leadership of Mohammadu Marwa ('Mai Tatsine'). With a unique mix of Islamic heresy, populist fundamentalism and witchcraft Marwa drew a vast following among the teenage boys who travel around for *karatu* every dry season. A number of them came from Niger, Chad and Cameroon, and it will never be known to what extent harvest failures played a part in sending them to the city. Marwa's short-lived republic was located in what had once been a colonial building-free zone between the wall of the Birni and the old caravan terminus of Fagge: an urban no-man's land marginal to the centres of power, both traditional and modern (figure 5.1). After the carnage was over, following army intervention, the deaths were variously estimated up to 3,000.

The expulsion of illegal aliens from Nigeria, early in 1983, caused the disappearance of the Buzaye, and in its first six months of office the military government which assumed power in January 1984 cleared away the street economy. This was justified partly in terms of returning labour to the land and partly for the improvement of urban sanitation, but its relevance to urban security was obvious, even while successors to Mai Tatsine were generating further outbreaks in Kaduna, Maiduguri and Yola. These developments notwithstanding, the persistence of *cin rani* is ensured by the continuing weakness of the structure of opportunities in rural areas.

MOBILITY IN THE REGION

The escalation in mobility that occurred throughout the hinterland of urban Kano in 1973–74 was investigated, to begin with, through the knowledge of

Mobility in the region

the heads of 631 villages whose experiences of drought and food shortages have been reviewed in chapter 3. Quantitative answers were, of course, out of the question. Since village heads are responsible for the allocation of land to newcomers, and the re-allocation of land belonging to those who have left, they were in a position to report better on permanent than on seasonal movements. It is possible that errors of judgement were made; that an answer might relate only to a part of the village area most closely observed by the head; that behaviour deliberately concealed (e.g. by tax defaulters) was inaccurately reported. However, the results allow an approximation to be attempted of the pattern of mobility at the macro-scale.

Out-migration

Even before Kakaduma, it was not unknown to uproot and move in response to hardship, the most commonly stated reason. But during the famine, the number of villages reporting out-migration (understood to be permanent) jumped from 26 to 43 per cent, and a further 51 per cent reported empty houses, whose owners were still expected to return.

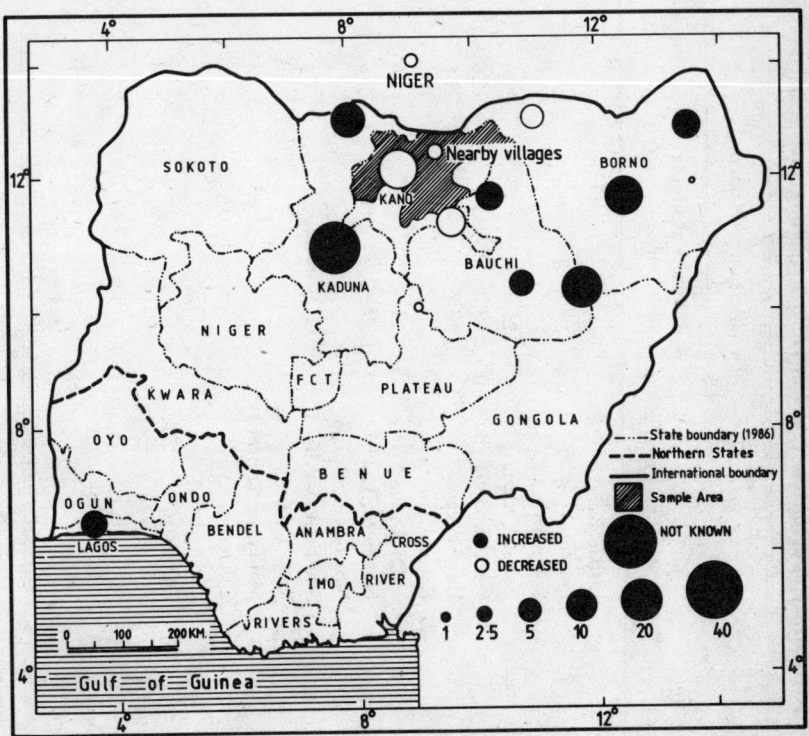

Figure 5.4 Out-migration: frequency of destinations cited by village heads in the sample area, 1973–74 (showing increase or decrease relative to earlier out-migration), per cent

Table 5.3 *Frequency of destinations cited in villages reporting out-migration, 1973–74 (per cent)*

Region	Destination group	Increasing frequency		Decreasing frequency	
		Before Kakaduma (n = 140)	During Kakaduma (n = 483)	Before Kakaduma (n = 140)	During Kakaduma (n = 483)
Kano state	Urban Kano			20.1	18.2
	Northern Kano			10.5	2.6
	Southern Kano			14.7	10.8
Northern states	Zaria, Kaduna	26.2	29.6		
	Gombe, Bima (Bauchi)	19.2	22.6		
	Borno, 'east'	17.7	23.6		
	Katsina, Daura (N Kaduna)	10.6	11.8		
	Northern Bauchi	9.9	10.6		
	North-west Borno	6.2		6.3	5.4
	Southern Bauchi		7.2		
	Jos			4.2	2.5
	Maiduguri (Borno)			1.4	0.6
	Others			3.5	2.2
Southern states	Lagos	3.5	6.0		
	Others			1.4	2.2
International (Niger, Chad)					
	'Everywhere'	7.1	28.2	2.8	1.4
	Concealed	6.4	8.7		

Note: Destinations attributed to temporary migrants are included.

Table 5.4 *Frequency of origins cited in villages reporting in-migration, 1973–74 (per cent)*

Region	Origin group	Increasing frequency		Decreasing frequency	
		Before Kakaduma ($n=134$)	During Kakaduma ($n=71$)	Before Kakaduma ($n=134$)	During Kakaduma ($n=71$)
Kano state	Nearby villages	19.4	28.2		
	North-eastern Kano	17.1	21.0		
	Kano city area			9.6	8.4
	Others			7.2	5.6
Northern states	Katsina, Daura (N Kaduna)			17.8	13.0
	Borno, 'east'	3.6	4.2		
	Sokoto	2.9	4.2	6.3	1.4
	Others			0.7	0
Southern states					
International (Niger)				35.7	26.7
Not identified				2.9	2.8

The destinations given by the village heads are set out in table 5.3. Their relative frequencies are a poor proxy for the volume of actual migration streams, but do reflect the informants' perceptions of the distribution of alternative opportunities in space. Some accuracy was lost through the habit of using the name of a major city to describe its rural hinterland as well; thus rural and urban destinations could not always be distinguished. However, table 5.3 and figure 5.4 show that a decline occurred in the popularity of destinations in areas adversely affected by drought – Kano State (even urban Kano), north-west Borno, Niger and Chad – while some major urban destinations further south were mentioned with greater frequency: Zaria, Kaduna and Lagos. An increased frequency of some destination groups east and west of Kano – especially Bima (in Bauchi), Borno and the east (associated with the direction of Mecca, and eschatological expectations which revive at times of disaster) – was certainly not based on accurate information on the food supply, which was as difficult there as in Kano. The great increase in the category 'not known' reflects the village heads' poor knowledge of journeys as yet uncompleted.

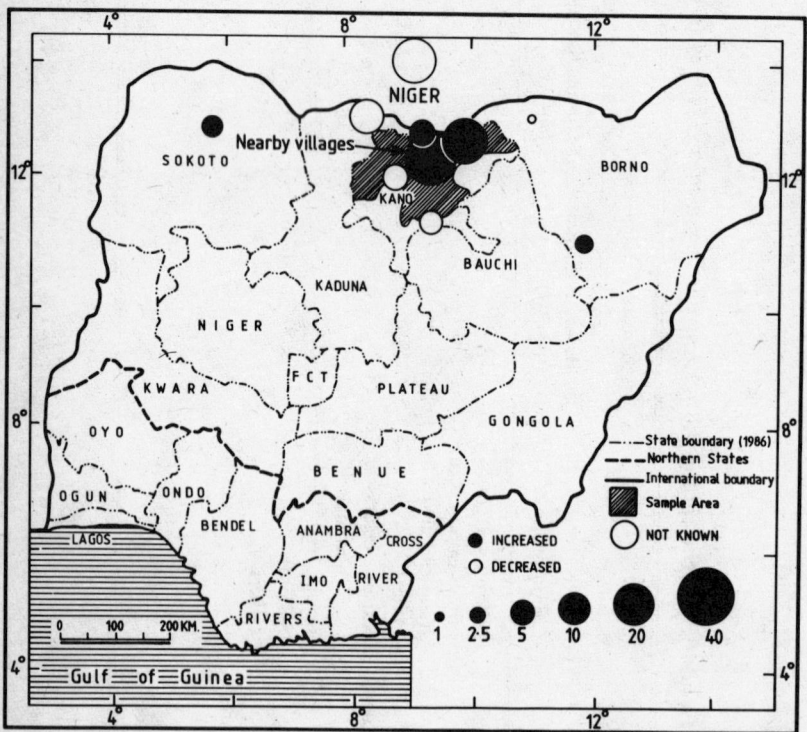

Figure 5.5 In-migration: frequency of origins cited by village heads in the sample area, 1973–74 (showing increase or decrease relative to earlier in-migration), per cent

Mobility in the region

In-migration

While Kakaduma nearly doubled the incidence of permanent out-migration, from 26 to 43 per cent of villages, it reduced that of in-migration from 21 to 11. It may be conjectured that an approximate balance may have been replaced by net out-migration. Most of the villages reported families looking for land or grazing. Table 5.4 shows their places of origin (figure 5.5).

The first two rows in table 5.4, which report movements within the drought-affected zone of Kano State, suggest a loosening of territorial ties in a time of disaster, as more families decided to try their luck in another village where, perhaps, they had heard that conditions were better, or had previously contracted kinship ties through marriage. The same would be true of those who moved in from Niger, who could remain within the same ethnic community (Hausa, Fulani or Manga) when crossing the border. A southerly movement from Niger has long been a recognised feature of the demography of Kano state (McDonnell, 1964; Mortimore, 1968). Its reduction in 1973–74 may be explained by food aid over there; also, certain areas in Niger suffered a rainfall deficiency in 1973 that was less severe than in Nigeria.[8]

Before Kakaduma, the incidence of in-migration was determined by land

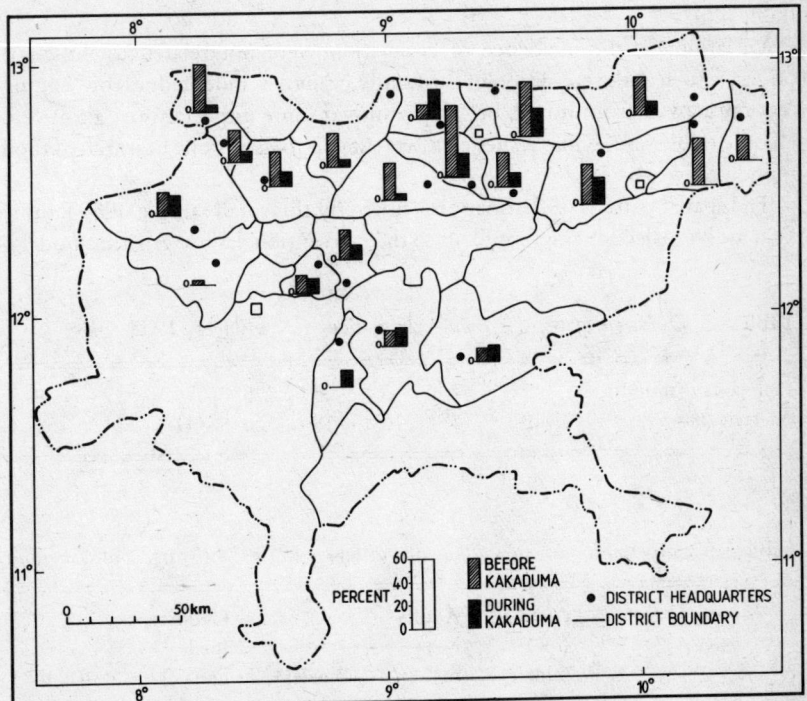

Figure 5.6 In-migration: percentage of village heads reporting, by district (Kano State)

supply: it was most frequently reported from villages having a land surplus, and was more common in the northern districts, where population densities were lower than in the districts of central Kano State. During Kakaduma, however, this pattern was reversed (figure 5.6). All but one of the northern districts reported a decreased incidence, ranging from 16 to 40 per cent of villages, but in the three southern districts the incidence increased from 8 to 15 per cent.

Circulation

Time-series data would be necessary to measure oscillations or establish trends in dry-season circulation. Since the village heads were not in a position to provide useful information, this review is restricted to the data collected in the five villages (figure 3.3), where 47 per cent of family heads reported that some members of their families had been away during the dry season of 1973–74.

A bi-modal pattern occurred in the length of their trips (table 5.5), the most common being seven months – related to the expected duration of the rains (five months) – and only rarely exceeded.[9] The mean incidence of circulation in participating families was three persons (ranging from 2.1 to 4.3 among the five villages); this had risen substantially in the second year of drought, and women, children and even the aged had joined the regular *masu cin rani* (table 5.6).

Variations in the incidence of circulation were not related to indices of poverty, such as the condition of the family granaries, indebtedness or begging. Yet once away from home, begging assumed some importance as a source of income (table 5.7), while 90 per cent were stated to have gone in search of food, money or work.

The spatial pattern of destinations chosen by this small sample of 59 families (206 individuals) does not conform to the conventional West African model of

Table 5.5 *Duration of trips: seasonal circulation, five villages, 1973–74*

Time away (months)	1	2	3	4	5	6	7	8
Per cent ($n = 59$)	10	14	16	18	6	11	24	1

Table 5.6 *Participation in circulation, five villages, 1973–74 (by age and sex group)*

	Aged (male, female)	Adults		Children (male, female)	Total
		males	females		
Per cent ($n = 59$)	3	32	24	41	100

Mobility in the region　　　　　　　　　　　　　　　　　　　　　129

interaction between the savanna and coastal regions (figure 5.7). Only two families travelled south of Kaduna. Kano, Zaria and Kaduna, which collectively comprise the urban–industrial core of northern Nigeria, attracted the three western villages. These were predominantly of Hausa or Fulani ethnicity; but the Fulani tended to avoid Kano, the culture-hearth of Hausaland. In such urban destinations, *k'wadago*, begging, trading and services were the major sources of income.

The two eastern villages, on the other hand, favoured eastern destinations – the Nguru and Gashua area, Gumel, and locations in Niger. In

Table 5.7 *Sources of income during seasonal circulation, five villages, 1973–74*

	K'wadago	Begging	Mat-/rope-making	Firewood selling	Relatives	Others	Total
Per cent ($n = 59$)	29	29	21	5	5	11	100

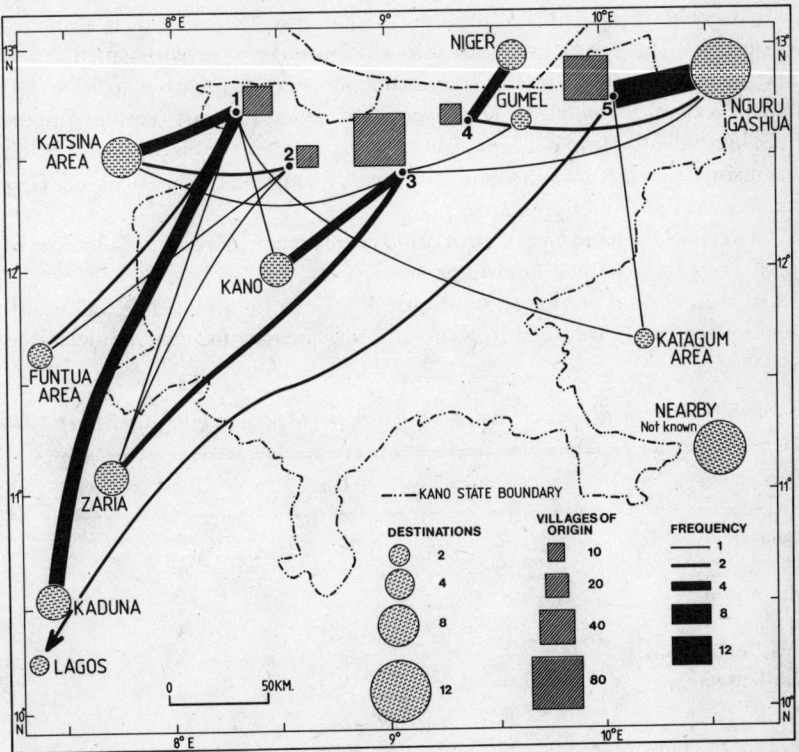

Figure 5.7 Spatial preferences in seasonal circulation: five villages, 1973–74

the first, the frequent citation of mat- and rope-making suggested limited opportunities for urban employment. All these areas were as hard hit by drought as were the source villages. Their popularity was a consequence of the eastward orientation of the Manga towards the Kanuri homeland of Borno, and of kinship links crossing the international frontier. The Manga do not see Kano, in particular, as 'their' city. Thus ethnic affinities entered into the determination of the spatial pattern of circulation, complicating the usual relations of core and periphery.[10]

MOBILITY IN ONE VILLAGE

Incidence and pattern

In March 1976, after a poor harvest in the preceding year, 44 per cent of male family heads were absent from Dagaceri (table 5.8). This understates the true incidence of seasonal circulation, because young men who were not family heads were excluded.

Sharp differences may be seen between the two ethnic communities. The incidence was twice as high among the Hausa as among the Manga. The Manga quoted economic pursuits – labouring and mat-making – as their principal occupations. But the Hausa, with their commitment to *karatu*, named this as their main occupation. In this, the distinction between teacher (*malam*) and student was not always clear. The *malam* is entitled to alms and receives support from his students. He may sell embroidered caps, charms or his prayers.[11] But it was insisted that there was no gain to be made from *karatu*, other than relieving pressure on the family granaries.

Thereafter, the incidence of circulation correlated negatively with the size of the harvest (as described in chapter 4).

Dagaceri appears as village 5 in figure 5.7. Accordingly, in 1976, although some of the Hausa travelled to Kano and to Damagaram (their homeland in

Table 5.8 *Male family heads temporarily absent from Dagaceri village in March 1976*

Occupation	Manga	Hausa	Total (per cent)
Karatu	2	32	50
K'wadago	18	3	31
Mat-making	6	—	9
Trading	—	3	4
Social obligations	—	2	3
Not known	—	2	3
Total	26	42	100
Per cent of family heads	29	65	44

Niger),[12] the Manga travelled east, one third to the Hadejia–Nguru area and another third further east, to Maiduguri, the shores of Lake Chad, and elsewhere in Kanuri Borno. Opportunities sought in these areas, during the past three decades, included: labouring for middlemen in the natron trade at Baga on Lake Chad; fishing in the receding floodwaters of the Yobe River system; dry-season sharecropping on irrigated gardens in the Biu and Gombe areas; retailing groundnuts in the same area; and contract labouring on the rice farms of the Hadejia–Yobe River valley at Gashua.[13] Such opportunities tended to change from year to year. Finally, adaptive experimentation led the Manga *masu cin rani* of Dagaceri all the way to the federal pot of gold – Lagos.

The Lagos trade

A handful of Manga men ventured as far as Lagos during Kakaduma, and again in 1976. Then the number swelled. In the eighties it was commonplace for twenty young men to embark on the trip. In view of the distance (1,000 km), and the number of intervening opportunities available in Kano, Zaria, Kaduna or Jos (to mention major urban areas only), the growth of the Lagos trade calls for some explanation. Three questions of particular interest are: the means of entry, the organisation of the trade, and its place in household economic strategy.

Entry was achieved via a bridgehead established by two Manga migrants in Lagos who each set up as a trading landlord (K *fatoma*),[14] in 1970 or 1971. They were used by most of the Dagaceri men to obtain accommodation, working capital, or both. Even if the need for accommodation could be foregone, by sharing the pavement with a fellow Kanuri night-guard, a sum of money was still required to purchase a job as a labourer or guard, or to get into the goat trade. It was the last that became the hallmark of Dagaceri's *masu cin rani*.

Goats are imported to Lagos from the north of Nigeria and from Niger by the truckload, especially at religious festivals. On arrival, each consignment of up to 140 goats is distributed through one of about twenty *fatomai* who control the goat market at Alaba, on the western outskirts of Lagos. The *fatoma* provides feed and accommodation for both goats and selling agents. Agents may either sell the goats on commission for the *fatoma* or, operating with borrowed capital, buy the goats themselves and hawk them around the town. Goats have to be sold individually; thus the need is for bulk-breaking, distribution and handling skills, which fits Hausa or Kanuri agents well for a marketing niche that resident Yoruba are unable to fill. The allocation of roles is thus the result of the special requirements of the trade in live animals, and is based on culturally transmitted skills and attitudes. Language is not a problem though none of the Dagaceri men understand Yoruba. Such a system may be compared with the waybill system that fed Onitsha market before the civil war, in which the acquisition, transporting and distribution of foodstuffs were entirely in the hands of Ibo (Onyemelukwe, 1974). It resembles the cattle trade controlled through the Hausa *maigida* system in Ibadan (Cohen, 1969).

Later, responding to profit margins with increasing entrepreneurial confidence, some of the men from Dagaceri arranged to purchase and transport their own goats, employing younger townsmen as selling agents. Irrespective of inflation, a goat or ram gains 50 per cent in value between purchase in Kwubsa market (near Dagaceri) and sale to the consumer in Lagos. With transport charges at 10 per cent of the purchase price and a risk of loss (through theft, injury or exposure) of perhaps 10 per cent on average,[15] deduction of the *fatomas'* commission still left room for handsome profits. A ban on street hawking, imposed in 1984, and the ensuing requirement to pay a market fee of ₦5 (irrespective of the number of animals sold), failed to extinguish enthusiasm.

The goat trade is considered to be more remunerative than any alternative source of income, and relieves the household of the necessity to sell part of its food grain in place of the vanquished groundnut. The goat trade, therefore, does not compete with farming, but rather helps to consolidate the subsistence priority in agriculture.[16] Migrants bring back from Lagos not only money, but clothing, foodstuffs and other goods, which may be resold in the village. Indeed, the possibility of entry into general trade is seen as a major attraction of the goat business. Alternatively, funds have been invested in the purchase of ploughs, work bulls, and in hiring farm labour. The goat traders are always present for the first weeding no matter when Id el Kabir or Id el Fitr may fall, and none of them has yet been tempted to settle in Lagos: they go 'in order to have enough, so that it will be easier for them when they get old'. Accumulation is still the safest form of insurance.

Mobility in one family

Born in the village around 1921, A (see chapter 4, p. 115) departed with a donkey for Mecca when he was 16, in a party of seven from Dagaceri. He spent the next 14 years in the Sudan, mostly working on the Gezira cotton scheme, and marrying there. After his return, around 1951, he embarked on regular *cin rani* to destinations within a few days' journey: working as a labourer, burning and clearing on the flood plain rice farms and, later, making mats from the abundant *kaba* in the Hadejia valley at Guri, Jajimaji and Gashua to sell to Hausa traders from further south. With six dependants (his mother, elder sister, wife and three young children), he needed extra income. During this period, the number embarking on *cin rani* from the village each year fluctuated inversely with the yield of groundnuts, but had an underlying upward trend. His movements continued until the season of 1977–78, when his son was old enough to go in his place.

K, son of A, and already married at his father's expense, first travelled to Lagos in the season of 1972–73, following an unprofitable trip to Biu to trade in vegetables. At Lagos, he borrowed from a *fatoma* to trade in goats on commission.[17] On hearing of the second failure of the rains in 1973, and of 'everyone leaving the village', he brought financial help to his father and returned immediately for another trip.

By 1980, he was borrowing capital in the village and arranging his own consignments of goats, fifteen to twenty per trip. He returned to the village for every weeding season, and was in Lagos for the religious festivals. He had financed his second and third serial marriages (at about ₦250 each) – he had no sons – and contributed substantially to food, clothing, and other expenses of a household which consisted, in 1985, of his father, his father's wife, a young brother, his own wife and a girl relative. His success had made him something of a role model for aspiring Lagos traders in the village.

Incomers in the eighties

At Zinder, the capital of Damagaram in Niger, the rainfall was only 52 per cent of the long-term mean in 1981, 56 per cent in 1983, and 52 per cent in 1984. Following three major harvest failures (the worst in living memory), some 22 families from Damagaram were added to the existing Hausa community at the west end of Dagaceri, in the months of October, November and December 1984. 'Everyone is looking for his relatives', and ties of kinship or ethnicity were supplemented by reports of a better millet crop at Dagaceri. Building plots were allocated on the authority of the Dagace, and it was expected that abandoned fallow farms would be allocated before the rains of 1985. Hunger necessarily overrode considerations of prior community claim. The new arrivals, who came with their wives and children, meanwhile sought income as fuelwood cutters,[18] fodder collectors, and water carriers; and women and children appeared begging in the streets, for the first time in memory.

One case history illuminates the role of spatial mobility in the cycle of impoverishment, survival and recovery, and re-emphasises the regional ties that transcend the former colonial and current national boundaries.[19] S was born in Kanya (a few kilometres north of Zinder) and had joined the Dagaceri community of *almajirai* when a young man. After returning to Kanya to inherit his father's farms, his family had grown to two wives and eight children aged from eight to 21 years.

The harvest failures of the early eighties left his household weakened by hunger, and stripped of almost all assets. The sole remaining marketable possessions were a donkey, and some ten used flour bags S had collected. Land could not be sold, being jointly owned. To sell the donkey would have sacrificed the option of movement. All other animals had either starved or been sold. The family had lived on famine foods, with very little grain, for more than a year. Even *cin rani* was not considered possible because a man's family might no longer be alive when he returned – possibly empty handed. Deaths from starvation were said to be occurring among those poor who lacked better-off relatives.

So the flour bags were sold for a pitiful ₦7 (the naira circulates north of the border), a further sum of ₦20 was borrowed, and the family field – S (who had a history of treatment for tuberculosis and was afflicted with guinea worm in one leg), his wives, and the smaller children on the donkey. The route to Dagaceri passes through Magaria district and normally takes seven or eight days – 12, on this occasion, with the children. The money was first used to buy small quantities of food; then the wives and children begged: finally arriving at Dagaceri destitute and weak.

S was collecting fodder in February 1985, which yielded insufficient income to feed his large family, but he received help, and their condition was by then improving. He did not want to return to Kanya, house and farm notwithstanding, and said that the village was already deserted by its poorer inhabitants. Significantly, he had rejected the possibility of staying at Jamburji, where a Niger government official fed the family and offered help, because the social environment was unfamiliar. Possibly, also, the expectation of obtaining land at Dagaceri was influential; or its somewhat better millet harvest. By February 1985, the children's energies were returning on a diet of (thinned) *tuwo*, though S's health was poor. By January 1986, the guinea worm had left him but not the symptoms of tuberculosis. Destitute, weakened and dependent, and no longer young, he had a long road to travel to rehabilitation. But he had a house, a farm and, most important, sons.

TERRITORY AND MOBILITY

Why did two years' harvest failure, following several years of food shortages, generate so little permanent redistribution of the population? Notwithstanding the evidence for increased mobility during Kakaduma, two facts are noteworthy. Firstly, circulation was preferred to migration, and when migration did occur it was largely between rural areas. Secondly, the incidence of out-migration remained low, 57 per cent of villages reporting none at all, and Dagaceri losing hardly any of its Manga households. It should be remembered that hunger was occurring, even though few died of it. Abandoned houses – even whole villages – were reported widely in drought-affected areas, and a massive convergence of the hungry on the major urban centres was being predicted. Yet by the harvest season of 1974, all had returned to normal, as far as the eye could see.

Circulation is perceived as a preferable alternative to out-migration, for several reasons. Firstly, circulation between home and (usually) urban income-earning opportunities is infinitely flexible with regard to the commitment of time and money. Visits may be long or short, and overhead costs can be minimised by sleeping in the streets. Such flexibility meets the requirements of a rural economy characterised by violent and unpredictable fluctuations in output.

Secondly, it is adaptable to the changing requirements of the life cycle. Young men are the most free to travel, while fulfilling their obligations on the family farm in the growing season. Later in life, increasing responsibilities will keep them at home, except in critical situations. Their sons will go in their places.

Thirdly, it is flexible with respect to destination. Having invested little more than the bus fare in his move, the *mai cin rani* is free to respond to changes in the spatial distribution of opportunities as he sees them. The development of the state capitals, new industrial investments, agricultural projects, and infrastructure led to diversifying destinations in the seventies, the 'discovery' of Lagos by the Dagaceri community being an example.

Fourthly, access to land and labour resources, and to rights of community membership, in the village may be retained, while exploiting the usually insecure possibilities for earning income in the city. Without educational qualifications, such opportunities offer few, if any, longer-term prospects, and urban unemployment, if not growing throughout the seventies, certainly took an upward leap in the eighties.

The tenacity of this hold on the land, no matter how low its productivity, acts as a brake on the process of urbanisation. Kano State was about 11 per cent urbanised in 1962, and perhaps 17 per cent by 1980 (Mortimore, 1974). In place of a massive permanent transfer of people to the cities is substituted an 'urbanisation of the countryside', whereby more and more rural people endeavour to have something of both worlds. Voluntary relinquishment of all rights to land, and with them membership of rural society, seems likely to be undertaken only by a small minority.

All circulation involves risk; Manga natron traders once used to travel together for security. In 1978–79 it was considered a matter for satisfaction that the Dagaceri contingent had returned from Lagos without the loss of any of its members. For all merchant venturers, the purpose of these trips is accumulation which, it is hoped, will render future ones unnecessary.

6

TWO DRY DECADES

From the sky you send rain on the hills   Psalm 104:13

The foregoing analysis of the nature of adaptive response to meteorological drought, and to the food shortages that have been closely associated with it in time, provokes the question of its future recurrence. To portray human communities solely as adaptors to exogenous events does not go far enough. However impressive the adaptive capabilities of such populations, the absence of any knowledge of future rainfall restricts individual choice to *ad hoc* decisions from year to year. This annual rhythm may obscure from view the possibility of longer-term trends, including that of ecological degradation (or desertification). Droughts are themselves contributory factors to such degradation. But so are anthropogenic factors, and increasing credence is being given to the view that land use may be linked with rainfall by means of 'feedback' mechanisms. For these societies, the roles of victim and agent are not easily distinguished. At this point, an examination of the climatological evidence is necessary, firstly as a pointer to the future significance of social adaptation, and secondly as a preliminary to a more systematic discussion of some evidence relating to desertification.

This chapter aims to review the related subjects of meteorological and hydrological drought. The most important questions concerning meteorological drought, from the standpoint of the present study, are (a) its persistence, (b) the possibilities for predicting its occurrence, and (c) the existence of feedback mechanisms. The first two of these questions relate primarily to identifying appropriate adaptations, while the third relates to the agency of human communities in causing drought. Meteorological drought in Africa is the subject of a controversial, proliferating and specialised literature, and no attempt can be made here to provide more than a layman's summary of these questions (see Hare, 1985; Farmer and Wigley, 1985). With respect to hydrological drought, which is dependent on meteorological drought but is also influenced by other determinants, an attempt will be made to review the evidence available for northern Nigeria.

THE PERSISTENCE OF METEOROLOGICAL DROUGHT

The climax of the Sahelian Drought in the years 1972 and 1973 provoked a number of analyses which sought to understand the social and economic disaster mainly in terms of rainfall failure (Dalby and Harrison Church, 1973; Davy, 1974). These years were soon found to have immediate antecedents as far back as 1968. But the spatial distribution of rainfall departures from normal in West Africa shows that it is misleading to think of seven years of uniform drought for the entire Sahelo-Sudanian Zone. And drought did not end in 1974. As the seventies proceeded, it became clear that a run of dry years was occurring, and the end of this run was still not in sight in 1985 (figure 6.2).

With each additional dry year the persistence of drought becomes more noticeable and its statistical significance increases. No dry spell of similar length can be shown from the rainfall records except, possibly, that of the 1860s and 1905–15 – when records were scarce (Mason, 1976); a similar conclusion emerges from the analysis of rainfall changes in the central Sudan (Trilsbach and Hulme, 1984). Few authorities, however, were initially prepared to see a permanent discontinuity or trend, preferring to regard it as another fluctuation of the kind that has occurred ever since the Quaternary (Grove, 1977a; Schove, 1977; Nicholson, 1978).

Analysis of West African drought has progressed from the selective use of rainfall records from a small number of stations to increasingly rigorous examination of ever larger samples.

West Africa

Rainfall records for West African stations are short (mostly 70 years or less), the principal exceptions being coastal stations. The period 1931–60, which forms a major part of most inland series, and was used to define 'normal' rainfall prior to 1970, was a period of low drought occurrence when compared with the decades before and after. For example, for Kano the mean for 1931–60 was 868 mm, but that for 1906–85 was only 822 mm. The difficulty of establishing an appropriate base for evaluating rainfall departures from normal after 1960 is increased by the fact that the number of rainfall recording-stations was very small indeed before 1920. Attempts to project rainfall trends further back still – into the nineteenth century – must depend heavily on indirect evidence. Yet the identification of precursors of the recent droughts is clearly important. From the standpoint of this study, the period 1931–60 did define 'normal' expectations for it fell within the lifespan of a large proportion of the population, in a period characterised by considerable expansion of agriculture and settlement.

An analysis published in 1976 concluded that recent years, drought notwithstanding, fell within statistical expectations based on the period 1931–60 (Bunting et al., 1976). But as one dry year followed another, the evidence for non-random persistence of drought accumulated. Tanaka et al. (1975), using a data set for the years 1941–73, showed that the rainfall in August declined after 1965 in the Sahelian Zone (between Long. 10°W and 25°E), and significant changes occurred in the latitude of maximum rainfall, and in the northward gradient from that position. Extending the analysis to all months of the year, and including 1974 and 1975, Kidson (1977) concluded that at Lat.

15°N a downward trend in rainfall was apparent prior to 1970, and that this trend, of about 7.7 mm/year, was equivalent to a southward displacement of the desert margin, climatically defined, of around 5 km/year. Such a trend was not apparent further south at Lat. 10°N. Kraus (1977) concluded that subtropical African drought had exhibited statistically significant persistence, but his conclusions were challenged by Katz (1978). Nicholson (1979, 1980), using records from 419 stations, constructed series of standardised annual rainfall departures from normal for 1901–75 for the West African subtropics, separating four zones: Sahelo-Saharan, Sahel, Sudan and Sudano-Guinean. There was a marked coherence of rainfall variation throughout the semiarid region south of the Sahara and the persistence of drought was noticeable, notwithstanding earlier findings to the contrary. But Ogallo (1979), using a smaller sample of 69 stations in West Africa, concluded nevertheless that no significant trends or periodicities were evident. In some stations *south* of the Sahel, there was even an increasing rainfall tendency prior to 1974.

Motha *et al.* (1980) analysed records from 813 stations from Senegal to the Central African Republic, for the years 1968–75, and concluded that the prolonged drought had still not been completely 'alleviated' by the end of 1975. Nicholson (1983) extended her series for the four major zones south of the Sahara to the year 1980, and showed that, in all four, there was a marked decrease in annual rainfall in the period 1960–80. Lamb (1982, 1983) constructed an index for the period 1941–82, which showed the yearly average of the normalised rainfall departures in April–October for 14 to 20 West African stations between Lat. 11°N and 18°N, and west of Long. 9°E. This series showed negative departures for every year from 1970 to 1982, and confirmed that the subSaharan drought which commenced in 1968 persisted strongly until 1982. Dennett *et al.* (1985) concluded, after analysing a data series from 50 Sahelian stations (starting in the 1940s), that both a persistence of drought and a downward trend in rainfall had been clearly demonstrated. A similar conclusion emerged from a longer series (starting in 1905) for five stations.

Progressively more elaborate analysis of instrumental records has confirmed what inhabitants of semi-arid West Africa have experienced. And none of these studies were able to take full account of the renewed intensification of drought that occurred in the eighties. Thus, 'the drought in Africa which began in 1968 has not yet ended in 1983, especially in the Sahel' (World Climate Programme, 1983:3). Less cautiously, Lamb (1974) has warned: 'it would be very injudicious to take the easier years as anything more than short-term interruptions of the drift towards increasing drought in the regions nearest the Sahara' (see also Lamb, 1977).

Nigeria and Niger

In Nigeria the occurrence of major droughts in 1972 and 1973 attracted attention initially as an isolated phenomenon with established antecedents, especially in 1913. On the basis of rainfall records for Kano and Sokoto, Kowal and Kassam (1975) estimated the expectation of rainfall as low as that of 1973 as once in 100–34 years, very similar to Landsberg's (1975) conclusion based on the records of Dakar. The pattern of departures from normal showed that the worst affected area of the country was northern Kano and north-western Borno States – the location of this study – where negative departures exceeded 40 per cent (figure 6.1), while in the south-west of the country there were positive departures of up to 10 per cent (Oguntoyinbo and Richards, 1977; Oguntoyinbo, 1981; Ayoade, 1977). The shorter length of the growing season, uneven

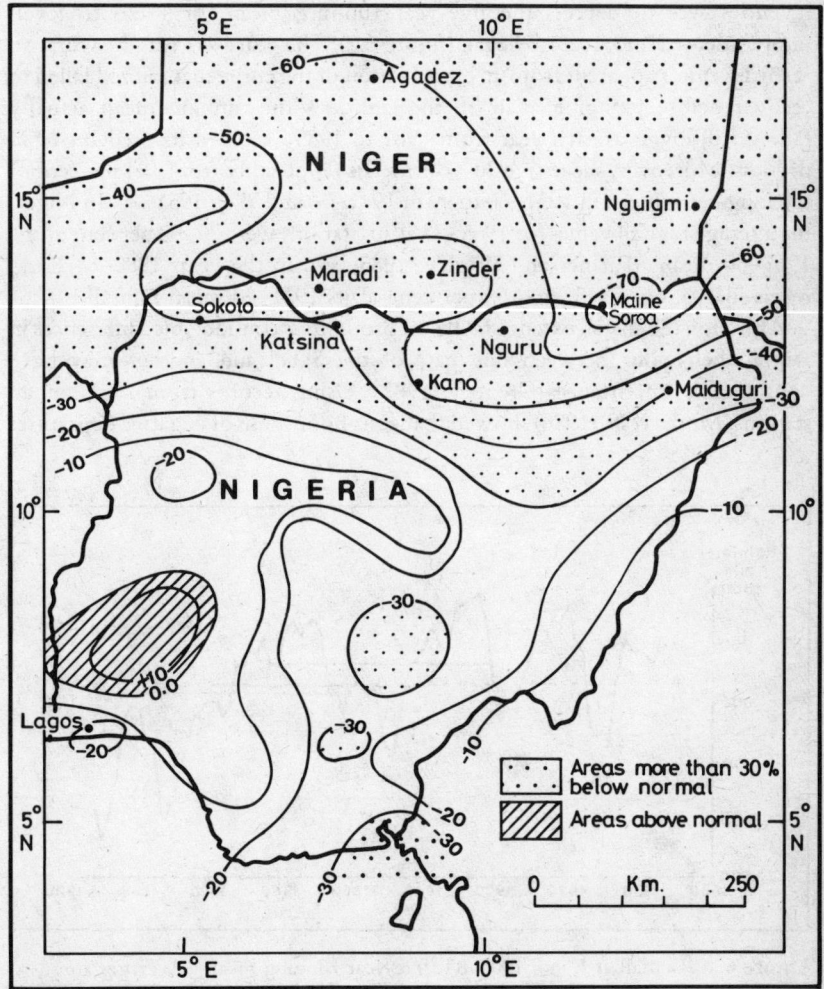

Figure 6.1 Rainfall departures from the mean in 1973 (per cent) (after Oguntoyinbo and Richards, 1978; Derrienic et al., 1976)

distribution of rainfall in time and space, and a southward shift in the mean axis of maximum weather activity were all significant (Adefolalu, 1983).

Using the lower quintile as a defining parameter, Derrienic et al. (1976: 166) showed that, in Niger, 34 rainfall stations recorded drought on average five times in six years, 1968–73, the incidence ranging from 17 stations in 1969 to 33 stations in 1973. Although central and southern Nigeria did not experience significant negative departures before 1972, the records of northern stations conformed more with those across the border in Niger. This long run of dry years reduced the state of preparedness in which systems of animal and crop husbandry confronted the full onslaught of drought in 1972 and 1973.

In the longer term, the significance of these negative departures in the seventies was to depress the five-year running mean for Kano to levels unprecedented since records began (figure 6.2). The year 1973 was the worst on record; yet it is apparent from an examination of the curve that rainfall failed to recover before plunging again in the eighties – the running mean actually falling still lower as each year from 1982 to 1985 exceeded its predecessor in dryness. Moreover, successive 30-year means fell steadily from 104 per cent of the long-term mean in 1931–60 to 99 in 1941–70 and 97 in 1951–80. In Niger, the average rainfall values for 1965–83 at six stations were 70–90 per cent of the 1931–60 means (Farmer and Wigley, 1985: 35). In the years 1980–84 Kano received 105, 66, 73, 57 and 55 per cent of its 1931–60 mean rainfall.[1]

Although the spatial extent of African droughts is considerable, differences in rainfall behaviour between one part of the Sahel and another may have considerable importance (Gregory, 1982). Using records from 28 Nigerian stations, Motha et al. (1980) showed that continuous runs of negative departures

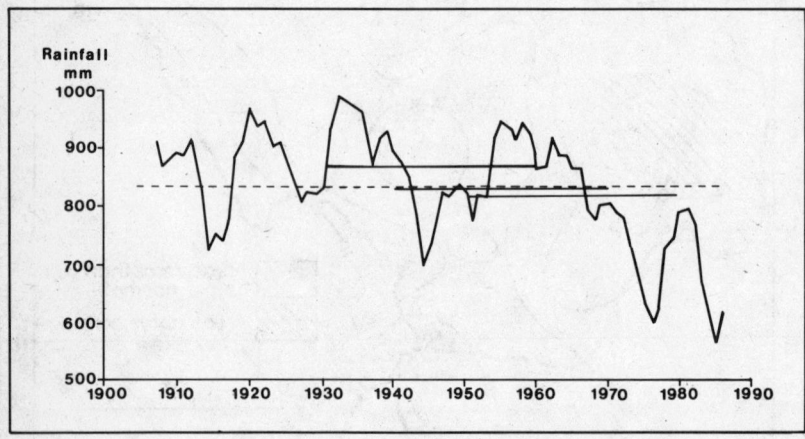

Figure 6.2 Rainfall at Kano, 1906–85 (five-year running mean). Averages are shown for 1906–85 (839 mm), for 1931–60 (868 mm), 1941–70 (836 mm) and 1951–80 (813 mm)

from normal (1927–76) occurred in the north of the country in the years 1940–44, 1947–49 and 1968–76 (the end of the period of analysis). Although these 'drought periods' were also detected in the centre of the country, they did not occur in the south-west or the south-east; yet in certain years – 1971–73 and 1976 – rainfall was below average throughout Nigeria. Leow and Ologe (1981) compared mean rainfall for the double-decades 1918–40, 1941–60 and 1961–78 at ten stations in northern Nigeria. They found that north of Lat. 11°N a significant decline occurred in the last of the three double-decades, but that south of this line, the decline was not statistically significant.

A strong relationship exists between the rainfall patterns in West Africa and the movements of the Inter-Tropical Discontinuity (or Convergence) in the northern hemisphere summer, and since the coast of the Atlantic Ocean (which is the source of the moisture) runs from west to east, there is a correlation between latitude and the amount of annual rainfall, the date of the start of the rains, and the end of the rains (Kowal and Adeoye, 1973; table 3.1). The probability of dry spells of seven days or more also varies systematically with latitude (Stern et al., 1981). Ilesanmi (1973) proposed that a displacement of one degree in latitude of the ITD is equivalent to a change of 175 mm in annual rainfall in northern Nigeria. The greatest effect of the ITD occurs at 8°–9° of latitude south of its surface position, in the zone of maximum convective instability. Motha et al. (1980) confirmed a relatively strong relationship between the position of the ITD and regional rainfall patterns both north and south.

It has often been assumed, therefore, that the southward displacement of the isohyets which may be discerned at the regional level in dry years is to be explained in terms of the failure of the ITD to penetrate as far north as usual (see chapter 3, pp. 42–4). Durand (1977) found a southward displacement of the 300 mm isohyet from its normal position, in 1968–74, of about 100 km in central Niger, and 200 km in Mauritania and northern Senegal (cf. figure 3.1). However, according to Nicholson (1982: 27), diagnostic analysis shows a southern displacement of the ITD to be the least likely explanation for drought. The different spatial patterns exhibited by rainfall anomalies in different years argue for a greater complexity of factors.

Such complexity becomes inescapable when the monthly distribution of rainfall is taken into account. In Kano in 1984, this completely reversed synoptic expectations (Adebayo and Mohammed, 1987). The August rainfall of that year was only 17 per cent of normal, and the lowest ever recorded (figure 6.3); while the total rainfall was only 55 per cent of the 1931–60 mean. Such a pattern, which was observed widely in northern Nigeria, mimicked the twin-peak regime more characteristic of the south of the country (Ireland, 1962).[2] If the relationship between the movements of the ITD and rainfall, as postulated above, holds true, then a low annual rainfall should be associated with a late start, an early end and a pronounced peak in July or August, as in 1983. In the following year, 1985, the August peak was again suppressed, but the actual

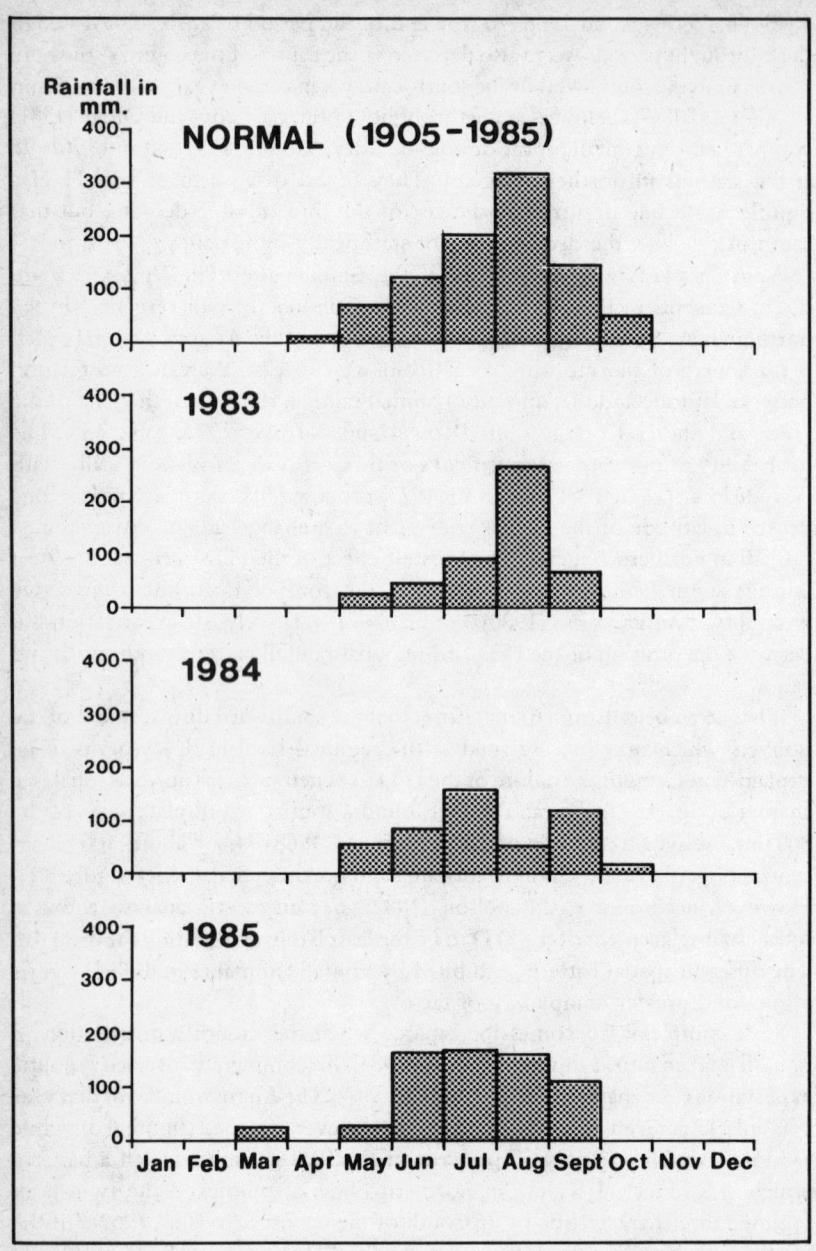

Figure 6.3 Monthly distribution of rainfall at Kano Airport (after Adebayo and Mohammed, 1987)

Table 6.1 *Monthly distribution of rainfall at Kano (June–September), 1906–85*

Time period	June	July	August	September
1906–85 average (mm)	115.3	201.1	294.7	126.9
1 1906–25	102.3	211.2	329.3	139.7
2 1926–45	119.7	197.6	305.3	135.6
3 1946–65	130.8	220.8	313.7	129.2
4 1966–85	108.4	174.9	230.4	103.3
1906–85 average (per cent)	100	100	100	100
2 1906–25	88.7	105.0	111.7	110.1
2 1926–45	103.8	98.3	103.6	106.8
3 1946–65	113.4	109.8	106.4	101.8
4 1966–85	94.0	87.0	78.0	81.4
5 Percentage change from period 3 to 4	−17	−21	−27	−20

amount was satisfactory, the total for the year was 75 per cent of the 1931–60 mean, and agricultural expectations were fulfilled.[3]

An elementary analysis of the monthly distribution of rainfall draws attention to several trends which appear rather well established. Table 6.1 shows that in the 80 years 1906–85 rainfall in the critical growing months of July–September was lower in the two decades 1966–85 than in any previous double-decade for every month except June; that expressed as a percentage of the monthly mean, August rainfall in the last double-decade was reduced by more than that of other months, followed closely by September, and June the least; and that the greatest percentage reduction in mean rainfall between the 1946–65 and 1966–85 double-decades was in August. The experience of 1984 is thus seen to be but an extreme case in the trend that has become established over two decades, confirming what has been reported elsewhere (Tanaka *et al.*, 1975; Dennett *et al.*, 1985; Hutchinson, 1985).[4]

Such a trend has the most serious implications for rain-fed agriculture, whose entire rhythm is based on the expected August peak. Late-maturing crops in particular are adversely affected by early termination of the rains. It appears from the data in table 6.1 that a reduction in total rainfall in the last two decades has not meant simply the substitution of a regime previously associated with latitudes farther north, but also a significant change in the configuration of the rainy season. Of five proximate rainfall stations north of Kano, all but one recorded an even greater diminution of the August rainfall in 1966–85 (table 6.2) than did Kano (table 6.1).

The persistence of drought (Nicholson, 1982), and the changing configuration of monthly rainfall distribution, can thus be shown from the instrumental

Table 6.2 *August rainfall 1966–85: departure from the 1931–60 mean*

Station	1931–60 (mm)	1966–85 (mm)	Departure (per cent)
Kano	311	230	−26
Maradi	260	169	−35
Nguru	233[a]	159	−32
Zinder	232	158	−32
Maine Soroa	176	123[b]	−30
Nguigmi	141	109	−23

[a] 1942–60
[b] 18 years only
For locations, see figure 6.1

record. However, it needs to be remembered that farmers and livestock breeders do not respond to trends, but to departures above or below expectations for individual growing seasons and parts of the season. Each season is a separate accounting unit and to a considerable extent decoupled from those before and after it (see chapter 4). For this reason analysis in terms of quartile or quintile distributions about the median approaches more closely the perceptual viewpoint of rural West Africans (Derrienic et al., 1976).[5]

PREDICTION

Until recently the prospects for predicting West African rainfall seemed to be as remote as they were desirable (see Winstanley, 1974; Glantz, 1976b, 1977b). The two main approaches are through projecting trends or periodicities that have been established from the past records and through the understanding and simulation of dynamic weather processes. The second approach offers the best practical prospect for forecasting annual rainfall; but seasonal forecasting is another matter.

Trends or periodicities

Evidence such as pollens found in deep-sea cores from Atlantic Ocean sediments, and in the Chad Basin (Rossignol-Strick and Duzer, 1980; Maley, 1981) indicate that the Sahara Desert has both expanded and contracted relative to its present position during the past 25,000 years. Such fluctuations account for the existence of stabilised dunes over large areas of West Africa at present inhabited by agricultural and pastoral communities, and for the recovery of archaeological material of Neolithic age from sites at present uninhabited in the southern Sahara. On a shorter time-scale, fluctuations continued in historical times (Mauny, 1961). There was a relatively wet period from the mid-sixteenth to the eighteenth centuries; and major droughts in the mid-eighteenth and early nineteenth centuries (Nicholson, 1978). Such fluctuations have to be seen in the context of global atmospheric circulation (Flohn and Nicholson, 1980), in which the annual movements of the ITD across West Africa form but a part.

Whether such fluctuations, and the shorter-term variability of rainfall which has been observed in the present century, form part of a long-term downward trend is a matter of controversy. Winstanley (1985) has challenged the concept of climatic stability with an analysis of rainfall data from 1850 to the present. According to this analysis, a continuous decline can be detected in the subSaharan zone of Africa. Taking 1941–70 as normal, this decline was from 120–30 per cent in the period 1854–93 to only 80 per cent in the decade starting in 1974. Conversely, in the tropical zone of Africa rainfall increased from 67 per cent of normal (1931–54) in 1874–93 to 125 per cent in the last decade. Explorers' accounts are claimed to support this interpretation; the well-known droughts of the 1860s were not as dry as those of recent years. If this view is correct,[6] expectations that rainfall fluctuations form part of a stable climatic regime may be 'wishful thinking'.

Nevertheless, the bulk of expert opinion is much more cautious with regard to trends. Farmer and Wigley (1985: 78) do not see any significant evidence for 'century time scale trends' and since such trends, if continued, must end in zero there are grounds for doubting their physical realism. 'Trends cannot be extrapolated forward in time, as climatic fluctuations occur abruptly' (Nicholson, 1982: 26).

The search for periodicities in rainfall – such as the proposed 30-year drought cycle – has also been inconclusive (Farmer and Wigley, 1985: 71–6, 101–2); although periodicities may be shown to have statistical significance, the records are too short for more than a small number of completed cycles to have occurred. Their amplitude is too small in proportion to year-to-year and month-to-month fluctuations for their determination to have practical usefulness in agricultural planning (Davy et al., 1976: 268). 'As these time series exhibit apparently random changes in phase, they have no predictive value, and are unlikely to be useful unless we can gain some insight into the underlying physical mechanisms' (World Climate Programme, 1983: 8). This appears to be a fair comment on Faure and Gac's (1981) projection of Senegal River discharge data, which suggested that the dry cycle would end in 1985 and be replaced by wet conditions by 1992. Meanwhile, 'there is a very real *possibility* of continuing drought in the Sahel' (Farmer and Wigley, 1985: 13–14, 107). Planners have been advised to base their expectations on the rainfall of the last twenty years (Dennett et al., 1985).

Dynamic processes

It has been argued that drought is caused by displacement of climatic belts, and in particular of the subtropical high-pressure zone, towards the Equator. Such a displacement would strengthen the north-east trades and impede the penetration of the ITD in summer. Winstanley (1973, a, b) suggested that such a change occurred in the early seventies in the northern hemisphere. Lamb (1978) found evidence that dry conditions in the Sahel from 1911 to 1972 were associated with anomalous southward displacements of the equatorial low-pressure trough in the Atlantic, the zone of maximum sea surface temperatures, and the southward penetration of the north-east trades. But Miles and Folland (1974) drew different conclusions for an overlapping period, 1900–73. And Nicholson (1980) identified and mapped precipitation anomaly types in Africa, and concluded that climatic fluctuations do not primarily take the form of a north–south displacement of the desert and other climatic zones, but rather an expansion or contraction of the entire desert belt.

The circulation of the global atmosphere is imperfectly understood, and pending progress in this field, the most promising area of research lies in identifying 'teleconnections' with climatic or oceanic variables that can be shown to be linked, via other factors in the circulation system, with the occurrence of drought in West Africa (Nicholson, 1982; Hare, 1984a). Such a role has been proposed for sea surface temperatures in the tropical Atlantic. Lough (1980, 1986) explored the possibility of such links, and concluded that *some* rainfall anomalies in the Sahel between 1948 and 1972 were related to sea surface temperature anomalies, with sea-level pressure as the linking mechanism.

Folland *et al.* (1986) have now demonstrated that warm conditions in the oceans of the southern hemisphere (relative to the northern) during the period from 1901 to 1984 tended to be associated with dry epochs or years in the Sahel rainfall zone (as defined by Nicholson), and cool conditions with wet. The correlation is more significant in respect of the global oceans than for the Atlantic alone.[7] Modelling experiments, using data for 1984 (an exceptionally dry year) and 1950 (an exceptionally wet one) reproduced the rainfall and wind conditions of those years quite well on the basis of the worldwide pattern of differences in sea surface temperatures (Owen and Folland, 1987). On the basis of this work, a forecasting technique has been developed which enabled the Meteorological Office of the United Kingdom to issue a forecast for Sahelian rainfall in 1986 (Parker *et al.*, 1987). The forecast was for about 71 per cent of the 1951–80 average; and observed rainfall was about 66 per cent. This significant development offers a possible key to forecasting annual rainfall in major regions of West Africa, which must now be understood to be intimately related to global atmospheric circulation.

It has also been proposed that changes in global atmospheric temperatures, which may be attributed to pollution, including increased CO_2, might shift the positions of the dry margins by several hundred kilometres (World Climate Programme, 1984b:29f.). But the CO_2 hypothesis is unproven, and while some global warming has occurred, its effects on tropical African rainfall are not understood (Farmer and Wigley, 1985:11–13).

Seasonal forecasting

Farmers and livestock breeders depend not merely on the amount of rainfall received annually but on the distribution within the rainy season. This distribution varies from place to place. The mechanisms controlling rain formation at given times and places are inadequately observed, owing to the scarcity of synoptic stations. Maley (1981:503–16) divides Sahelian rainfall into two types: fine rain (characterised by small droplets) associated with monsoon weather in the heart of the rainy season, and heavy rain (large drops) associated with disturbance lines in the vicinity of the advancing ITD. Such a distinction may have significance in the sedimentological record, and is not without importance for erosion, infiltration and crop development, especially when the quantity of rain is subnormal. Omotosho (1985) estimated that line squalls contribute 48 per cent, thunderstorms 39 and monsoonal rain 13 per cent to total annual rainfall at Ilorin, Nigeria. Such distinctions, however, are not always possible using standard station observations. The trajectories of individual rain formations and their precipitation histories are also a neglected area of research. They cannot be reconstructed until the network of weather stations is improved and rainfall, windspeed and cloud formations are recorded at short intervals. Yet the discontinuities in rainfall that occur over quite short distances are of enormous agricultural and ecological significance (Davy et al., 1976: 255–64; Hulme, 1984). Weather forecasting within the season will not be possible without improved synoptic capability.

Satellite data have been proposed for a role in seasonal forecasting, but since they can monitor rainfall only through proxy variables, such as cloud cover or vegetation development, their usefulness seems rather limited.[8]

Probability estimates of total rainfall or the length of the rainy season may, perhaps, be based on rainfall behaviour in the early part of the season (Bunting et al., 1974). For example Bradley (1973) reported a significant correlation between cumulative rainfall to the end of May and total annual rainfall at Samaru, Nigeria. But no dependence between rainfall at different times within a single season emerged from a study of daily records from Kano, Say and Zinder (Dennett et al., 1983). Models combining hydrological and rainfall data, and based on the soil conditions of specific areas, may be capable of predicting agricultural output or pasture development later in the season. Probability statements, calculated from aggregate rainfall and hydrological data, are available for the subregions of the Sudano–Sahelian Zone of West Africa (Davy et al., 1976; Sivakumar et al., n.d.). Such probability statements are potentially valuable for agricultural planning at the regional level, and to governments and agencies involved in remedial action, for whom drought is defined administratively as a large-scale food shortage (Sandford, 1978). But they cannot be effectively incorporated into the plans of small-scale farmers and livestock breeders.[9]

FEEDBACK

The tropical atmosphere is highly sensitive to surface characteristics and it is therefore quite plausible that changes in the temperature, wetness, reflectivity or other characteristics may affect rainfall. Such changes may be brought about by human activity, although it has not yet been demonstrated that they do occur on a scale sufficient to modify regional climates. Several theories have proposed linkages between landuse management and rainfall. Should any of them find confirmation, anthropogenic influences will assume increased significance, and *Homo Sahel* becomes responsible, at least in part, for his own droughts.

Albedo

The case for the 'albedo effect' on Sahelian rainfall has been put as follows:

> Dry, sandy or rocky soil has a much higher albedo [or reflectance value] than soil covered by vegetation. Hence desert regions reflect more solar radiation to space than their surroundings, all else being equal. Desert surfaces are hotter than surrounding regions and the air above them less cloudy. Hence deserts emit more terrestrial radiation to space. The net result is that the desert is a radiative sink of heat relative to its surroundings... Since the ground stores little heat, it is the air that loses heat relatively. In order to maintain thermal equilibrium, the air must descend and compress adiabatically. Since the relative humidity then decreases, the desert enhances its own dryness, i.e. it feeds back upon itself. A bio-geophysical feedback mechanism of this kind could led to instabilities or metastabilities in border regions, to advances or retreats of the borders themselves, which might conceivably be set off or maintained by anthropogenic influences.
>
> (Charney, 1975)

In a model of this process, an increase in albedo from 0.14 to 0.35 (representative of vegetated and desert surfaces respectively) between June and August predicted a drop in precipitation of up to 40 per cent in each week, and a similar reduction in convective cloud (Charney, 1975; Charney and Stone, 1975).

The albedo hypothesis was criticised for ignoring the effects of evaporation (Ripley, 1976a, b) but its authors, in a re-run of their model, produced evidence that 'with or without evaporation, the increase of albedo causes a net decrease of radiative flux into the ground and therefore a net decrease of convective cloud and precipitation' (Charney and Stone, 1976; Charney *et al.*, 1977).

Albedo values increase as vegetation diminishes northwards. Representative values have been obtained for Nigeria which range from 0.12 for mangrove swamps to 0.22 for dry savanna woodland (Oguntoyinbo, 1974). They also vary seasonally. For example, values obtained for Sudan savanna woodland ranged from 0.18 in July (rainy season) to 0.25 in March (at the height of the dry season). In view of such spatio-temporal variations, and the nature of weather processes, it may be questioned whether the conditions under which the model was run were realistic.

There is no doubt that albedo values increased during the dry years between 1967 and 1973 in West Africa; and as the drought progressed, the reflectance of the surface in the wet season approached that of the normal dry season (Norton et al., 1979). Comparative measurements in semi-arid regions elsewhere indicate that anthropogenic pressures can increase albedo by nearly 0.2 (Otterman, 1981), but such effects are locally reversible by (for example) excluding certain areas from grazing. Definitive empirical support for the albedo hypothesis is still awaited.

Soil moisture

Alternatively, it has been argued that the albedo of bare soil is linearly related to its moisture content, which also affects evaporation. A dynamic model based on West African conditions gave more rainfall over moist land surfaces; and in the desert, even though convective activity occurred, little rainfall resulted (Walker and Rowntree, 1977). If moist ground was substituted for desert, the rainfall belt reached Lat. 30°N after ten days' run of the model. Thus it was suggested that rainfall anomalies occurring early in the wet season might tend to persist, and the effects of drought in one year might, by reducing transpiration, reduce rainfall in the next; removal of the vegetation might have the same effect.

The inception of the rains takes place when soil moisture is at its yearly minimum. However, the rains progress northwards behind the ITD between March and July, and presumably the soil moisture in the area of precipitation is of less importance than that in its 'catchment area' to the south and west, whence evaporation may influence the humidity of the monsoonal air before it becomes subject to convective activity. But the proportions of water vapour derived from evaporation locally and by importation from the ocean are not known (World Climate Programme, 1983: 13). The erratic pattern of observed falls and dry spells seems to call for a more complex explanation of daily rainfall. As for the seasonal distribution, the proportionately greater significance of negative rainfall anomalies in August and September does not seem to be consistent with the model's dependence on anomalies in the early part of the season. Empirical support is again scarce, and the collection of soil moisture data over huge areas presents considerable problems, but the authors were nevertheless optimistic about the possibilities for forecasting rainfall for periods of 1–2 days or 2–3 weeks or more.[10]

Suspended dust

The presence of suspended dust inhibits surface heating and may cause temperature inversions. Macleod (1976) suggested that the south-westerly

150 Two dry decades

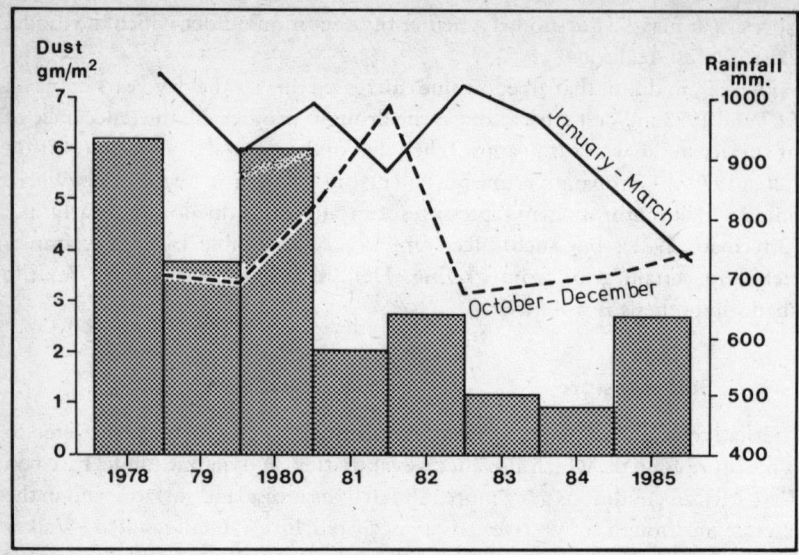

Figure 6.4 Rainfall and mean weekly dust deposition at Kano, 1978–85. Annual rainfall is shown by the columns, and trends in dust deposition by the lines. Deposition data by courtesy of Harmattan Research Group, Bayero University, Kano

monsoonal air tends to have high and uniform temperatures, and the presence of suspended dust adds to its stability, inhibiting convection, cloud formation and rainfall along the ITD. However, it is rare for the south-westerly monsoonal air to contain significant quantities of suspended dust. Aerosols associated with the north-easterly winds of the dry season are quickly deposited when the northward transgression of the ITD occurs, and visibility improves conspicuously.[11]

This theory raises the question of the link, if any, between land use and the Harmattan dust plumes of West Africa. The source of Harmattan dust is the depression of north-central Chad (McTainsh, 1980; McTainsh and Walker, 1982), a lightly vegetated area at the best of times; and there is no doubt that additional material is mobilised downwind. Whether the intensity of the Harmattan in a given dry season is related to the rainfall of the previous year (through the level of protection afforded the surface by its vegetation), or by synoptic conditions in the source area at the time of mobilisation, has not yet been determined. Also, the impact of land use, as distinct from rainfall, on the vegetative cover from one year to the next is largely a matter for speculation. It has been suggested that the intensity of the Harmattan provides an index of the progress of desertification (McTainsh, 1985, 1986). Such intensity may be measured in terms of visibility conditions, deposition rates, or the length of dust spells. But measurements at Kano (which lies in the path of the great majority of dust plumes) show little significant trend in the mean weekly rate of deposition during the eight years 1978–79 to 1985–86 (figure 6.4).

During the first three months of the dry season (October–December), the influence of locally mobilised dust is minimal owing to the presence of crops and other vegetation. At this time, the Harmattan should consist predominantly. of dust imported from its catchment area. No relationship can be discerned between rainfall and dust deposition. During the later months (January–March), the surface is often quite bare and locally formed dust clouds are a common sight. Yet the relationship between dust deposition at this season and rainfall in the preceding year is, if anything, stronger – and positive rather than negative, as would have been expected.

An attempt to link dust mobilisation with a decline in the yields of major crops in the western Sudan (Ibrahim, 1978) has been disputed on methodological grounds (Olsson, 1983). At present, therefore, no firm basis appears to exist for linking suspended dust in the atmosphere with drought or desertification.

According to Hare (1983, 1984a) the possibility of feedback mechanisms should be taken seriously although their status is essentially hypothetical. It appears from the weight of the evidence that their role is, at most, secondary to that of changes in the general atmospheric circulation. The evidence that rainfall anomalies may and often do transcend climatic zones, and of the influence of sea surface temperatures, suggests that changes in the general circulation are ultimately to blame.[12] But if confirmed, a linkage between the removal of vegetation and rainfall deficiency–translating West Africans from victims to procreators of drought–has far-reaching implications.

HYDROLOGICAL DROUGHT

The impact of rainfall variations on surface- and groundwater regimes, and the occurrence of hydrological drought, is mediated by land use, as well as by geological and soil conditions. Positive or negative trends are necessarily specific to time and place, and may vary considerably between river basins and geological formations. The storage of water in the regolith delays the full impact of seasonal or longer-term changes in rainfall. There is also some controversy concerning the effects of anthropogenic modifications to the vegetation. To relate desertification to trends in surface- and groundwater regimes is not, therefore, a straightforward matter and in Nigeria such an attempt faces the added difficulty of a severe scarcity of hydrological data, especially with respect to groundwater. In the semi-arid zone of Nigeria, the possibility of hydrological drought is an addition to the intrinsic constraints facing the development of water resources, which have been identified by Olofin (1985) as low and variable rainfall, and high rainfall intensity and rates of evapotranspiration.

Before 1970

In 1955, rises in the level of the groundwater table were reported in the Potiskum area (Carter and Barber, 1958; Barber and Dousse, 1964; Du Preez and Barber, 1965: 49). In some locations a rise in level of as much as 70 m was reported. This area is underlain by the Kerri Kerri Formation, a series of sands, grits, silts and clays up to 200 m thick, and occupies a watershed location between the Chad and Niger-Benue drainage basins (see figure 2.1). The water table occurs at depths down to 200 m, but pronounced ridges, revealed by mapping, occur under the main rivers. These are due to seepage during river flow, and relatively low permeability of the sediments. Such groundwater ridges are common under northern Nigerian rivers. The reports of rising water tables came from the vicinity of rivers while, between the ridges, the water level fell or remained stable. An inventory of wells confirmed these trends in 1964.

The hypothesis was advanced that deforestation and cultivation facilitated infiltration of rainfall, reduced evaporation, and increased surface run-off.[13] Increased, and more perennial, streamflow resulted, which fed the groundwater table by influent seepage, leading to the lateral and upward extension of the ground water ridges. Support for this hypothesis was provided by estimates of annual losses by transpiration (based on a nearby forest reserve) in two river basins, and of the net gain in groundwater storage during the period 1955–64. These estimates showed that a reduction in transpiration of 15–25 per cent could account for the observed increase in groundwater storage. Such a reduction was considered probable on the basis of reports of extensive woodland clearance in the area during the previous forty years.

Similarly, rises in the groundwater table were reported at about the same time on the Cretaceous sediments of Daura and Katsina Emirates (Jones, 1960; Du Preez and Barber, 1965: 42). The same explanation was invoked. In this basin (the Sokoto-Rima), a contemporary government memorandum expatiated on the theme of 'the dying rivers of the Northern Region of Nigeria' (Ministry of Agriculture, n.d.). Increased run-off was choking river valleys with sand; perennial flow and even seasonal surface flow was reduced or halted altogether, as in the Maradi River, a north-bank tributary of the Sokoto River. This view ignored the advantages of increased subsurface flow and reduced depths to the water table, and proceeded with gloomy foreboding to predict diminishing water supply, increased erosion by water or by wind, and reduced yields, even large-scale depopulation. The author suggested that diminished transpiration would reduce atmospheric humidity and therefore rainfall, anticipating the soil moisture feedback theory by many years. The end result would be desert conditions.[14]

Later work in Borno surveyed the depth to water in wells and boreholes on four routes radial from Maiduguri between 1966 and 1970 (Offodile, 1971). The fluctuations were greatest in sandy soils and least in impermeable clays, owing to variations in the rate of infiltration. Other factors causing fluctuations were believed to be climatic conditions and evapotranspiration, but there was no evidence of long-term decline in the water table. Eastern Borno is underlain by artesian aquifers, and these were exploited intensively from the sixties onwards. Some loss of pressure was reported from boreholes in the lower Middle Zone aquifer, whose source and rate of recharge, probably from outside the region, were imperfectly known. Such 'mining'–consumption in excess of recharge – was, and still is, a source of concern (Olowu and Uzoma, 1965: 19), notwithstanding the large quantity of water still in storage.

Before the onset of drier conditions in the seventies, then, there was little evidence of any deterioration in groundwater supply on the sedimentary formations of northern Nigeria. There was, on the contrary, some evidence that land-use changes associated with intensified exploitation could lead to changes in surface water regimes and improvements in subsurface water, at least locally.

The large area of Basement Complex rocks which separates the sedimentary basins of the Sokoto-Rima Rivers and the Chad system was a poor source of groundwater, with impermeable weathered mantle, and poorly developed jointing systems in the underlying rocks (according to Du Preez and Barber, 1965: 29). Wells yielded rather poorly and often went dry. General trends were hard to discern since hydrogeological conditions vary from place to place.

After 1970

Notwithstanding the importance of rural water supply to livelihoods, and considerable public investment in wells and boreholes, little systematic effort has been made to monitor trends since the droughts of 1972 and 1973 impelled public attention towards the possibility of shortage.[15] Kano State is exceptional in this respect. The state-owned Water Resources and Engineering Construction Agency has collected and published both hydrological data on surface water resources and well-depth records of groundwater. Attempts were made in the years 1973, 1977, 1979 and 1983 to map the height of the groundwater table above sea level in that part of the State underlain by the Chad Formation (WRECA, 1985, figure 6.5).

The surveys did not take account of seasonal variations in the depth of water. There were no data on the heights of well heads, so these had to be estimated by interpolation from the nearest available heights on published maps, or by the use of a barometer.[16] Data were available for 300 wells in 1973, 160 in 1977, and 84 in 1979 and 1983. Standardised contour maps were prepared for each year.

The area mapped forms part of the Chad drainage basin, but the drainage system is only partially integrated. The major river, the Kano-Hadejia system, collects a number of tributaries from the Basement Complex in the south-west, and then divides into two main branches flowing north-east through a flood plain having many of the characteristics of an inland delta: channel meanders, ox-bow lakes and cut-offs, levees, backswamps and extensive flooding. To the

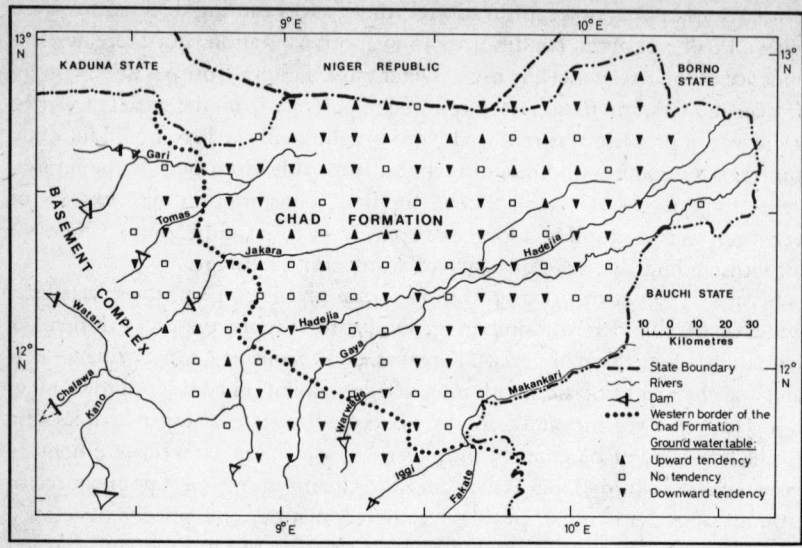

Figure 6.5 Trends in the depth to water table in part of Kano State, 1973–83 (after WRECA, 1985)

north of this system, the Jakara and the Tomas–Gari Rivers, which also follow the regional slope off the Basement to the north-east, disappear into the sand. Even in the Hadejia system, losses from seepage and evaporation accounted for an average of 20 per cent of run-off between Wudil and Gashua (a direct distance of about 300 km) in the years 1963–77 (WRECA, 1985: 11–12).

These data, however approximate, show that downward trends were more common than upward or stable ones in the decade spanned by the surveys, and that the position grew perceptibly worse during the period (table 6.3).

Table 6.3 *Trends in the groundwater table in part of Kano State, 1973–83*

Period	Upward	Stable or fluctuating	Downward	Total
1973–77	73	34	40	147
1977–79	52	39	56	147
1979–83	8	37	102	147
1973–83	36	40	71	147

These figures were obtained by superimposing a grid of 147 equally spaced points on the area covered by all four contour maps (WRECA, 1985) of the groundwater table, in order to compensate for the irregular distribution of the wells.

See figure 6.5.

In 1973–74, 25 per cent of the 631 village heads interviewed in Kano State had reported that half of the wells in their villages were dry; 11 per cent reported that nearly all were dry. Deepening work was carried out very widely. This indicates that in the short term, depths to water in sandy soils may respond quickly to rainfall deficiency. The improvement recorded in the period 1973–77 reflects the recovery which followed the droughts of 1972 and 1973. Unfortunately the data for 1973 cannot be compared with any obtained before that drought.

Were these trends directly dependent on rainfall? Only in part. The importance of river basin management is shown, not this time by the removal of vegetation, but by the effects of dam construction on the headwaters of nearly all the rivers in the zone (figure 6.5). In 1973, a prominent ridge of groundwater followed the Kano-Hadejia River downstream, but after the improvement of the Tiga Dam in 1974 (Olofin, 1982) this ridge shrank, and negative trends were observed in many valley locations. After the dam had filled, a small perennial flow was assured and the ridge stabilised, according to the data, 10 m lower. Similarly, a groundwater ridge followed the Tomas-Gari system north-eastwards; in 1958 this ridge had been mapped at least 20 km beyond the normal termination of flow (Du Preez and Richards, 1958; Du Preez and Barber, 1965: 53). By 1979 this ridge seems to have vanished.[17] On figure 6.5, 31 of 71 points registering downward trends during the decade as a whole were close to rivers whose regimes had been modified by the construction of dams.

In Kano State, therefore, it seems barely feasible to separate the effects of meteorological drought from those of basin management – both land use, and hydraulic engineering.

LAND USE AND DROUGHT

Judgement has to be suspended on the linkages between land use on the one hand and the occurrence of meteorological drought. The 'feedback' theories that propose a causal relationship between anthropogenic modifications of surface characteristics and rainfall deficiencies lack empirical support and it seems unlikely, at a time of widespread deterioration in weather reporting systems, that adequate data for their validation will be available for some years. Meanwhile, the possibility that African droughts will eventually be explained in terms of the behaviour of the global atmospheric circulation seems to be strong.

The systems of land use that have developed adaptive capabilities during the course of many centuries should be strengthened, rather than replaced. Present and future development should continue to take account of the salient characteristics of the climate. For the Sahel, these have been identified as: low and variable rainfall, a prevalence of dry years, the extreme magnitude of variability, the rapidity with which new persistent conditions can become established, the great geographical extent of the major episodes of abnormal rainfall, and the 'spottiness' of rainfall even in normal years (Nicholson, 1982: iii).

On the ground, the linkages between meteorological and hydrological drought have been obscured by anthropogenic factors, in particular the clearance of woodland for agriculture and the construction of dams on seasonal rivers. The only area for which data on trends in the depth to the water table have been collected systematically gives conflicting, but mainly negative, signals. This area is underlain by sedimentary formations. Elsewhere on the sedimentaries, evidence of significant if localised upward trends in groundwater levels was obtained before the present period of persistent drought began around 1970. The level of Lake Chad (which receives only 10 per cent of its inflow from Nigerian rivers) peaked in 1963 and fell dramatically in the seventies; its failure to recover mirrors the persisting meteorological drought, as do the shorter-term fluctuations in the regimes of perennial and seasonal rivers.[18] On the igneous Basement Complex, which underlies most of central northern Nigeria, there is evidence of deteriorating groundwater conditions north of Lat. 11°30'N in hundreds of dry or deepened wells, and reports of seasonal watercourses that no longer flow, even in areas where no significant change in land or water use has taken place for decades.

On balance, therefore, we may conclude that hydrological drought (deteriorating surface and groundwater conditions) in northern Nigeria is associated with persistent meteorological drought in the region, but that anthropogenic factors affect the relationship at the level of the individual river basin or locality.

7

SHIFTING SANDS

> ...like abandoned land in the desert   Jeremiah 17:6

From the perspective of adaptive response, desertification differs from drought not only in its time perspective (a long-run process as opposed to a short-run event), but also in the extent to which it can be considered as exogenous to the human system. The question of adaptation becomes subsumed in the larger one of causation. To what extent are the communities that inhabit areas prone to desertification themselves responsible for the degradation that threatens their livelihood?

It was argued in chapter 1 that a simplified definition of desertification as 'the degradation of ecosystems in arid or semi-arid regions' has theoretical and practical advantages. The standpoint of the present chapter is therefore ecological. Its objective is to review field evidence for ecological degradation, in relation to land use, in four of the subsystems proposed in table 1.1: the woodland, grassland, soil and morphodynamic subsystems (the hydrological subsystem was discussed in chapter 6). The evidence is ambiguous if not, in places, contradictory. Such an outcome should be expected in a complex, multivariate problem involving most aspects of the relationship between a society and its environment. Nevertheless it is only by accumulating judgements based on empirical studies that environmentalist slogans can be replaced by more balanced evaluations. This case study gives ground for questioning conventional wisdom that emphasises the role of 'over-exploitation' at the expense of that of meteorological drought.

It was also argued in chapter 1 that any attempt to measure the rate and extent of desertification should be specific to time and place. The selection of a study area exhibiting evidence of advanced desertification has the advantage of providing a standard by which other areas may be conveniently measured. The most obvious evidence of desertification is often found in the morphodynamic subsystem, and moving sand conforms very closely with most people's perception of the process. An area possessing such characteristics is the Manga Grasslands.

THE MANGA GRASSLANDS

The Manga Grasslands (figure 7.1), largely treeless and densely populated with moving sand dunes, provide a dramatic impression of 'Saharisation'.[1] They are

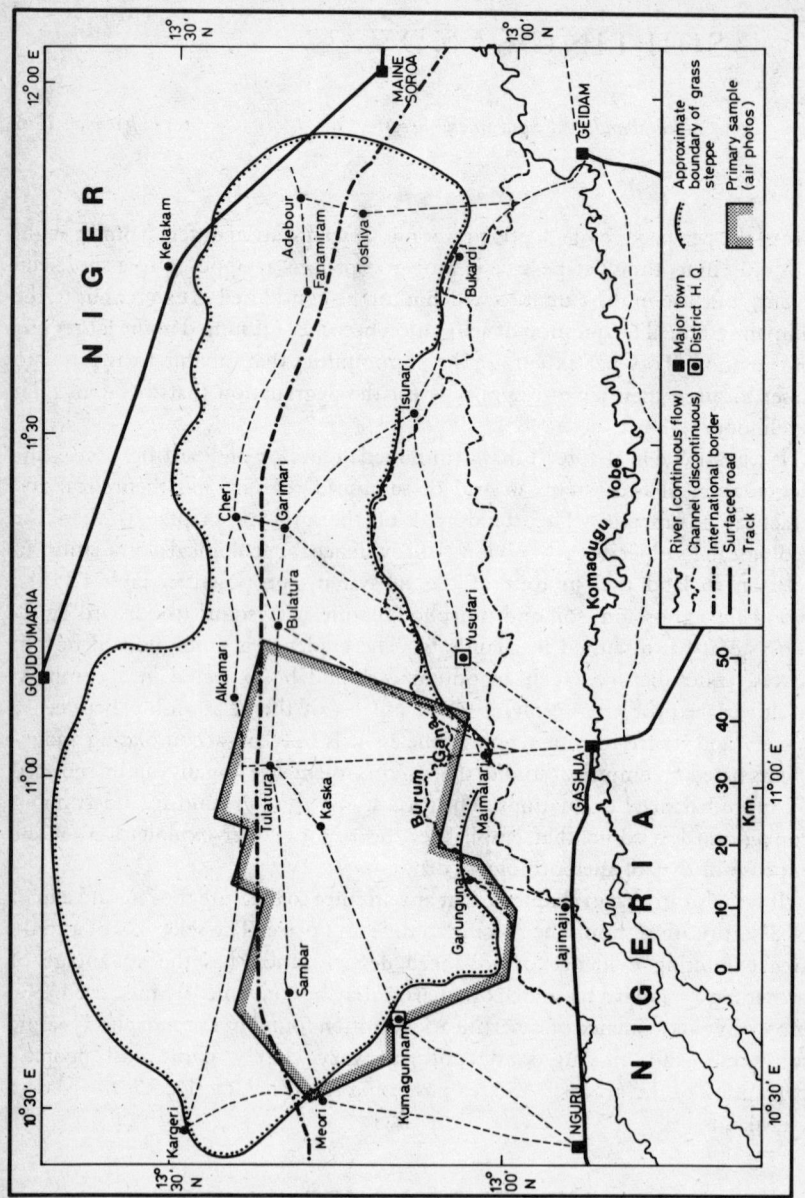

Figure 7.1 The Manga Grasslands

The Manga Grasslands

7 The Manga Grasslands in the early rains (late July 1981). In the foreground, *Cenchrus biflorus* grassland on a stable dune formation; in the middle distance, a *tafki* lake surrounded by dum woodland (*Hyphaene thebaica*) with farmland on the lake shore on the far right; in the far distance, remobilised dunes associated with the village site. Shade and economic trees grow in the village (left), and an aged individual of *Acacia albida* on the dunes. (Looking south to the village of Ligaridi Babba, figure 7.5)

occupied by sedentary Manga farmer–stockowners and transhumant Ful'be stockowner–farmers, with nomadic stockowners visiting the area from time to time (Poncet, 1973). The two groups, as in Dagaceri, have come to terms over the sharing of resources (not without friction) on the basis of their complementary specialisations in farming and stockbreeding (see Horowitz, 1972: Bovin, 1985).

This distinctive region is entered abruptly from the Sudanian woodlands and farmlands to the south and west, and is superseded northwards by wooded savanna or shrub steppe, which continues for 200 km before the living dunes of Ténéré are reached.[2] The Grasslands display an irregular hummocky landscape of stabilised dunes, 300–50 m above sea level, in which quite small relief features may be visible from some distance owing to the general absence of trees. These slopes, usually of less than 10°, are clothed with grass whose changing colour indicates the progress of the seasons, and its luxuriance the abundance of the previous season's rainfall.

Traversing this seemingly uninterrupted surface, the observer cannot go far before encountering a circular or elliptical depression, whose sides drop abruptly, at angles ranging from 12° to 24°, to a flat floor. This floor is (or once was) densely wooded with the dum palm (*Hyphaene thebaica*) in sole stands, often encircling a natron lake, or its dry bed. With a relative relief of 15–20 m, enough to conceal a fully grown palm, such a depression may be invisible a few hundred metres away. Its white or grey lake bed and dark soils, and the dark green foliage, contrast vividly with the orange soils and grass-covered dunes that surround it.

This contrast in terrain is mirrored in two dominant systems of land use. The upland (H *tudu*) is grazed extensively, with occasional fields of millet in years when the rains are adequate. It also provides hilltop sites for Manga villages and Ful'be camps. The depression (H *kwari*), however, is intensively used. Its lower slopes are the preferred sites for rain-fed agriculture. Shallow (but saline) groundwater permits dry-season farming in small plots around the lake (H *tafki*). Cattle are brought in to graze on the lake-bed pastures, and to water from shallow wells. The dum palm is exploited, fruit trees are cultivated, and salt and natron (or 'potash', H *kanwa*) workings are found amongst the trees. Intermediate in characteristics between the *tudu* and the deep *kwari* is the more extensive shallow *kwari* which lacks both dense woodland and lakes. Its soils are heavier and more moisture-retentive than those of the *tudu*, and these sites are favoured for rain-fed farming. Control of the resources of the *kwari* has always been fundamental to the Manga political economy.

Moving sand formations have been noticed in the area for at least fifty years, but increased attention has been given them, within Nigeria, since the droughts of 1972–74 (for example, Mustapha, 1977; *Daily Times*, 2 June 1982). Funds of the Federal Government's Arid Zone Afforestation Committee have been invested in tree nurseries and shelter-belt plantations.

The sands of the Grasslands form part of a huge system of fixed dunefields that were formed under arid conditions from alluvial deposits left behind by drainage systems in the Lake Chad basin (Grove and Pullan, 1964; Prirard, 1966; Maley, 1981). As the climate became drier and the level of the lake fell, the rivers cut down through their thick sandy deposits, and under the influence of a new regime with reduced discharge, the character of their deposits changed from coarse sand to finer silts and clays. The Komadugu Yobe, which flows into Lake Chad, is the last vestige of a formerly extensive river system.[3] Other branches have died: the north-flowing Gari and Tomas rivers (in Kano State) disappear into the sand, and the Burum Gana, which forms a southern boundary of the Grasslands, no longer has continuous surface flow. The depressions of the Grasslands are broken sections of still more moribund channels. The presence of natural woodland not only to the south, east and west, but also to the north indicates that the climate is not prohibitive to tree growth. Probably, the Manga Grasslands are the last portion of the subSaharan dunefields to have been stabilised under vegetation. Under conditions of temporarily increasing aridity, they are, therefore, the first to be remobilised.

ECOLOGICAL CHANGE, 1950-69

Methods

The measurement of degradation over time requires longitudinally compatible data, and the most important source of such data is conventional air photography, which became widely used in northern Nigeria from 1949. By comparison, earth resources data obtained from satellites have insufficient scale resolution for micro-scale studies, and rather shallow time depth.[4] Side-looking airborne radar imagery, which is available in Nigeria for one date (1976), is insufficiently sensitive to surface variations of the kind relevant to this investigation. Black-and-white air photography, however, with a history of nearly four decades, permits medium-scale monitoring of ecological changes that occurred between the different dates of photography.[5]

Two sets of air photography are available for the Manga Grasslands, dated 10 October 1950 (1:30,000) and 7 November 1969 (1:40,000).[6] The interval between the two was one of economic expansion in northern Nigeria, including the probable growth of the livestock population, of farm production for the market, of populations, settlement and agricultural colonisation. Rainfall was above normal (1931–60) in most years. The photography allows an evaluation to be made of changes in land cover types in the Manga Grasslands before the impact of the droughts of the seventies and eighties.

The air photography varies in scale and in quality. The earlier set lacks the resolution of the later, but the gain in resolution is partly offset by a smaller scale. The two sets were flown on different flight patterns, the first in arcs which in this part of the country ran from WNW to SSE, and the second in conventional east–west flight lines. Owing to this problem, and the lack of ground control, coincident sampling was impossible. Independent primary and secondary samples were generated for each set of photographs in approximately coincident areas of 1,480 and 1,840 km². These comprise a block extending from the centre to the periphery of the south-west segment of the Grasslands (figure 7.1).[7] The dominant land cover type was interpreted for each of 3,300 blocks of 2.25 ha (for 1950) and 2,175 blocks of 4.0 ha (for 1969).

Results

The results, which are shown in table 7.1, should be treated with caution. Although both sets of photographs were taken in the early dry season, annual variations in the rainfall and in the dates of termination of the rainy season affect the soil moisture, groundwater, and vegetation, influencing tonal and textural characteristics. Owing to such variability on annual or even shorter time-scales, the inference of a trend from a change in class frequencies may not always be reliable. Furthermore, the results are subject to the limitations of visual interpretation on materials of different scale and quality.

Table 7.1 *Ecological change in the Manga Grasslands, 1950–69*

Terrain type	Land cover class	Frequency (per cent) 10 Oct. 1950	7 Nov. 1969
Upland (*tudu*)		67.8	71.8
	1 Grassland	66.3	70.3
	2 Woodland	0	0.1
	3 Cultivation	0.3	0.5
	4 Settlement	0.3	0.1
	5 Mobile sand	0.9	0.8
Lowland (*kwari*)		32.2	28.2
	6 Grassland (sparsely wooded)	2.9	3.7
	7 Open woodland	13.1	10.7
	8 Dense woodland	4.3	3.3
	9 Cultivation (lightly wooded)	9.9	7.9
	10 Lake beds	2.0	2.6
Total		100.0	100.0
Number of blocks		3,300	2,175

The terrain was divided into the two major classes: upland (*tudu*) and lowland (*kwari*). The first contained all the unwooded grassland, settlements and mobile sand; the second contained all the woodland, most of the cultivation, and the lake beds. The only significant change was an increase in grassland (class 1), which occurred at the expense of open woodland (class 7) in shallow *kwari* sites. The apparent increase in the extent of upland at the expense of lowland is an error of interpretation, for there is no other evidence that such a transfer occurred.[8] All the lowland categories of woodland (classes 7, 8 and 9) registered a decline. The frequency of mobile sand showed no change (but this evidence will be questioned below); neither did the percentage of sample locations falling within 250 km of a sand formation (15.2 in 1950 and 15.9 in 1969).

Interpretation

Evidence of ecological stability is unexpected, particularly if growing human and livestock populations are assumed. It accords better with rainfall behaviour, however. The year 1950 had 14 per cent more rainfall than normal (1927–76) in northern Nigeria as a whole (Motha *et al.*, 1980). While eight of the ten preceding years had negative departures, 16 of the following 19 years had positive departures. But in 1968 and 1969, rainfall deficits of 11 and 21 per cent proved to be harbingers of the drought-stricken seventies. That the situation was no better in 1969 than in 1950 is also significant. Increased pressure of land use evidently cancelled out any positive effects of a rainfall regime better than average.

Extending the study to the present, the four subsystems – woodland, grassland, soils and surface materials – will now be examined separately. To supplement the air photography, ground transects and profiles of sand formations were carried out between 1978 and 1986. In some places, surface photography provides a supplementary record. Interviews were conducted with Manga and Ful'be informants in a wide range of locations.[9]

WOODLAND

There is a large literature (referred to in chapter 1) linking deforestation in West Africa with desertification. Rarely, however, is the causal link given the benefit of empirical demonstration. Later writers on the subject have either dodged this question or interpreted deforestation as itself synonymous with desertification. Do the Manga Grasslands provide evidence of significant deforestation and, if so, is it meaningful to interpret such deforestation as a form of degradation?

Grassland may be man-made by cultivation and fallow cycles, and such places usually have a sparse scatter of old trees, or vigorous shrub regrowth. But the scarcity of trees of any age on the *tudu* suggests that cultivation was not the agency responsible for the origin of the grassland communities. Further evidence that the grassland is not man-made is provided where the boundary intersects the cultivated zones surrounding a series of major settlements.[10] Between 1950 and 1969 the mean distance from village to field increased significantly in these places as inner lands were exhausted. Nevertheless this process failed to affect the stability of the boundary of the Grasslands.

There is no evidence that the upland (*tudu*) soils have ever supported woodland; the oldest informants affirm their almost treeless status throughout living memory. The depth of the water table is from 10 to 15 m. *Acacia albida* occurs occasionally, its very rarity assisting easy recall of its history. While some older specimens have been observed to die recently, in other places – particularly in the eastern Grasslands – spontaneous regeneration is taking place. Well-established village sites on the upland have a few *Balanites aegyptiaca*, *Adansonia digitata*, *Acacia albida* and introduced neem (*Azadirachta indica*). The last are planted, but the indigenous trees are usually spontaneous (the seeds of fodder trees being brought to the village by goats). The marginal conditions for tree growth are shown by the failure of several government shelter belts and, unsurprisingly, of at least one attempt to plant neem seedlings directly in moving dunes. In conditions of sandy free-draining soils and variable rainfall, the water budget in some years is simply insufficient to provide for the survival of plantations, unless irrigated.[11]

Lacking any evidence of deforestation, therefore, the treeless upland soils of the Grasslands must be supporting a natural grassland community, in the earliest stage of woodland colonisation. Such an ecosystem is youthful, if not ephemeral.

The lowland ecotype is subdivided into the deep *kwari*, where the water table reaches the surface for part or (occasionally) all of the year, and the shallow *kwari*, whose sand-filled floor has a depth to the water table of 3 to 9 m. In the shallow *kwari, Acacia albida, A. senegal* and *Hyphaene thebaica* form a sparse woodland, associated with farmland and fallows. Such woodland diminished in extent between 1950 and 1969 (class 7 in table 7.1), replaced by open cultivation or grassland.

In the deep *kwari* the position is different. The depth of the water table is less than 2 m in the dry season. When protected, woodland occurs in close stands entirely composed of *Hyphaene thebaica* growing 15 m tall, the ground covered with fallen trunks and branches. Protected it is, on the Niger side of the border, by heavy fines imposed on illegal woodcutting, and even on cutting for browse. On the Nigerian side, such dense woodland is now rare, and there is plenty of evidence (from local informants) that it has diminished in the vicinity of the larger villages.

Almost all construction work uses the trunks of dum palms, and most domestic energy is obtained from burning them. They are also burnt in the salt works that still operate in a few locations. In fact the *kwari* is the focus of competing and complementary land uses throughout the dry season: commercial irrigated or perennial agriculture, intensive livestock grazing, watering of herds, natron collection and storage, salt-making and timber cutting all occurring within a small area. Rights to these resources are jealously protected, and they attract migrant labourers and farmers in the dry season, who camp on top of the adjacent hills. Notwithstanding the antiquity of the salt industry (Lovejoy, 1986), there are no reports of any shortages of free timber.

Not surprisingly, some woodlands have been much reduced, although it should be added that other economic trees (date palm, mango, citrus) have been planted in their place, especially near villages.[12] It is sometimes suggested locally that the reduction of the woodland has contributed to the overblowing of the *kwari* banks by moving sand, but this seems improbable since the woodland's effect on the aerodynamic properties of the adjacent upland must be little. Overwhelmed date palms on *kwari* banks are, however, a common sight near some villages.

On balance, therefore, the woodland ecosystem gives contradictory signals. The *tudu* never was wooded, but there is evidence of some spontaneous regeneration, and localised protection or planting, notwithstanding conditions marginal for tree growth. Diminution of woodland has occurred in both shallow and the deep *kwari* sites especially on the Nigerian side of the border. The existence of any linkage with the reactivation of moving dunes is unproven. However, the implications of woodcutting in excess of regeneration, amply testified in some places, are serious enough for the future of the woodland itself. In Niger, Wata (1979) traced the southward recession of the gum arabic tree (*Acacia senegal*) in the Manga Grasslands, equating it directly with desertification.

GRASSLAND

The grassland community is dominated in the 1980s by the annual *Cenchrus biflorus*, with *Aristida* spp. in second place, and also *Ctenium elegans* and *Eragrostis tremula*. Under good rainfall conditions they form a continuous cover, usually less than one metre tall. *Cenchrus* may recolonise the lower slopes of moving dunes in the wetter years, but is swiftly killed by moving sand in the drier, when bare patches appear, and individual plants are more widely spaced. On the windward side of active dunefields, *Leptadenia pyrotechnica* forms a dispersed shrub community, separated by deflation patches; its association with dune boundaries that have remained stable for many years suggests that its growth is a response to conditions marginal for grass, and may contribute to the eventual stabilisation of these dunes. In the vicinity of settlements, where grassland has been eliminated by trampling, the pernicious *Calatropis procera* appears. In the *tafki*, *Phragmites vulgaris* provides valuable fodder.

The Grasslands have been occupied by semi-sedentary Ful'be groups for at least half a century. Their camps are occupied intermittently for many years and, near them, pressure on the rangeland is partly compensated by the addition of considerable quantities of manure. During the dry season the herds may be split, the younger men taking one part south to the riverine areas near Nguru and Gashua, and the remainder staying in the Grasslands. At the same time, herds from farther north may enter, the numbers depending on how bad conditions are in their home (rainy-season) pastures. Their rights to grazing are not questioned by the resident groups, who also farm extensively during the rains.

Many resident herds never leave the area, and in a good year there is adequate dry forage throughout, notwithstanding the dramatic reduction that occurs between October and April in the moisture and crude protein content of such grasses (Wickens and White, 1979). In a bad year, the decision to move to pastures further south may not be an easy one to take. Most herders are ignorant of rangeland conditions south of the Komadugu Yobe, of conditions for their animals' health and of the situation regarding access to water. The decision to move the resident herd may be postponed until the livestock are too weak for the journey, and instead they are fed on the tough leaves of the dum palm. Such difficulties have been reported among pastoralists elsewhere in the Sahel (Barral and Benoit, 1977). Obligatory transhumance, as recommended by the Land Resources Division (1972: map 13), is not a realistic management option.

In 1983, the rising value of fodder provoked the sudden eruption of fodder selling, as in Dagaceri – a commercial opportunity for the Manga population at a time of labour surplus. The short, dry annual grasses are easily raked up and stored against the seasonal rise in price. In recent years there has also been a slow penetration of industrial feeds (such as wheat offals), through the market system.

The size of the resident livestock population is locally perceived to depend directly on the condition of the rangeland and on the incidence of disease. The rangeland in turn depends on the rainfall. Cattle mortality on a large scale is remembered for the first time in 1972–73. The herds were reconstituted in part during the seventies – among the Wodaabe of the agropastoral zone in central Niger, such reconstitution has been estimated at over 60 per cent (Bonfiglioni, 1985). But poor grazing in the late seventies culminated in heavy mortality in 1980 in the Grasslands. In the eighties, grazing conditions reported in the dry season reflected the rainfall of the preceding season quite closely (table 7.2). The years 1983 and 1984 were considered even worse than 1972 and 1973. In addition, rinderpest struck. Not surprisingly, therefore, informants agree that the resident cattle in the mid-eighties were fewer than those of twenty years earlier.[13]

However, the density of the livestock population, and of cattle in particular, fluctuates violently from month to month, depending on the size of the influx from further north, which depends in turn on grazing conditions there. In July 1983 (when the rains were failing) cattle densities of over 80/km² were observed, from two to eight times those estimated for a normal wet season (10–40 km²: Land Resources Division, 1972: map 8). In the *kwari*, where there was some fresh grass, day-time cattle densities of 800/km² were seen. In the Gouré and Goudoumaria areas of Niger, grazing conditions were reported to be worse in 1984–85 than at any time in living memory, after three years of very

Table 7.2 *Rainfall at Maine Soroa, grain yields and grazings in the Manga Grasslands*[a]

| Year | Rainfall: per cent of normal (1941–60) | | | | Reported harvest | Reported dry-season grazing |
	June	July	August	September		
1978	84	194	112	82	?	not good
1979	100	162	32	44	good	not good
1980	144	81	45	63	rather poor	bad
1981	128	158	95	37	excellent	good
1982	44	45	111	13	poor	poor
1983	0	43	61	45	very poor	disastrous
1984	0	102	19	50	disastrous	better
1985	109	74	67	n.a.	good	very good

[a] Variability, even between adjacent villages, is high; the records for Maine are not necessarily representative.

low rainfall. The wave of incoming herds marched through the Grasslands between December 1984 and March 1985, annihilating the poor pasture growth that already bore the marks of low rainfall in three successive years.

Under such conditions, it might be expected that the reduced plant cover would expose the soil to aeolian action, a sight that is common enough in the late dry and early rainy seasons. But the surviving roots and stems of the grasses provide some protection, even when dead. And the regenerative capacity of annual grasses is not impaired. Following good rainfall in 1985, a luxuriant grassland – the best observed in eight years – sprang to life, and was still intact by the following February, well in excess of the needs of the resident herds. All the incomers had gone.

According to the Land Resources Division (1972: vol. 4, 22f.), the grassland community was dominated by *Andropogon gayanus*, var. *tridentatus*, and *Aristida* spp., with *Cymbopogon schoenanthus*, *Ctenium elegans* and *Diheteropogon hagerupii*. *Andropogon* spp. was here at the northern limit of its distribution.[14] The Land Resources Division's study was carried out before the onset of drier conditions in the seventies. Later, the vegetation and land use map of the Federal Department of Forestry (1977) identified the same dominant species in an inaccurately designated 'wooded shrub' formation. This map was based on side-looking airborne radar imagery of November 1976. However, few ground transects were carried out in this area (Parry and Trevitt, 1979). Therefore it may be assumed that the classification follows the Land Resources study and does not, as the date might imply, indicate conditions after the drought of 1972–74.[15]

Informants in the area do state that *Andropogon gayanus* (H *Gamba*), which is an easily recognised, tall perennial, favoured by cattle, was more common in the area before 1972–74, though whether dominant is not clear. Its disappearance is dated to that time, and attributed to the decline in rainfall, although in such marginal conditions preferential grazing by cattle could have accelerated its demise. In the eighties, the grassland community resembles more closely a more northerly community of 1972 described by Rippstein et al. (1972: 172–4) in an area adjoining, but not co-extensive with, the Grasslands in Niger. This was dominated by *Aristida* spp., with *Eragrostis tremula* and *Cenchrus biflorus*. Substitution of the annual grass *C. biflorus* for perennial species is consistent with the onset of drier conditions, even if accelerated by grazing pressure on the perennials. Annual grasses do not remain succulent for as long as the perennials, but they are nevertheless considered excellent year-round grazing; such a substitution does not necessarily carry negative implications for forage yield (Rippstein et al., 1972: 223).[16]

Rainfall variations, therefore, are primarily responsible for changes in the condition of the grass, both from year to year, and in the longer term. When its condition is worst, it is subject to the greatest pressure, owing to the constraints that limit transhumance among the resident herds, and the influx of animals from the north. But the capacity of annual grasses to regenerate vigorously after intensive grazing and poor rainfall has been demonstrated beyond doubt.

SOILS

The dune soils of the Manga Grasslands have been described as weakly developed sandy soils of aeolian deposition, having little or no profile development, and little colour change along the catena from crest to base, indications of their recent origin (Land Resources Division, 1972: vol. 1). They are very low in fertility. On the other hand, the halomorphic soils of the depressions contain a higher clay/silt fraction and are white or grey in colour, but also incorporate much sand washed from upslope. Although benefiting from a shallow water table, they are affected by salt.

The upland soils occupy from 68 to 72 per cent of the area and lowland soils the remainder (table 7.1). The first are of importance principally for grazing, the second for cultivation and woodland. Further subdivisions, however, may be recognised on the basis of topographical location and the categories recognised by the users (table 7.3). The upland soils divide into orange grassland soils (M *kaisa*) and those of moving dunes (M *kabai*); the lowland soils into grey or black sands found in the depressions (M *tulo*), and the sticky, saline soils found in the beds of the seasonal *tafki* lakes (M *danko*).

In fact, the properties of these soils (at 15 cm depth), as shown in table 7.3, form a continuum from moving dunes at one extreme to the *tafki* floors at the other. On the basis of these properties, no real distinction can be sustained between moving and stable dune soils (nos. 1–4), which is not surprising, the chief differences being in site characteristics. The remaining samples (nos. 5–7) show an improvement in clay/silt ratios with lower topographical locations, and a sharp increase in potassium, calcium and magnesium content to levels which render them almost sterile (no. 7). The carbon content of all these soils is exceptionally low.[17] Control samples (nos. 8 and 9) from 10 km and 50 km beyond the boundary of the Grasslands do not differ significantly, in any of the properties measured, from the upland soils of the Grasslands.[18]

Rain-fed cultivation of millet, cowpeas and sorghum is practised in the arable depressions, which are the mainstay of the subsistence sector (nos. 5 and 6). In good years, millet cultivation is extended to the stable dune soils (nos. 3 and 4). In the *kwari*, the lower slopes are used for grain and the bottom soils (no. 6) carry intensive market gardening during the dry season, sometimes supplemented by small irrigation works using shallow wells. The range of market crops includes sugar, cassava, peppers, maize, cucurbits and melons. These soils may waterlog in the rains and are not usually planted with grain. Land is fallowed when exhausted, which partly accounts for marked changes in the distribution, but not the quantity, of arable land observed between 1950 and

Table 7.3 Soil properties (15 cm depth)

No.	Site	Vegetation	Type	Clay/silt fraction (%)	Sand (%)	Exchangeable (me/100 g)				pH	Organic carbon (%)
						Potassium	Calcium	Magnesium			
Manga Grasslands											
Upland											
1	Moving dune	none	kabai	4.6	95.4	0.06	0.65	0.21		8.91	0.1
2	Moving dune	none	kabai	4.6	95.4	0.07	1.00	0.29		7.93	0.16
3	Stable dune: upper slope (10°)	grassland; *Cenchrus* spp.	kaisa	3.0	97.0	0.02	0.80	0.26		7.00	0.09
4	Stable dune: lower slope (5°)	grassland; *Cenchrus* spp.	kaisa	6.2	93.8	0.02	0.90	0.16		7.30	0.12
Lowland											
5	Shallow *kwari*	arable	?tulo	8.6	91.4	0.10	2.95	1.47		7.41	0.16
6	Deep *kwari*	arable	tulo	14.6	85.4	0.09	11.25	1.98		7.55	0.18
7	Lake bed	none	danko	21.0	79.0	0.43	45.90	1.30		8.50	0.25
Control samples											
8	Stable dunes, 10 km south-west of boundary	woodland	—	2.6	97.4	0.04	1.20	0.50		7.51	0.15
9	Stable dunes, 50 km south-west of boundary	arable	—	7.6	92.4	0.05	1.00	0.60		7.00	0.09

Determinations by E.U. Essiet and S. Nuwanshong at the Department of Geography, Bayero University, Kano.

1969 (table 7.1). The stability, if not decline, of the area of arable land suggests a stable demographic pressure during the period.[19]

Conclusive evidence of soil degradation over a known period of time would be diminished fertility of cultivated soils, diminished crop yields, or soil profiles damaged by erosion. Without long-term monitoring of the soils, nothing conclusive can be said about the trends in soil fertility. With regard to the main arable soils (nos. 5 and 6), some informants report that yields are lower than in former times, and that soils become exhausted unless fallowed or fertilised. But it is impossible to separate the effects of soil degradation, if any, from those of rainfall. Records from Maine Soroa, the nearest rainfall station, confirm that the size of the reported harvest is primarily a function of the monthly distribution of rainfall (table 7.2). Neither can the profile be examined for evidence of loss of the upper horizon through wind erosion; profile development is barely perceptible in the upland soils. The soils of the depressions, whether under cultivation or not, might be expected to show signs of increasing concentrations of mineral elements to levels intolerable for crops; but notwithstanding the very high levels reached in the *tafki* floors, there were no reports of *tulo* soils (no. 6) going out of cultivation for this reason.

SURFACE MATERIALS

Using the air photography, direct measurements of sand formations in the area of the primary sample allow an analysis to be made of their patterns and dynamics during the two decades, 1950–69. Owing to the absence of later photography, it is not possible to extend this analysis to cover the following one and a half decades of drier conditions. Instead, field measurements were carried out in five locations, to provide a basis for case studies of sand formations over a period of 36 years.

Morphodynamics, 1950–69

The results of direct measurements of all moving sand formations (with the exception of small blowouts) in the sampled area are given in table 7.4. Such formations occurred in two characteristic locales, and these demonstrated opposing tendencies during the period under review. In the first – village perimeters – continuous systems of dunes and deflation pits developed, which might be up to 70 ha in size. In the second – open rangeland – the formations averaged little more than 1 ha in extent, but were far more numerous. The first type tended towards stability, but the second increased noticeably.[20]

The analysis shows that the first type (row 1) nearly doubled in frequency, but diminished substantially in average size, from 42 to 16 ha. The second type (row 2), however, increased twelvefold in frequency, increasing its share of all sand formations from 5 to 51 per cent, but maintained the same average size.

Table 7.4 *The development of moving sand formations, 1950–69*

Number and size	Number		Area (ha)			Average size (ha)	
	1950	1969	1950	1969		1950	1969
1 Village perimeter formations	13	23	551[a]	370[a]		42	16
2 Open rangeland formations	6	73	30	390		5	5
3 All formations	19	96	581	760		31	8
4 Change in area (%)				+31			
5 Percentage of area[b]			0.34	0.44			

Dynamics	Number in 1950	Condition in 1969			New in 1969		Number in 1969
		Larger	Smaller	Gone	Blowout[c]	New site[d]	
6 Village perimeter formations	13	0	7	6	7	9	23
7 Open rangeland formations	6	0	1	5	38	34	73
8 All formations	19	0	8	11	45	43	96

[a]The built area of the village is included, where surrounded on three or four sides.
[b]1,480 km² in 1950 and 1,840 km² in 1969.
[c]Developed on the site of blowouts visible in 1950.
[d]Developed on new sites

172   Shifting sands

The area of both types together (row 3) increased by 31 per cent, from 0.34 to 0.44 per cent of the study area.

Table 7.4 also investigates the dynamics of the formations. Of the village perimeter dunes (row 6) all 13 observed in 1950 had diminished in size, or disappeared, by 1969; but 16 new dunes had appeared at other villages. The frequency of new rangeland dunes (row 7), as already mentioned, was more striking. About half of the new dunes observed in 1969, of both types, were on the sites of small blowouts (such as deflation patches, eroded footpaths, and blown crests) that had been visible in 1950.

Thus, although the village perimeter dunes are more extensive, more spectacular, and even threaten the existence of certain settlements, they were in relative equilibrium from 1950 to 1969 when compared with the dunes in the rangeland which, in spatial terms, posed a greater threat. The fact that only eight substantial formations lasted throughout the period shows that, given the right conditions, mobile sand can be recolonised by vegetation. To explore these ambiguities further, a number of case histories will now be reported, using field traverses and profiles, and interviews with resident observers, to extend the evidence obtained from air photography to the present. Table 7.5 summarises the area measurements for these case histories.[21]

Table 7.5 *Sand mobilisation in selected localities*

Locality	Figure reference	Area of moving sand (ha)		
		1950	1969	1986
Village perimeter dunes				
1 Kaska	7.2	40	40	30
2 Ligaridi Karami	7.4	60	54	33
3 Ligaridi Babba	7.5	52	34	39
4 Bulari, Gaptari	7.7	0	0	c. 20 (1981)
5 Old Ngelsandi	7.8	6	2	2
New Ngelsandi	7.8	0	70	30
Lala'a	7.8	36	10	6
		1950	1969	1986
Rangeland dunes (4.8 km²)	7.9	0	35	99
Number of formations (>0.2 ha)		0	29	46
Percentage of area		0	7.3	20.0

8 The leading edge of the Kaska dune proceeds slowly through the village market place, a number of houses having already been submerged (July 1986). Bundles of grass fodder collected for sale may be seen on the left. Behind the forked tethering posts is the SW base for the transect shown in figures 7.2 and 7.3

Village perimeter dunes

The case histories of village perimeter dunes which follow will show that their dynamics were closely dependent on the growth and decline of human settlements.

(1) *Kaska*. At first sight, the dunes of Kaska (figure 7.2) assume disastrous proportions, and are indeed slowly overwhelming the Manga village. Kaska is situated on a slope of 4–6° overlooking a large *kwari* from the north-east, whence the prevailing wind blows throughout the dry season. According to elderly informants, there was no moving sand in the area before the arrival of the Europeans in the first decade of the century. But in the 1940s, the foundation of a market, which soon became the most important in the area, led to the expansion of the village and the construction of some mud brick *soro* houses – usually associated with trade – amongst the more usual thatched huts. By 1950, the village perimeter dunefield was established, as the air photography proves, and it changed little in size (about 40 ha) until 1969. By 1977, the dune reached its most fearsome proportions:

The highest of these active dunes was about 45–50 feet high, and together with a second one of about 30 feet forms a formidable creeping front which has already completely

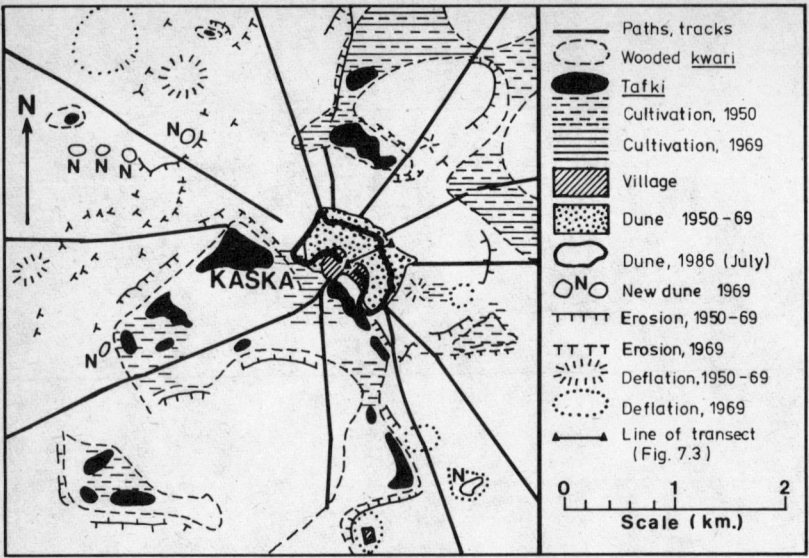

Figure 7.2 Kaska dunes (air photo interpretation and field survey)

buried not less than twenty houses and about a dozen trees. Evidence of already completely buried houses can be seen from their exposed dark flat tops. At the time of the visit we saw a 'Soro' in the process of being buried. The gate... was facing the east and there was a Kurna tree (*Ziziphus spina-christi*) about 10 ft in front of it. These sand dunes which have already completely buried the tree... at first piled up at the door and then continued to pour into it till at last the roof of the soro was pushed down... Audu on his part said before he finally vacated the room about four years ago, he used to sit and rest under the Kurna tree.

(Mustapha, 1977: 11–12)

By this time, a newer market at Tulatura on the international border was competing with Kaska, whose market has since declined in importance. The village was little larger in 1986 than it was in 1969, although the advancing dune had overwhelmed the centre, decapitating the *dendal* and distorting further house building sideways. While this westward advance continues, the area of moving sand on the eastern (windward) side appears to be stable or declining (Figure 7.2).

Along a transect of 420 m (figure 7.3), the dune formation may be divided into an expanding eastern zone of deflation, and a western one, where accumulation is dominant.[22] In the first, three deflation pits deepened by 1.5, 3.0 and 4.5 m between 1980 and 1983. In the second, three dune crests migrated south-westwards at rates of 5, 11 and 12 m/yr. The slower movement of the leading dune led to its absorption by the second in 1986. There was a steady

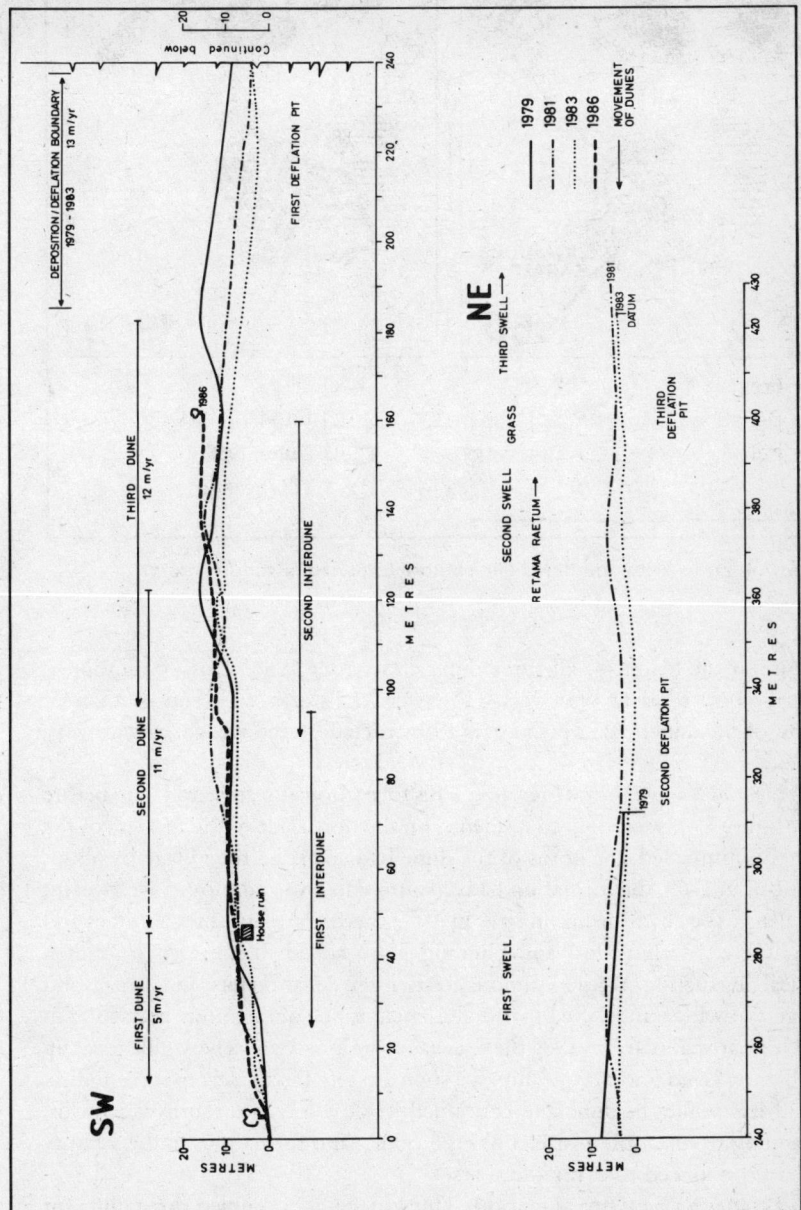

Figure 7.3 Profile of the Kaska dune, 1979–86 (field survey; transect for 1979 by G.H. McTainsh)

176 *Shifting sands*

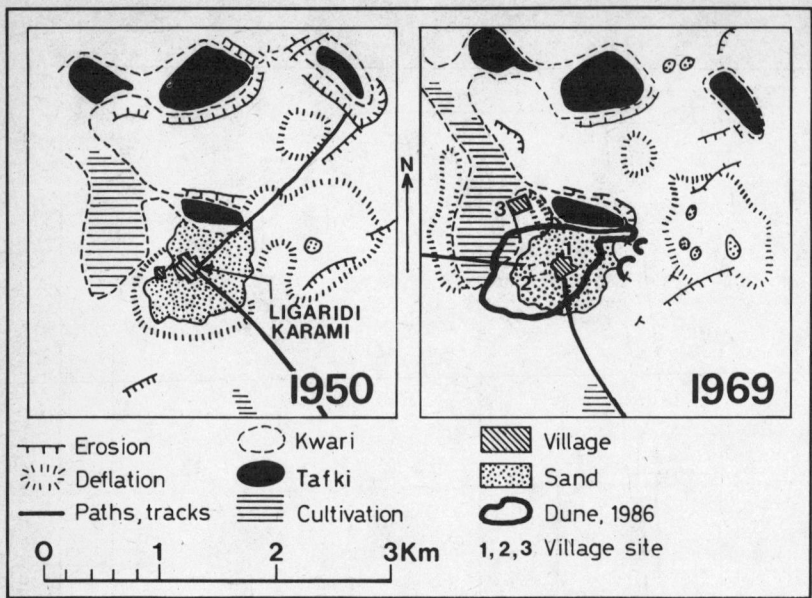

Figure 7.4 Ligaridi Karami dunes (air photo interpretation and field survey)

decrease in the height of all three dune crests after 1980. This diminution is confirmed by resident observers. However, the north-west and south-west 'horns' of the dunefield, which are not obstructed by the village, remain very active.

(2) *Ligaridi Karami* (Little Ligaridi). This formation also appeared long before 1950 (figure 7.4), according to residents, on the north side of the first site of the village. Unimpeded, the horns of the dune had encircled the village by about 1960. But in 1969, the formation had *diminished* in area and moved westwards; the village, too, had shrunk in size. In 1979 there were but three families still living there; swirling dust and blinding glare added discomfort to gradual physical destruction. With its sinuous crests, huge deflation pits, and contrasting orange and white sands, the Ligaridi dunefield is one of the most spectacular in the whole area. Finally, in 1984 the site was abandoned, and a new village set up in a shallow *kwari* north of the dunes; it soon attracted settlers from surrounding areas. Meanwhile the dunefield continued its slow advance south-westwards, continuing to diminish in size, although other sand formations in the vicinity took a pronounced turn for the worse.

(3) *Ligaridi Babba* (Great Ligaridi). This village has occupied three different sites during the lifetime of its dunefield (figure 7.5). At the first, deflation has exposed a spread of potsherds 30 m in diameter. By 1950, however, the village had moved to its second site, atop a substantial dune and overlooking a small *kwari* to the north-west, which supports both natron collection and farming in

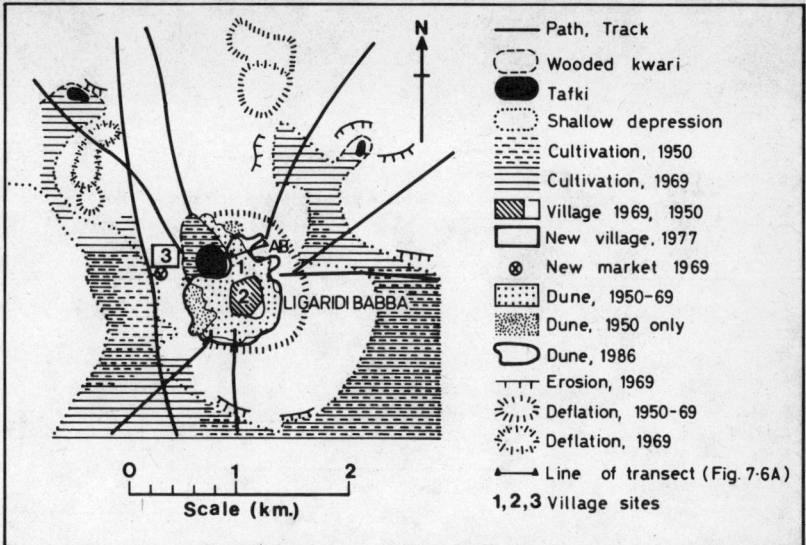

Figure 7.5 Ligaridi Babba dunes (air photo interpretation and field survey)

the dry season. Between 1950 and 1969, the village decreased in size, the dunefield diminished and extensive deflation scars, present in 1950, disappeared.

The decline of the village continued when, in 1977, a new market was set up on a site more accessible to motor vehicles, and adjacent to it, the third village. Now much larger than the old, the new layout has not snuffed all life out of its forebear, but situated as it is in a shallow *kwari*, where the heavier soils are less liable to movement, and with its market and motor route, new Ligaridi has its future better assured. Meanwhile, the dunefield surrounding the old site has enlarged slightly and is advancing south-westwards. A profile measured in 1981 (figure 7.6A) shows how the reactivation of a stable dune has raised the maximum slope from 12° to 26°, while overblowing is threatening the valuable natron lake and its foreshore.

(4) *Bulari and Gaptari*. A more recent example (figure 7.7), Bulari had no dunes or extensive deflation in 1950, or even in 1969. The grassland was, however, well marked with broken crestlines. The large *kwari* to the north of the settlement of Bulari is a major centre of natron working; there is dry-season farming in this *kwari*, and rain-fed cultivation in the shallow *kwari* to the south. These opportunities attracted a group of 30–5 migrant families from Bulatura to the east, who set up a second village (Gaptari) in 1969. They appropriated vacant lands in the shallow *kwari*. The rainfall deteriorated in the seventies. The dunes then appeared around both villages. Several of those observed in 1981 were only a few years old, and figure 7.6B shows how such a formation modifies the antecedent slopes, steepening the banks of the *kwari*, which is 5–10 m below the gently rolling topography of the grassland.

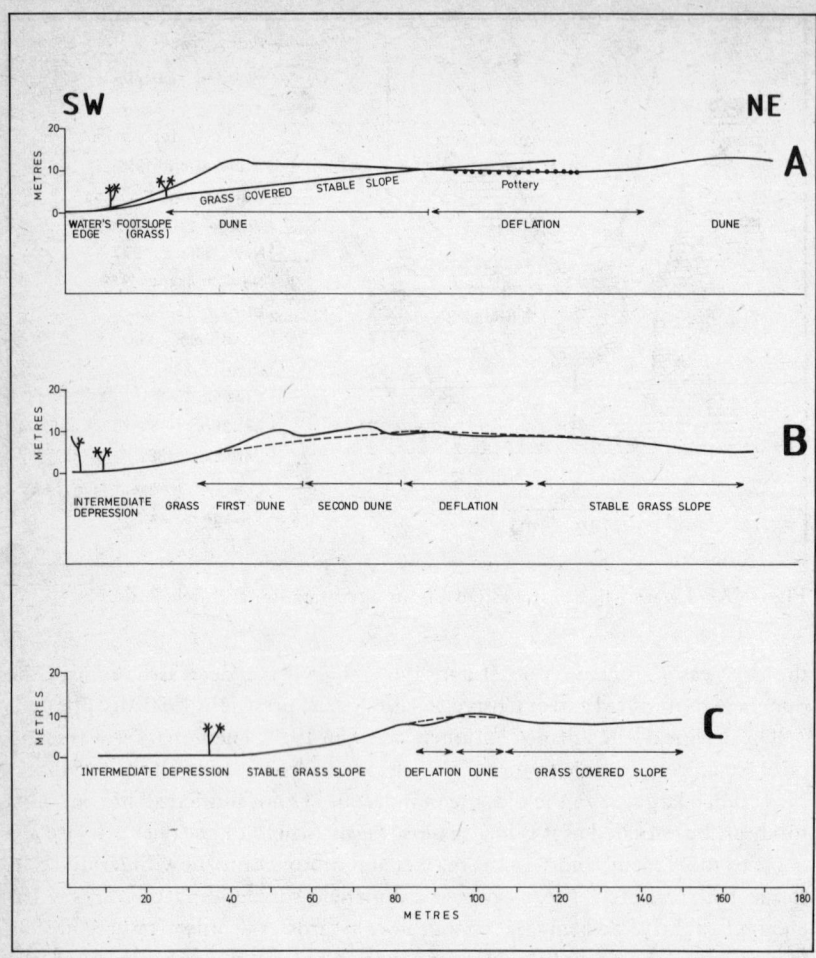

Figure 7.6 Profiles of dunes: A, Ligaridi Babba; B, Gaptari; C, a rangeland dune (field survey)

(5) *Ngelsandi and Lala'a*. Sedentary villages are not necessarily permanent. Shifting cultivation is characteristic of rain-fed farming and associated with movement of villages or groups of families. The next example shows how such movements may be related to the growth and decline of sand formations. Figure 7.8 shows a group of sites on either side of the international border. one of these has since been abandoned; one has grown and then declined during the period under review; and at the third, four different sites have been occupied in turn by members of the same families. Sand movement has been correspondingly complex.

Surface materials

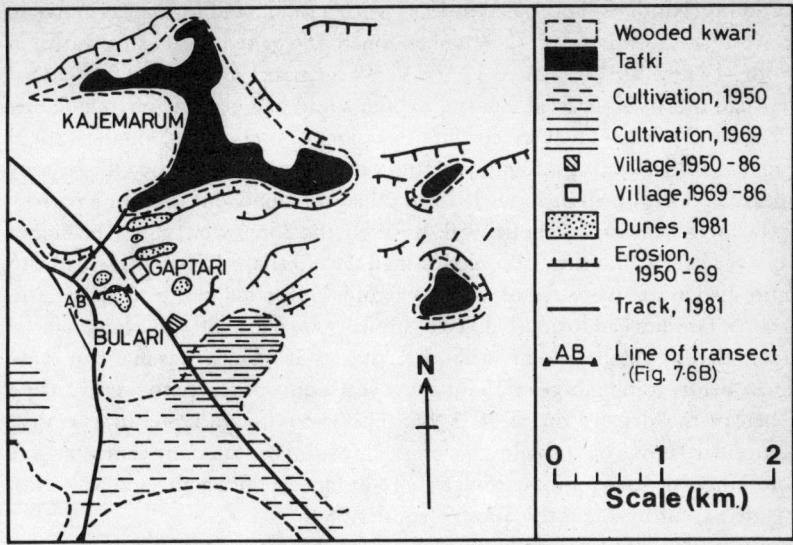

Figure 7.7 Bulari and Gaptari dunes (air photo interpretation and field survey)

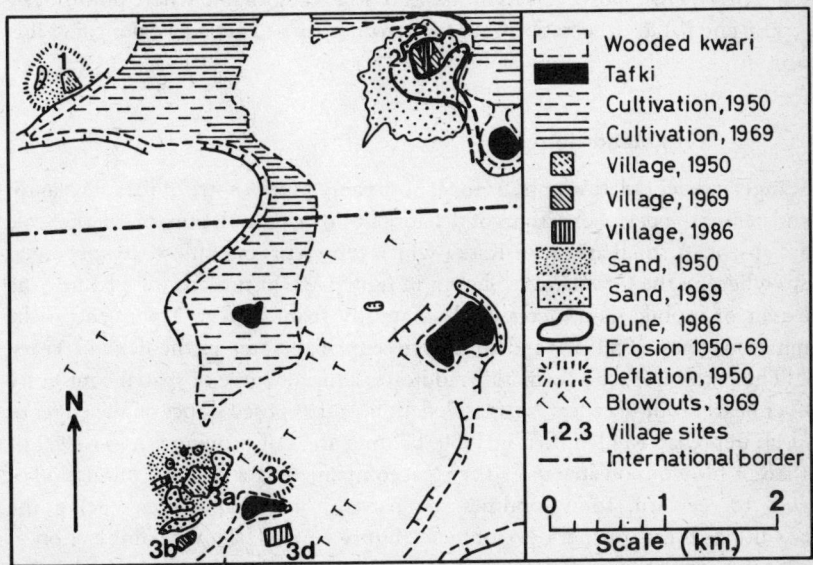

Figure 7.8 Ngelsandi and Lala'a dunes (1, Old Ngelsandi; 2, New Ngelsandi; 3, Lala'a; air photo interpretation and field survey)

Old Ngelsandi, still occupied in 1950, with a small sand formation active to the west (site 1 on figure 7.8), was abandoned one or two years afterwards, in favour of a new site (2) situated 3 km east. By the year 1969, the moving sand at site 1 had mostly re-grassed, but site 2 (Ngelsandi) had become surrounded by bare sand, 70 ha in extent. According to resident observers, no dunes were in evidence until then; beginning soon after on the south side of the village, they appeared on the north in about 1975, and had surrounded it (but for a narrow gap facing northeast) by about 1980. By 1986, the dunes were said to be higher than ever before, but they occupied a small area (30 ha).[23] The village had also diminished in size when ten of its thirty families migrated to site 3(Lala'a). This group of families had formerly lived at site 3a (which in 1950 was a large village surrounded by a dunefield of 36 ha), later at sites 3b (which was in use in 1969) and 3c, before joining Ngelsandi for a few years, only to return to site 3d during or before the dry season of 1985–86. The last was, in 1986, an extensive rectangular layout on a shallow *kwari* site. Doubtless, these movements were some amongst many more similar short-distance relocations, and were not impeded at all by the international frontier.[24]

In July 1986, site 1 was still occupied by about 70 farm trees (predominantly *Balanites aegyptiaca*) in stable grassland. Site 2 was almost surrounded by a fast developing dune formation. Site 3 retained one small dune, but the rest of its formerly extensive dunefield was colonised by *Leptadenia pyrotechnica* and *Cenchrus biflorus*. However, its limits were still recognisable where hummocky terrain and patchy vegetation gave way to undisturbed grassland along a visible scar.

Rangeland dunes

Rangeland degradation appears most noticeably in the centre of the Grasslands; and nearer the periphery, signs of deflation, or of accumulation, on a large scale are absent. A small area near Kaska which represents conditions as extreme as anywhere in the Grasslands is shown in figure 7.9. In this area of 4.8 km^2, the extent of mobile sand increased dramatically from zero to 7 per cent in the nineteen years, 1950–69, and to 20 per cent of the area in the next 17 years.

The pattern of sand formations indicates a high degree of spatial continuity over nearly four decades. Areas of deflation, and exposed slopes on the edges of *kwari* depressions, identified in 1950, became areas of moving sand in 1969 or 1986. A blowout on an exposed crest, accompanied by a small accumulation of sand to leeward, soon modifies the pattern of windflow, enhancing the conditions for its further development (figure 7.6C).[25] Thus a combination of exposed crestlines, and slopes steeper than average, in the antecedent topography provides localising site factors: on which the general causes of degradation, whether climatic or anthropogenic, are superimposed.

Surface materials

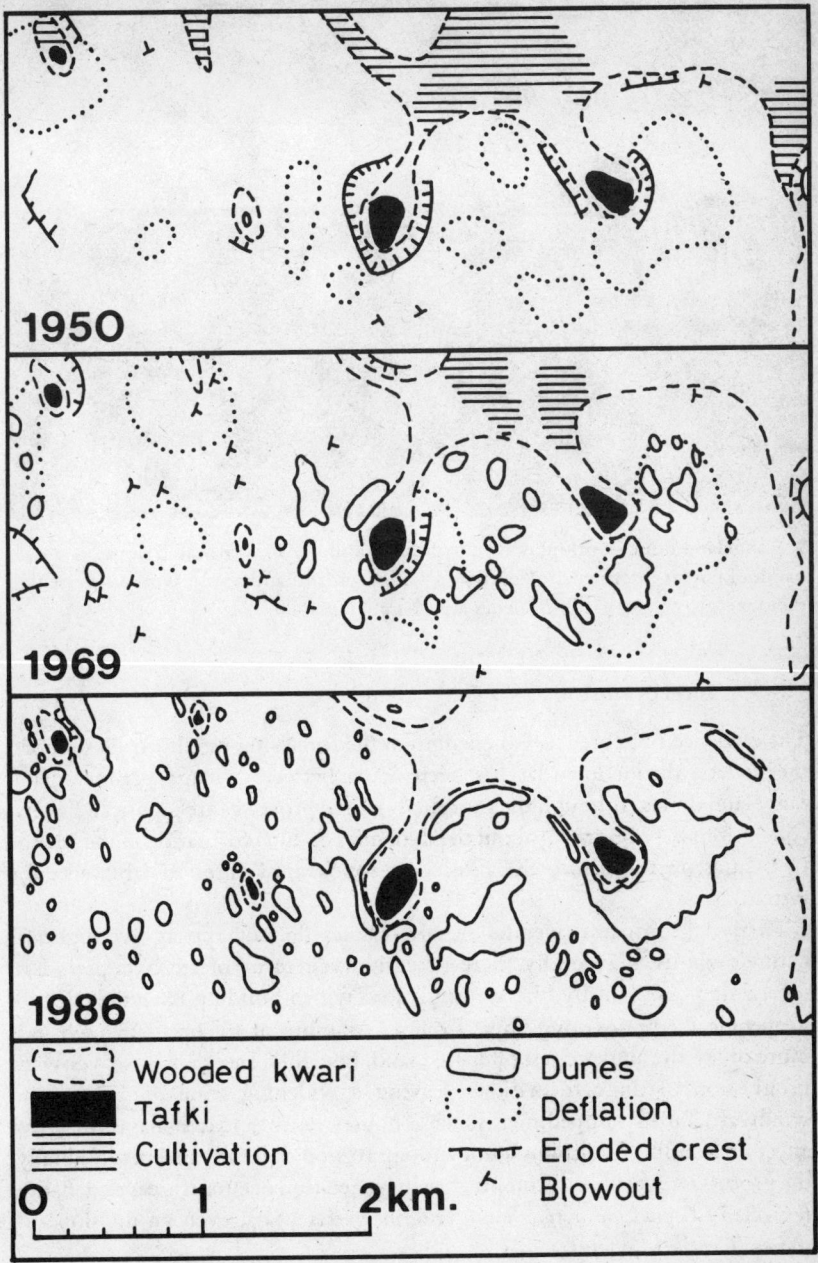

Figure 7.9 Rangeland dunes near Kaska (air photo interpretation, 1950 and 1969; field traverses, 1986)

9 Rangeland dune developing on the crest of a thinly grassed stable slope, overlooking an intermediate depression with farmland and sparse woodland. Behind the dune grows the hardy *Balanites aegyptiaca* (July 1980)

Interpretation

The evidence presented at some length in the foregoing case histories supports the genetic distinction that has been made between village perimeter and rangeland dunes, although the morphodynamic processes are identical. Several conclusions may be drawn about the dynamics of dune formation in the Manga Grasslands, in relation to the antecedent topography (itself a stabilised dune system).

Firstly, the process operates slowly. Dunes do not appear overnight, as nature's whimsies. Secondly, there is continuity, in terms of site, over periods of several decades. Thirdly, the accumulations which build up have the effect of steepening the pre-existing slopes from a maximum of 15° up to 35°, which is more or less the angle of rest of drifted sand. Fourthly, there is a slow westward progression of dune formations, leaving a widening zone of deflation to windward. Fifthly, impediments such as houses, fences and trees have a negative effect on windflow, although not enough to stop dune advancement. Finally, the process is reversible: recolonisation by vegetation occurs in areas of deflation (especially *Leptadenia pyrotechnica*) and, in wetter years, even on the slopes of dunes (*Cenchrus biflorus*).

Does the analysis provide any pointers to the anthropogenic factors in sand remobilisation?

Farming can be exonerated from direct blame. Dry-season cultivation, which occurs where the water table is within the rooting zone of the crops, protects the soil from wind action. Rain-fed cultivation is confined, in most years, to the shallow *kwari* where windflow is reduced by scattered trees and a lower topographical location, 5 to 15 m below the *tudu*. Although surface material may blow about the exposed fields (drifted white sand is called *cidibul*), there are *no* sand formations in the *kwari*. On rare occasions when millet is cultivated on the *tudu*, the fields are not maintained for long enough to provide opportunities for dune accumulation. After one or two years, weed and grass growth recolonises them.[26]

If farming can be exonerated, the same cannot be said of village-based grazing. Manga farming families own substantial numbers of small livestock. These are penned or tethered in the village at night, and taken out to graze each day by the women and children, exerting intense pressure on proximate rangeland. For example, of 35 households in the new village of Gaptari in 1981 (figure 7.7), ten owned goats and ten sheep; ten individuals owned horses, and two, camels; and there were a few cattle. All save the cattle were kept in the village. The demands of these livestock, added to those of the older village of Bulari, pressed hard on the trampled and nibbled rangeland close to the settlements.

In a Manga village, the customary plan of the *dendal* causes the village to face west, and the most convenient location for small livestock, and fowls, is the east end of the village, behind the chief's house. It is at this end that livestock most denude the vegetation. It is also from this direction that the dominant dry-season winds approach the village. Commonly, therefore, dunes begin at the east, and later progress around the north and south sides of the village, finally uniting at the west, if unimpeded.

With the increasing inconvenience of the site, households detach and move elsewhere, taking their livestock with them. The reduction of grazing activity gives an opportunity for grassland to recolonise the dunes, beginning to windward where *Leptadenia pyrotechnica* has already become established on the deflated surface. The homogeneity of the soil profile, and the ubiquity of blown seeds, facilitate such recolonisation. The livestock population thereby provides a functional link between the dynamics of settlements and living dunes.

For rangeland dunes, antecedent topography exerts the major influence on location. Localised grazing pressure could not be deduced from livestock numbers, even if known, because the Ful'be family and herd may occupy their camp and rainy-season pasture for only a part of the year. And Ful'be informants insist that the resident cattle population has not increased since 1972–74, yet they agree that the incidence of rangeland dunes has increased (figure 7.9). A functional link between grazing management and living dunes has not yet been demonstrated. But to lay some of the blame on the poor rainfall of the last two decades would be consistent with the change from dominant perennial to annual grasses.

On the other hand, rangeland dunes started forming during the period 1950–69, when rainfall was good, and before the floristic transformation

10 Cattle grazing around a semi-nomadic Ful'be camp (*ruga*); a major dune formation occupies the entire skyline. The very sparse cover afforded by the new grass in a year of poor rainfall is apparent (July 1986). The location of this photograph is at the centre of the area depicted in figure 7.9

occurred (tables 7.4, 7.5). Since an increase in livestock numbers is generally believed to have occurred at that time, an intensification of grazing pressure is a reasonable hypothesis. Lacking more decisive evidence, we may tentatively conclude that rangeland dunes began to develop under grazing pressure, but their growth in size and numbers accelerated, after 1969, under conditions of drought stress. That living dunes are nothing new is clear from the Report of the Anglo-French Forestry Commission (1937: 3), which enumerated four locations where moving sand was seen, including:

Small dunes near certain villages in the Manga country, particularly near Cheri (in Niger). These are part of old fixed dunes which have become live at certain points. Although the Commission has no evidence to show how these small live dunes came into being their proximity to villages leads to the supposition that man and stock are not unconnected with their creation.

The Commission toured the Niger–Nigeria border on both sides, but did not enter beyond the southern border of the Grasslands in Nigeria. Both Cheri and Fanamiram in Niger (figure 7.1), the only villages named in connection with moving sand, still have their dunes. How many more there were cannot be guessed.

OVER-EXPLOITATION OR UNDER-PRECIPITATION?

From the foregoing analysis, we may conclude as follows. Firstly, woodland (which is confined to lowland sites) was the only land cover type to show a significant change from 1950 to 1969, according to the interpretation of air photography; this change was negative, and there are indicators of the deteriorating status of woodland, where cutting exceeds regeneration. Secondly, upland grassland (which has always lacked trees) has undergone a change in floristic composition, since the early seventies. This change (an increase in annual species at the expense of perennials) does not imply a reduction in the dry forage available, whose violent annual fluctuations are due to rainfall variability. Thirdly, no conclusive evidence of soil degradation has been obtained, except for indications of the movement of surface materials. Fourthly, the evidence for such movement is contradictory, but the overall trend is one of pronounced deterioration. Village perimeter dune formations are dynamically dependent, via the small livestock sector, on the fluid Manga settlement system. Rangeland dunes have increased noticeably since 1950, and their location is governed by topographical site factors. Whereas their earlier development can be reasonably attributed to grazing pressure, their increased size and frequency in the last two decades must be due at least as much to deteriorating rainfall.

The conventional wisdom on the relations between land-use management and desertification is that 'overexploitation gives rise to degradation of vegetation, soil and water... Desertification is a self-accelerating process, feeding on itself' (UNEP, 1977c:3, quoted in full in chapter 1, p. 16; see also UNEP, 1977e).

The Anglo-French Forestry Commission (1937) reported the boundaries of the Manga Grasslands at three locations (west, east and south), and they are found in exactly the same places today. The Commission reported virtually treeless grassland communities; and remarked that the country on the Niger side of the border was 'apparently uninhabited except for temporary villages of salt makers in depressions' (p. 17).

Whether this last observation was accurate or not[27] (and the Commission did not enter the Grasslands on the Nigerian side of the border), the dense scatter of Manga villages found today implies an increase in the level of 'exploitation' which is consistent with what is known about human and livestock populations in the Sudan and Sahel zones generally. But there is evidence that this increase has levelled off. The aggregate surface area of Manga villages in the area of the primary sample of air photo interpretation did not increase between 1950 and 1969. Neither did the cultivated area. And from the known effects of drought and rinderpest, and the evidence of interviews, an increase in the resident cattle population since the late sixties can be ruled out.

The stability of the borders of the Grasslands, their largely treeless flora, and the known existence of moving dunes since 1937 shows that the basic characteristics of the area have some continuity in time. Most of the evidence for ecological degradation derives from the last two decades, corresponding to a decline in the rainfall.

Wherein lies the case for 'over-exploitation'? Such a concept can, of course, be defined only with reference to the idea of optimal carrying capacity, and such an idea can in turn have meaning only in relation to expected rainfall. And rainfall was below expectations more often than not in the seventies and eighties. In practice, governed by land users' perception of rainfall variability, surpluses from good years (stored crops, and livestock holdings) cushion the bad. But when drought recurs with exceptional frequency, pastoral and agricultural systems make demands that cannot be met. Farmers, denied their subsistence, ransack the ecosystem for alternative income – such as woodcutting and grass-bundling – and pastoralists attack the woodland for forage. Any farmer or stockowner in the Grasslands would call the problem under-precipitation, not over-exploitation.

8

INTERPRETATION

You know how to interpret the look of the earth and the sky   Luke 12:56

Recognition of the interaction of social with natural environmental factors in the genesis of famine is not new. For example, three early studies dedicated to major continental areas emphasised the effects of political and administrative factors in compounding those of drought. In India (Dutt, 1900) it was land taxation, in China (Mallory, 1926) it was administrative breakdown, and in Russia (Fisher, 1927) it was government food levies. In West Africa, historical research is uncovering similar complexity (Cissoko, 1968; Fuglestad, 1974, 1983; Salifou, 1975; Baier, 1980). The dry African savannas have been prone to famine both north and south of the Equator (Renner, 1926; Brooke, 1967; Maddox, 1986). Quite early in the colonial period, this susceptibility was linked to the process of desertification which, under various names, had been alleged by many observers. Huntingdon's (1915) ideas on the effects of climatic change on civilisations had a pervasive influence (Grove, 1977b), but most field studies in Africa were more willing to recognise a multiplicity of social and environmental factors behind ecological degradation.[1] Neither can the droughts of the last two decades be isolated from the kaleidoscopic patterns of social and economic change that were occurring, and in particular (as far as Nigeria was concerned) from the oil boom, which was at its height in the seventies. The distortions brought about by the oil revenues depressed agriculture and increased social and spatial inequality (Collier, 1983; Zartman, 1984; Watts, 1987). These interactions add to the difficulty of locating an appropriate theoretical framework for understanding the relations of drought, famine and desertification, especially since they transcend the conventional boundary between social and natural sciences (Richards, 1987).

There is a growing literature on the theoretical interpretation of the ecology of African land use (for a review, see Richards, 1983). My commitment to teaching in an African institution during the seventies and eighties ill equips me to embark on a critique of this debate, one of whose salient characteristics is a conspicuous shortage of African participation. Indeed, the debate is only partly audible to Africans owing to the structural inequities of the international publishing industry and the contemporary 'book famine' in African libraries starved of foreign exchange. My objective, as stated in the Introduction, is primarily empirical rather than theoretical. The findings reported in chapters

3–7 have been presented with a minimum of interpretative comment. They are not free from ambiguity or contradictions and may even be used to support conflicting theoretical positions. Some degree of agnosticism seems to me to be important because the pressure to develop general models – intrinsic to Western science – bear in this case on a flimsy foundation of 'in-depth' studies specific to time and place. Some influential contributions to the desertification debate, for example, appear to be based on quite modest field research and limited exposure to African perceptions – including those prevailing in governmental structures or educational institutions within Africa (and which it is unwise to ignore or dismiss). The only ultimate justification, in my view, for drawing on the resources and goodwill of African governments, institutions and citizens for research among the poor is to try to contribute to a better understanding of their needs, and to the discussion of policy alternatives. Without such an explicit ethical imperative, which would probably be agreed by almost all writers, research on the poor seems as morally cynical as it is intellectually stimulating. It is for these reasons – the need to be specific and the need to be practical – that the preceding chapters have deliberately adopted a theoretically naive position.

Nevertheless, at this point it becomes necessary to draw out the main conclusions of the study and to direct these towards certain areas relevant to public policy, national or international. Such conclusions impinge on the theoretical debate in several ways, but I shall discuss only three major issues, that arise directly from the evidence reported here. Others – the role of the state, for example, and the nature of 'green revolution' technology – are set aside not because they have no importance but because the results reported here have only a limited bearing on them. The three I shall discuss in more detail, drawing on other West African studies reported in the literature, are adaptive response, population mobility, and ecological degradation. Afterwards (in chapter 9) I shall take up the significance of these issues in terms of policy objectives.

ADAPTIVE CAPABILITY

Individuals or social groups have access to resources, including food, according to the rules of the political economy in question, and at a time of famine the entitlements of certain groups may collapse, owing to a rise in the price of food, or the loss of employment, for example. In Sen's (1981: 45–6) formulation of the 'entitlement approach' to poverty and famines, a distinction is made between a person's endowment (land, labour and other resources he owns) and what he can obtain in exchange for his endowment in a market economy: his 'exchange entitlement mapping'. The range of options available to such a person has been more simply termed his 'opportunity structure', divided into opportunities at home and those away from home; the need to exploit the latter type accounts for the significance of spatial mobility in semi-arid rural economies (see Hart, 1974; for a Nigerian application, Mortimore, 1982).

Adaptive behaviour during famine goes beyond the normal sphere of production and exchange, however. As chapters 3 and 4 have shown, it encompasses the mobilisation of social networks, the broadening of the definition of food, and the application of religious supplication.

In comparative context

Notwithstanding the devastation of the Sudano-Sahelian zone of West Africa in the seventies (Bernus and Savonnet, 1973; Bouquet, 1974; Pitte, 1975; Derrienic et al., 1976; Derrick, 1977; Toupet, 1977; Vermeer, 1981), and a widely quoted estimate of 100,000 deaths from hunger, the demographic expansion which had become established during the sixties continued unabated in the seventies, apparently attesting to the vigour of the adaptive mechanisms (Caldwell, 1975, 1977).[2] At the micro-scale, this conclusion seemed to be borne out by a study of one village in Niger (Faulkingham and Thorbahn, 1975). Even in Mauritania, the excess mortality amongst nomads was limited to an estimated 44,000 in a population of 1.2 million (Greene, 1974, 1975). According to Garcia and Escudero (1982: 89), the famine aggravated an already precarious health and nutritional situation, but a contemporary survey in Burkina Faso had suggested that the nutritional status of the settled population was not unusual for such an environment (Seaman et al., 1973). By comparison with such countries, it may be surmised that the opportunities for adaptive response in northern Nigeria – where no demographic effects have been demonstrated, and little is known of morbidity or nutrition at the time – were comparatively rich.

The droughts of 1972 and 1973 (chapter 3) when taken together were the worst in the memories of most Nigerians. The lack of rainfall destroyed part of subsistence farm output in both years, and the second was worse than the first. The resulting food shortage was widespread. Commercial output was also destroyed. In the absence of income, savings (especially livestock) were liquidated. A dearth of grazing and fodder caused widespread animal mortality. The market, on which people now depended for food imported from outside the region, was subject to inflation.

Adaptive response was determined on the one hand by the extent of loss, and on the other by the alternatives perceived. Economic adaptation can be characterised at three levels. Firstly, direct adaptation to drought was possible within the technical limits of farming and livestock husbandry systems. Secondly, individuals or families could adapt to poverty. Kinship, clientage or other networks might be mobilised, or alternative sources of income sought, or assets liquidated. Thirdly, adapting directly to hunger, alternative foods could be obtained by exploiting the diversity of the ecosystem. Material forms of response, however, did not subsume the wealth of adaptive behaviour: a social theology of drought and hunger underpinned individual action.

The adaptive behaviour of nomadic pastoralists in several West African countries has been the subject of a number of studies (for example, Horowitz, 1976; Gallais *et al.*, 1977; Bernus, 1977b; Boutrais, 1977). Some of the strategies employed – such as the alteration of migratory grazing circuits, shifting emphasis to fast-breeding small livestock (for example, Holy, 1980), changing grazing practices and methods of land management, communal borrowing of breeding stock, and changing relations with farming – were also available to the transhumant stockbreeders of the semi-arid zone. The droughts made some nomads sedentary, in particular those forced by the loss of their herds to turn to farming. And some went further to take up migrant labour – such as Wo'daa'be herders in southern Niger; according to one survey, 65 per cent of households were resorting to labour migration as a response to poverty (White, 1984; Bonfiglioni, 1985).

Sedentary populations considerably outnumber nomadic and semi-nomadic people in the semi-arid zone. Elements of the adaptive structure explored in chapter 3 have been reported in other studies in Nigeria and Niger: Abdu, Ahmed, Aliyu, Daudu, Kura, Mormoni and Ndaks (quoted in Van Apeldoorn, 1981: 57–72), Laya (1975), Faulkingham (1977), Oguntoyinbo and Richards (1978), Sidikou (1977) and Watts (1983a). Before the drought of 1972 occurred, Dupree and Roder (1974; Roder, 1976) had investigated anticipatory adjustments in the Kainji area. In East Africa, studies initiated in the early seventies have drawn many parallels. Hankins (1974) reported that adjustments to drought among farmers in Sukumaland (Tanzania) were strongly developed – such as using drought-resistant grain varieties, extra weeding, and moisture-retaining cultivation practices (cf. Bein, 1980). In the same country, other samples of farmers listed borrowing, storage, the use of substitute foods and migration as appropriate responses to drought (Berry *et al.*, 1971; Mkunduge, 1973). In a study in the Usambara mountains, Heijnen and Kates (1974) classified 18 adjustments to drought in a typology as follows: accepting (self-insuring against) loss; sharing loss with other members of the community; eliminating moisture waste in farming; changing crops; praying (attempting to affect the source of moisture); changing the location of farming; irrigation; and timing planting to optimise the use of moisture. In eastern Kenya, Wisner and Mbithi (1974; Mbithi and Wisner, 1972) measured the frequency of preferred adjustments: buying food, selling labour, planting in wet places, planting early, selling livestock, obtaining help from the government. Various attempts were made in these studies to identify structure or ordering in the patterns of response.[3] In the Ingessena Hills of the Sudan, a study in the eighties showed that the broad configuration of adaptive behaviour continued unchanged: people sold animals, handicraft production, labour; or bought grain on credit; or migrated; or gathered wild fruit and dug for edible roots or tubers, when their crops failed (Jedrej, 1985).

Several of the East African studies recognised the potential influence of ecological differentiation on adaptive opportunities (cf. Porter, 1965), and tried to take account of it by sampling farmers from different positions along a 'moisture gradient' from wet to dry. It is fairly easy to find examples of spatial variations in the structure of opportunities that can be explained in ecological terms. The villagers of Dagaceri have no land within reach that is suitable for dry-season farming, and have to be more mobile than the farming communities of the Manga Grasslands with their *kwari* farms. However, a man–environment paradigm is insufficient on its own. Since the distribution of wealth also varies in space, urban concentrations of investment and government revenues (for example) influence the availability of alternative opportunities for farming populations according to their location. In the sixties, inhabitants of the Kano Close-Settled Zone benefited from market and employment opportunities in urban Kano while those in the Sokoto close-settled zone, whose urban development was then rather restricted, had to travel further afield (Goddard, Mortimore and Norman, 1975).

On a much more general level, Burton, Kates and White (1978) attempted to model individual choice in response to the entire range of 'natural hazards'. Alternatives were supposed to be ordered in lexicographic fashion and judgements made between successively restricted options. While it is clear that, for the most part, individuals make decisions governed by their perception of alternatives (cf. Mortimore, 1982), and by considerations of cost, risk and conflicting obligations, the attempt made by these authors to classify the populations of 20 study areas into four patterns of responsive behaviour – absorption, acceptance, loss reduction and radical change – introduces an inappropriate evolutionary perspective. Applied to whole societies, it turns out to be a branch of modernisation theory: the social response to hazards is determined by income levels and by the transition from folk through industrial to post-industrial technology and organisation. Nigerian farmers, it seems, are assigned to the 'acceptance' category on the basis of the claim that: 'they have crossed the threshold of awareness of the hazard, regard its effects as significant, and judge that there is little that people can do to reduce its impacts. They tolerate the prospective loss without taking counter-measures. Their response is passive: the chief adjustments adopted are to evacuate or to seek help' (Burton, Kates and White, 1978: 108). But it is difficult to imagine a more misleading characterisation of Nigerian farmers. Even the authors of the cited study (Dupree and Roder, 1974) detailed an armoury of alternative opportunities available to their small sample – relatives, labour selling, firewood, crafts, grass collecting, fishing – just like others.

Gross characterisation of whole societies on the basis of a fundamentally technological variant is too crude a tool for interpreting the plurality of Third World societies and the multiplicity of factors that determine individual perception of opportunities. The concept of the 'natural hazard' as an independent variable has been questioned, in particular in relation to drought. The validity of such a catholic category (drought is only one of many 'natural hazards') is open to doubt when the diversity of adaptive behaviour is considered in relation to its immediate objectives. Such behaviour is only partly directed to meteorological drought (adaptations to farming or livestock production systems). Most of it is aimed at alleviating poverty and hunger, which are as uneven in their social impact as they are intimately related to social organisation. 'Response to famine' is a more useful paradigm than 'response to natural hazards' because it allows this distinction to be taken into account.

A comparative or cross-cultural approach to adaptive behaviour can establish common patterns where they exist (White, 1974). However, the more urgent priority is to place famines in their social and historical context, since this is the dimension most overlooked when emergencies occur. Droughts, as meteorological events, impinge on social relations, dampening or amplifying tendencies towards change. Adaptive behaviour, therefore, makes full sense only in the light of such historical, social and geographical specifics. The shallow historical depth of our knowledge of drought and famine in West Africa is an impediment to a deeper understanding of contemporary events (cf. Baier, 1980: 3). The transformation of a relatively self-sufficient precolonial economy was far from complete when the droughts of the seventies and eighties struck. Market penetration was proceeding, population was multiplying, but the tenurial links between the people and their land remained substantially undisturbed. In the light of these considerations, the important question is how fast, and in what direction, the opportunities for adapting to drought, poverty and hunger are changing, and what the significance of such changes is.

In historical context

Adaptive capability varies with the pattern of social differentiation. 'In the ancient world, as in the modern, famine was always essentially a class famine' (Gapp, 1935: 261). This ancient fact (see Keys *et al.*, 1950: 7) has also been demonstrated in northern Nigeria, where groups identified on the basis of economic indicators show an association between poverty and vulnerability to food shortage (Hill, 1972; Watts, 1983a). Rural people are well aware of this association. Hunger and poverty are historical correlates (Hill, 1977; 1982). For Ethiopia, Wood (1976) has constructed a model linking drought response to economic status. All this is no surprise. In any society faced with disaster, no matter what the cause, the rich are better insured in the size of their food reserves (Garcia and Spitz, 1986: 3–32), preparedness in the long term, and diversity of short-term options available (cf. Sorokin, 1942: 14).[4]

If adaptive capability depends on socio-economic status – among other factors – then it follows that changes through time in the relative capabilities of socio-economic groups are tantamount to changes in social relations. It also follows that low capability in the poorest economic groups, in accelerating the adoption of irreversible or relatively irreversible responses – like selling land – may further weaken their capability, and widen existing inequality within the community. This so-called 'ratchet' effect so far lacks convincing empirical demonstration from the longitudinal micro-studies of interpersonal

Adaptive capability

transactions that alone would settle the question whether mechanisms for economic recovery, which do exist, are ever effective enough to balance it in the medium term. Nevertheless, the issue has provoked a controversy which has some importance for the understanding of longer-term trends in rural society in this zone of Africa. This focuses initially on the question whether adaptive capability (the sum of available 'coping', 'response', or 'survival mechanisms' or 'strategies', or 'adaptive flexibility') has significantly diminished or 'lessened' (Bowden et al., 1981) for the rural population since the imposition of colonial rule at the beginning of the present century. Or, as expressed by Wisner (1977b), whether enlarged choice for the few has been countered by narrowing options for the many.

Kates et al. (1981) tried to compare the impact of drought in the Sahel and Sudan Zones in the years 1910–15 (which included the major famine, K'ak'alaba, in northern Nigeria in 1914) and 1968–74 (which included Kakaduma in 1973–74). The practicability of such a comparison is open to question owing to the dearth of evidence relating to the earlier period, particularly estimates of its impact in quantitative terms; and the significance of the exercise may also be questioned on the ground that colonial intervention had already had opportunity to disrupt precolonial relations of production by 1914. Nevertheless, Kates argued on the basis of reduced *pro rata* mortality and morbidity that for the majority of the population (excluding the pastoral nomads of the northern Sahel) there was a lessening of 'vulnerability' between the two periods. Also, food was relatively less scarce, and migration relatively less widespread, in 1968–74. This implies an improvement in social welfare, notwithstanding a greatly increased population.

This conclusion was vigorously contested by Watts (in Kates et al., 1981: 89–90) following earlier studies in the region (Copans, 1975; Raynaut, 1977). Raynaut argued that the precolonial subsistence system had been disrupted profoundly in its social aspects – a disorganisation of the principles of storage and redistribution within social groups, and of reciprocity and solidarity within such groups; the splitting of productive social groups and emergence of wage labour; and of land selling.[5] Famine cannot be understood except in terms of the penetration of capitalism (Ball, 1976; Piché and Gregory, 1977; Shenton and Watts, 1979). Watts (1983a) extended the hypothesis to encompass the gradual erosion of a 'moral economy', as distributive networks, patronage, the local grain trade and storage systems disintegrated, commodity production penetrated the rural economy, and the population became increasingly subject to indebtedness, exploitation by grain traders, taxation and the burden of marriage costs. 'Droughts and periods of food shortage were catalysts for increasing rural inequities' (Watts, in Kates et al., 1981: 89–90).

The present study has adopted a longitudinal perspective but the period of empirical observation was only 13 years. Does the evidence reported here contribute in any way to a resolution of this controversy? It is clear that contrary trends can be identified in each category of response.

In adapting to agricultural drought, there is significant experimental innovation, but the scope for adaptation is severely restricted by constraints beyond the farmers' control, such as the shortness of the rainy season. These constraints have given some the wrong impression that farmers tend to use an inflexible package of technical inputs from year to year, irrespective of the rainfall. In fact there is a tension between sticking to well-tried techniques, adapted over the longer term to the conditions of the environment, and risking modifications that offer a flexible response to the conditions of a particular season. No case can be made for a significant deterioration in this type of adaptive flexibility, except for the loss of the groundnut; there is evidence of considerable effort (chapter 4) being expended on improving it.

In adapting to hunger, the alternative foods available from intensified exploitation of the ecosystem were sought out in time-honoured fashion, but resort was not necessary to the most dangerous famine foods whose use was remembered or documented in earlier famines. On the other hand, increased dependence on the market was noticeable.

In adapting to poverty, the contrariety is discernible amongst responses internal to the community and also among those having an external orientation. The surveys carried out in 1974, for example, picked up a consensus on the view that patronage had weakened since former times: 'the rich no longer help the poor'. On the other hand, no evidence was found that kin-based ties had weakened significantly; and the experience of Dagaceri shows that new opportunities for earning income are not necessarily disruptive of the existing system of kinship relations. Although Dagaceri is certainly not representative of Hausaland, it would be dangerous to assume that family ties anywhere are on the verge of drowning in a tide of individualism.[6]

Responses having an external orientation – market-oriented opportunities and those involving mobility – also fail to show a consistent pattern. While the processing of local raw materials for the market (particularly mat- and rope-making) has suffered from unfavourable price trends, new opportunities have opened up elsewhere with the growth of the informal sector in the economy as a whole. The disappearance of the groundnut as a revenue earner has resulted in the diversification of alternative income-earning activity, and in Dagaceri such diversity increased during the period 1974–86.

This study, therefore, lends no support to the view that adaptive response to food shortage has been subject to a reduction in flexibility and choice – or increased vulnerability to the impact of drought – although the period of direct observation was relatively short. While some categories declined in importance, others increased. There is no denying, however, that survival is being bought at the price of intensified integration into the market system.[7] Such integration, accomplished if necessary by mobility, offers the diversification considered to be essential.

In the longer term, disintegration of the precolonial economic system is an obvious fact.[8] Increased dependence on commodity production, on the market and on mobility has diminished the self-sufficiency of rural communities, the growing importance of wage labour has altered the social relations of production, and such factors as taxation, the demands of a growing population, and the effects of compulsory education have contributed to the erosion of autonomy in the use of resources. However, the implications of such changes for welfare are not as clear as some authors seem to imply. Slavery – whatever its welfare implications – was asymmetrically distributed in society. To contrast the situation of the poor today with a precolonial economy in which distributive networks, patronage, trading and grain storage systems are assumed to have been more benign may not be comparing like with like. The disintegration of the old is occurring alongside the emergence of the new[9] and it is too early to be sure what final pattern this new, market-integrated economy will assume. It is doubtful whether those who participate in it would prefer to return to the relative isolation of the past. In Dagaceri, the long-term objective of *masu cin rani* is to enter into local trade, in order to enjoy the opportunity of accumulation without the obligation of frequent long-distance travel. Rural perceptions of welfare are pragmatically related to the economy of modern Nigeria.

The larger significance of such changes lies in the bearing they have on the view that social change is proceeding inexorably towards the creation of a landless proletariat of commodity producers (see Watts, 1983a: 453–9). The tenacity with which a subsistence priority is maintained is a principal difficulty confronting such a view. In Dagaceri this priority was just as firm in 1986 as it was before the first major harvest failure of 1972. In order to secure this priority, rights to land use must be maintained; and the significance of a famine lies partly in the possibility that, in order to survive, the poor may sell these rights and subsequently be unable to recover them: hence the preference for pledging land (often reported inaccurately as selling) rather than outright sale.

Land selling is not new to northern Nigeria and conclusive evidence of its increase is hard to come by. Land that is sold is normally manured land under annual cultivation and, as improved land, it may be bought even in places where unimproved woodland is available, owing to the costs of clearing such land. But land selling is most common in densely populated districts where free land is scarce. In such areas, a significant proportion of fields at any time can be shown to have been acquired by purchase. More common than selling, however, is the 'involution' of rights over small and subdivided holdings by individual usufructuary transactions of some complexity, as reported by Ross (1987) in Hurumi in the Kano Close-Settled Zone.

Certainly land selling occurred during Kakaduma (chapter 3), but it was one of the least common forms of response, and it had been taking place before. Significant dispossession of the land rights of the lower income group by the better off was reported in one village in Daura Emirate, but it would be inappropriate to draw general conclusions from such a study until confirmed elsewhere.[10] On the Kano River Irrigation Project, after ten years' operation under conditions strongly favouring larger-scale farmers, a study of land transactions (Orogun, 1986) provided little confirmation of the expectation that smallholders were being turned landless, although in the first five years of irrigation inequalities in the distribution of land worsened (Baba, in press; see also Wallace, 1980). In more peripheral areas, notwithstanding the powerful logic of land concentration – given the penetration of the countryside by capital and by the market, the absence of registered title to customary holdings, and the stress introduced by famine – the link between the producers and their land is resisting breakage.

Watts (1983a: 460–1) argues that 'technical and biological conditions mitigate against[sic] the emergence of a local agrarian capitalist class or the intervention of productive (international) capital based on dryland and upland agriculture. No convincing technological package has yet emerged that makes it profitable for producers to direct capital into dryland farming.' This is incorrect. Urban and rural capital has been invested since the seventies in upland farming in Sokoto, Kano and Borno States, both in extensive holdings acquired under statutory title in hitherto uncultivated woodland, and in consolidated holdings of improved land, for example near Sokoto (Labaran, 1986). Notwithstanding seasonal periodicities in labour demand, the very uncertainties that bedevil subsistence production in the semi-arid zone offer the prospect of windfall gains (see figure 4.4).

From what little is known about land transactions in semi-arid Nigeria, worsening inequality in the distribution of land appears to be more serious a problem than landlessness. The policy implications of this will be taken up in chapter 9.

Given the vigour of adaptive behaviour in the face of ecological and economic stress, it is hard to see the peasantry of northern Kano State, and of Dagaceri, under immediate threat of any 'final act of historical elimination' (Watts, 1983a: 513; cf. Meillassoux, 1974). Stress on adaptive capability, of course, runs counter to a deterministic view of social change: the individual is conceived less as the object of impersonal forces, and more as an experimenter, innovator and risk-bearer.[11] An adaptive framework need not imply a dualistic view of man and nature (Watts, 1983b) if the scope of adaptation is broadened to include economic and political events, reflecting the perceptive viewpoint of the subjects; neither does it reduce human behaviour to a mere 'stimulus–response' model if the anticipatory character of human cognition is recognised (Bennett, 1976, 1978; see chapter 1). As a guide to policy priorities, it has considerable advantages.[12]

MOBILITY

The resilience of the rural productive systems has a dual basis in secure access to land and in spatial mobility. The contradiction in this duality is resolved through preferring short-term circulation to long-term migration. By means of such mobility the necessary economic diversification may be achieved, whereby the productive deficiencies of the home farm can be made good by exploiting alternatives elsewhere. In this sense, the mobility of livestock breeders and farmers are analogous.

The movement of labour between the Dry Zone of West Africa and the Coastal Zone, which attracted attention quite early in the colonial era, was first interpreted in relation to seasonal drought, which was understood to create a labour surplus that could be transferred to areas of commercial agriculture, mining and industrial and urban development (Labouret, 1941: 222–34; Rouch, 1956; Prothero, 1959). Observers of these movements subsequently polarised between those who saw them, on the one hand, as a benign factor transfer bringing a regional subsidy to the poorer Dry Zone in the form of earnings, remittances, innovations and temporary relief to food stocks (Berg, 1965; Mabogunje, 1972); and on the other hand, as the consequence of unequal colonial investment (which favoured coastal regions) and the transfer of productive labour as a net loss to the Dry Zone, a malignant form of exploitation (Amin, 1974). Although an exception is sometimes made of Nigeria, on account of its size and relatively even distribution of investment, the chief difference is the fact that relatively few migrants happen to cross an international frontier.

Labour circulation (which implies a return to a permanent residence after every trip) is only a part of a complex system of population mobility that is far-reaching both in space and in time (Gould and Prothero, 1975; Chapman and Prothero, 1985: 1–11). Circulatory movements of less than a year, usually on a seasonal basis, occurred in precolonial times, although the colonial intervention brought about a process of change (Swindell, 1984). Migratory movements of a year or more, involving a permanent change of residence, also occur today, and their origins may be traced ultimately to the southward drift of prehistoric populations, responding to desiccation along the southern perimeter of the Sahara. Such redistribution of the rural population, as farmers go in search of new land, has been traced in the census record, although these movements have been overtaken in magnitude by rural–urban migration in later-twentieth-century Nigeria (Mabogunje, 1970a; Green, 1974).

While the conditions of colonial rule (and the spatial distribution of investments in export agriculture in particular) modified and extended circulation, it did not evolve 'from the impact of alien or western influences upon indigenous circumstances' (Chapman and Prothero, 1985: 8). Even in the Manga Grasslands, an illusion of remoteness is broken by the ancient trade in salt and natron, which gave work both to migrant labourers and to seasonal trading caravans from other places in the region. Men from Dagaceri participated in this trade as late as the 1950s.

Decisions to move are made in the context of a structure of opportunities divided into home and away categories (Mortimore, 1982). Even in a subsistence crisis, the costs of mobility have to be considered by a potential mover: transport costs, forgone social obligations, alternative arrangements for farm labour, or to support immediate dependants while away from the village. It is the elimination of alternative opportunities at home that pushes the balance in favour of mobility: the loss of marketable crops or animals, depressed prices for rural manufactures, trade goods, services, or labour. Circulation implies, however, retaining an option in the home structure, as well as access to land. Individual participation implies that kin act as guardians of this option. When the family moves, as occurred in 1973–74, the link is correspondingly tenuous. So irrevocable out-migration by whole families, though recorded, is unusual.

'The notion can...be advanced that Third World societies may be increasingly characterised as bi-local populations, relatively stable in their demographic composition, but composed of individuals in constant motion between village and nonvillage places' (Goldstein, 1978: 15). The behaviour of rural populations in northern Nigeria lends little support to the conception of the migrant as progressively committed to an urban way of life, moving through a system till finally becoming a fully-fledged urbanite (Mabogunje, 1970b). Rather, the territorial separation of obligations and activities between the home or natal place and the wider world of alternative opportunities supports the thesis that 'the rapid growth of urban places...has been a function of a dramatic increase in numbers of people who circulate into, through, and out of them, and is far less indicative of a rise in total urban commitment on the part of those who are in towns' (Chapman and Prothero, 1985: 9).

An urban-centred view (which dominates many studies of rural–urban mobility) bears little resemblance to the villagers' perception. For the inhabitants of Dagaceri, the city is the cultural periphery: remote, often linguistically and ethnically alien, economically insecure, politically unpredictable, and sometimes physically dangerous – a place for adventurers, and not the faint-hearted. 'Peoples' mobility behaviour can...be viewed as a system whose locus is the village, the local subgroup, the extended or nuclear family' (Chapman and Prothero, 1985: 8). Given the threat which agricultural droughts have administered to the subsistence priority – and deterministic models which explain West African mobility in terms of climatic stress – the continuing immobility of the population (or the low incidence of permanent out-migration) is as striking as the rising incidence of circulation.

While recognising that circulation, as it has been described, demonstrates rationality, flexibility, experimentation, and entrepreneurial innovation, it should be stressed that those who participate do so on disadvantageous terms. Peripheral to the national centres of capital formation, the mover, when he arrives at the centre, is doubly marginalised by being confined to the informal sector. Moreover, he participates for only a part of the year. He finances his own considerable travel, yet provides urban services at rock bottom prices. In sleeping rough and in every way eliminating overheads, movers 'suffer'. Only continuing rural poverty can make such an unequal participation appear attractive.

Disadvantageous or not, circulation shows no evidence of being a transitional phase in a process of modernisation as postulated by Zelinsky (1971).[13] The maintenance of a bi-local residence pattern must act as a brake on the emergence of an urban proletariat of full-time wage earners. The dogged retention of rights to land in the village likewise impedes the concentration of land holdings and the appearance of landless labour in the countryside. It cannot be stated with any confidence that these barriers will prove capable of resisting indefinitely the pressures placed upon them. But that their continuation conforms with the wishes of most inhabitants of the zone does not seem to be in doubt.[14]

ECOLOGICAL DEGRADATION

Desertification (the degradation of ecosystems in arid or semi-arid regions) is distinguishable from drought, as social impoverishment is distinguishable from famine, by its longer time-scale. Human communities are caught in a web of self-reinforcing causative and responsive behaviour. While meteorological drought contributes to it, the phenomenon is certainly more complex and embraces human activity at levels from the local to the international. Some such activity is a conscious *response* to the occurrence of drought. Some human activity may *cause* drought, if any of the feedback hypotheses are to be believed: 'We seem to have arrived at a critical moment in the history of mankind's relation with climate ... For the first time we may be on the threshold of man-induced climatic changes. Yet one must maintain a balanced view' (Hare, 1984a: 21; 1984b). And there is little doubt that some human activity contributes to ecological degradation.

Many statements in the literature are, however, either weakly connected (shopping lists of 'causes' unrelated to one another) or unduly hypothetical (plausible arguments unsupported by case-specific data). Glantz (1980: 81) offers an apparently unfinished catalogue:

... *Some* of the generally accepted causes of desertification of the Sahelian rangelands *include* an overtaxed rangeland carrying capacity (too many cattle on a dwindling vegetative resource base), human population increases within and to the south of the Sahel, misuse in some instances and non-use in others of climatological and meteorological information, ever-expanding herds, cultivation in meteorologically marginal areas, reduced fallow time, deep well construction, technology transfer, political and cultural rivalries, low level of international and regional planning, systems and [*sic*] land tenure, competing political and economic ideologies, *among others. All of this is now well known* [italics added].

The underlying causes are a commitment to the ethic of 'man-over-nature', the limitations of bureaucracies, and inappropriate technology. Even accepting such conclusive statements as proven beyond doubt, the systemic ramifications of desertification, as so presented, discourage anything better than piecemeal technical solutions.

But such conclusiveness is not warranted. For the purposes of the present study, the ecosystem in a small area of about 1,500 km² was dis-aggregated into the four subsystems woodland, grassland, soils and surface materials (chapter 7). The results of selective (if not fragmentary) fieldwork resemble an inventory of ignorance as far as many aspects of ecological change are concerned. The ambiguities that have emerged from this micro-study do not encourage the kind of confidence that characterises many statements in the literature. Nevertheless, the evidence does converge on some tentative conclusions. These will now be discussed in relation to the four main groups of hypotheses that have entered the crowded debate on desertification.

The first of these is the climatic hypothesis. Based on the evidence for reduced rainfall, desertification is explained in terms of persistent meteorological and hydrological drought. The second is 'environmental misuse', which opposes a scientific evaluation of environmental capability to indigenous land-use practices that are considered to be antithetical to conservation. An extension of this view seeks to explain drought partly as a 'feedback' from anthropogenic activity (chapter 6). The third is 'population pressure', usually considered as the main factor explaining misuse of the environment. Finally, structural theories put the blame on forces of socio-economic change impelled by external factors in the political economy of the region. Different authors emphasise different factors or attempt a balance. For example Grainger (1982) uses a two-tier structure, the direct causes (land-use practices) being influenced in turn by changes in population, climate and socio-economic conditions. Chains of cause and effect, however, are easier to hypothesise than to demonstrate empirically.

The climatic hypothesis

The regional evidence relating to the climatic hypothesis, reviewed in chapter 6, shows that the persistence of meteorological drought is now an established fact of the last two decades. This means that reduced rainfall is plausible as a factor contributing to ecological change. A review of the feedback hypotheses implicating human communities in causing drought is, given present knowledge, inconclusive. Therefore, until proved otherwise, rainfall may be treated as an independent variable.

Hydrological drought is dependent partly on meteorological drought, subject to a time lag which is longest for deep groundwater. The evidence for drought in northern Nigerian groundwater regimes is, however, somewhat inconclusive owing to the effects of intervening variables, in particular the construction of dams and changes in land use. The tendency is for deterioration, and the limited amount of hydrological data is supported in this respect by impressionistic evidence (cf. Berry, 1984a: 44). But subsurface flow in the major river basins has not yet fallen low enough to threaten existing patterns of settlement and land use, even where artesian water is being 'mined'. Surface water, of course, is more directly related to the rainfall regime, and dry years pose major threats both to hydraulic engineering schemes and to small-scale users.

The significance of reduced rainfall is suggested by successively lower 30-year averages since 1960 (figure 6.2). More significant for agro-ecology, however, was the rainfall in shorter periods. In the five years 1980–84, Nguru received on average only 61 per cent, and Kano 71 per cent, of average rainfall for the period 1931–60. This was equivalent to temporarily transferring Nguru from the southern to the northern Sahel, and Kano from well inside the Sudan Zone to the Sudan–Sahel boundary, as defined by some authorities (Kowal and Kassam, 1978: 70, 108). Isohyetal transgressions of such magnitude must have some effect on vegetational communities, especially when exacerbated locally by a falling groundwater table.

In the study area, trees are scarce but mortality was reported among isolated specimens of *Acacia albida*. Mortality among trees has been observed widely in semi-arid West Africa, especially on sites where the depth to the water table is greatest (Bernus, 1977a: 79; Boudet, 1977, 1979). *Acacia senegal* was hard hit, both in West Africa (Wata, 1979) and in the Sudan (Lamprey, 1975), threatening the future of the gum arabic industry and provoking a programme in Niger to protect the remaining groves whose regeneration is threatened by clearance, burning and cutting (CTFT, 1975). This suggests that pressure from human communities may be as important as drought in killing trees. The effects of grazing on dry savanna communities may be pronounced (Bradley, 1977): in drought stress.[15] In the Kano Close-Settled Zone, the valuable trees *Acacia albida* and *Parkia biglobosa* suffered significant mortality during the drought years 1972 and 1973.

and *Parkia biglobosa* suffered significant mortality during the drought years 1972 and 1973.

Underlying the anthropogenic effects, however, was a tendency in some areas for the composition of woodland to change in favour of hardier species (Berman and Cisse, 1977). In Darfur, the survival of *Balanites aegyptiaca* in the drought of 1972 was conspicuous, by contrast to other species (Sirag, 1980). Unprotected woodland responds differently on sandy and colluvial soils. Surface sealing on colluvial soils in lower sites causes mortality and prevents regeneration, leading to the emergence of alternating strips of bare ground and woodland in the longer term (Boudet, 1972, 1976). These vegetation arcs have been called 'brousse tigré' from their striped appearance on air photographs, patterns reflecting the regularities of the old dune surface. The phenomenon, which has been observed for several decades, appears to be caused by drought.

The main grassland community in the study area changed between about 1970 and 1980 from a community dominated by the perennial *Andropogon gayanus*, var. *tridentatus*, and *Aristida mutabilis*, to one dominated by the annual *Cenchrus biflorus*. Since this change coincided with heavy livestock mortality, it is improbable that it was due mainly to selective grazing and most probable that it was a response to a drier climatic regime. The *Cenchrus* community is well favoured for forage, reproduces early and is highly resilient under conditions of periodic aridity and heavy grazing pressure (Gillet, 1975).

Several studies of vegetational change during and after the droughts of the early seventies show that *Cenchrus biflorus* became prominent in dry conditions on dry sites – even moving sand – in central Niger, north Burkina Faso, the Gourma region of Mali, and Senegal (Bernus, 1977a: 81–2; Toutain, 1977; Boudet, 1977, 1979; Breman and Cisse, 1977). It has replaced perennials, such as *Andropogon* spp., all the way from Lake Chad to Dakar.[16] Such changes might suggest that droughts brought about a southerly shift in vegetational boundaries (consistent with the idea of an encroaching Sahara). But such a hypothesis would be simplistic. Perennials are not confined to the wetter environments; the effects of grazing pressure on plant communities are localised within a certain range of water sources; and conspicuous temporal and spatial discontinuities are characteristic of the grassland communities. Furthermore, such changes cannot necessarily be categorised as a form of degradation. The dry-matter yield of annuals may be superior to that of perennials.

The climatic hypothesis, therefore, cannot be dismissed. But neither is it adequate on its own to account for all observed changes in the vegetation.

Environmental misuse

Five major hypotheses of environmental misuse have been levelled at indigenous land users by experts and officials concerned with desertification. These are: deforestation (often said to be 'indiscriminate' or 'unplanned'); cultivation (often called 'wasteful' or 'inappropriate'); grazing (often said to be 'overgrazing' as the result of 'overstocking'); burning (also 'indiscriminate'); and irrigation (although 'bad' irrigation methods – e.g. inadequate drainage – are more commonly identified in large-scale public projects than in small-scale private works). Misuse is a common thread linking many theories of desertification from Malthusian to Marxist, whether the land user is conceived as stupid, or subject to forces beyond his control (Campbell, 1974; Ball, 1978; Timberlake, 1985: 14). A brief survey of the first three of these hypotheses shows that the evidence, both from the present study and from other work in the semi-arid zone, is far more ambivalent than is generally admitted.

Deforestation in the semi-arid zone is but a part of a process estimated to be ridding tropical Africa of 3.7 million ha of forest each year (FAO–UNEP Tropical Forest Resources Assessment, 1982, quoted by Kassas, 1985). Against this 'sucidal deforestation' (Eckholm, 1977), only 126,000 ha are being planted each year. In Chad, woodcutting, agricultural clearance, lopping and burning are blamed for the deterioration of the woodland in the recent past (Depierre and Gillet, 1971). Agricultural clearance for the export crops, groundnuts and cotton, was the chief culprit according to the mainstream view in the forestry profession (Michon, 1973). In addition, there were demands for more and more wood fuel. According to Ibrahim (1978), 800,000 families in Kordofan and Darfur provinces of the Sudan consume about 155 million trees annually, and the Sudanese government estimates that nomads alone uproot a minimum of 548 million acacia shrubs each year for cooking. Similar alarms have been sounded in every country in the semi-arid zone. It is a short step from such estimates to a gloomy prognosis of a landscape without trees, fuel and (eventually) soil. Such a scenario has long been used energetically in support of reafforestation programmes and forestry reservation in West Africa (e.g. Chevalier, 1952).

In the Manga Grasslands, open woodland, dense woodland and farmed woodland all decreased between 1950 and 1969 (table 7.1). Alarmism is out of place, however. By 1986 wood for construction or fuel was still not considered to be scarce: it had not entered the market economy. Dum woodlands in the deep depressions, admittedly, are denser on the Niger side of the border where felling or lopping is illegal. But they occupy potentially valuable dry-season farmland, where economic trees of higher value are usually planted or protected. Open or sparse woodland still occupies the shallow depressions where most rain-fed cultivation, and fallow, is found. Self-seeded economic trees are always protected. Planted neem trees, and indigenous trees seeded by domestic animals, grow in every village.

Elsewhere in the semi-arid zone, woodland is more extensive. Those who condemn 'indiscriminate deforestation' often ignore the need to estimate natural regenerative capacity before concluding that off-take is excessive; the fact that much rural fuelwood is harvested from living trees rather than by clear felling; the fact that rootstocks are often left in the ground under shifting cultivation; and the fact that economic trees are given protection on farmland, particularly but not only on fields cultivated annually – including trees maintained for firewood. An indigenous base of values and techniques already exists for the 'farm forestry' now being advocated by development agencies.

The values of the new class of farmer–entrepreneurs are a different matter. Mechanised enterprises of up to several thousand hectares in size, found in growing numbers along Nigerian arterial roads, are usually preceded by clear felling. Like the agricultural development corporations of Senegal (Wane and Kone, 1980) they are responsible for deforestation, more than small farmers. They receive, however, little official criticism, notwithstanding the clouds of dust that follow the tractors at ploughing time and the sheet wash that redistributes surface soil in times of heavy rain.

Certainly large areas of woodland have been diminished to secondary shrub growth on fallows. But few, if any, data have been adduced to show that the fertility of these soils is significantly and adversely affected by such a change in status. On the contrary, 'in the drier northern savanna, where the tree cover is sparse or absent, fallowing under grass or grass–shrub mixtures is an effective means of restoring fertility' (Jones and Wild, 1975: 129).

Since the fuelwood requirements of quite dense rural populations can be met from harvesting trees at densities of about 15/ha, the threat to woodlands from this quarter can be attributed to urban, not rural, demand. Around larger Sahelian settlements, according to Mensching (1982), 'there has been a vast increase in circular destruction of the savannah tree-stock... The circles of destruction... are still widening everywhere'. An example may be the town of Bara, in the Sudan, where a stripped perimeter has resulted from woodcutting (Hammer, 1977). Yet Pullan (1974) has shown that protection, not destruction, of trees characterises permanent farmland, including the hinterlands of settlements, in many parts of West Africa. In the Kano Close-Settled Zone, the density of trees was found to be stable at about 15/ha between 1972 and 1981 (Cline-Cole et al., 1987) – meanwhile urban demand for fuelwood, which had previously been met from harvesting farm trees, had been largely redirected further afield. Indigenous forest management is more likely to be rational than reckless and, given the right conditions, a conservationary ethic emerges.

Cultivation inevitably exposes the soil to water and wind erosion, under any climatic regime. Millet and sorghum cultivation has been blamed for sand remobilisation – beginning with the tell-tale 'ripple-marks' and going on to the formation of dunes – in western and central parts of semi-arid Niger (Chamard and Courel, 1975; Mainguet and Cossus, 1980); Burkina Faso (Bougères, 1979; Chamard and Courel, 1979); western Sudan (Ibrahim, 1978) and in the Sahel generally (Mensching, 1980); cultivation is said to increase evaporation, drying the soil, encouraging crust development.

However, Mainguet et al. (1979) have described in some detail the forms of degradation associated with Hausa villages in the Maradi area of Niger. In an 'aureole' nearest the village, the effects of repeated cultivation and trampling are partly compensated by manure, domestic waste and rubbish; the well, market, school and granaries provide focal points for denudation, and the surface material is mobilised by the dry-season winds. In an outer zone of millet fields alternating with fallows, conditions are more stable, but trampling by animals is a disturbing factor. Evidences of aeolian degradation include the following. Deflation – the removal of fine soil particles by the wind resulting in surface concentrations of coarse particles – and craters around the feet of *Acacia albida* trees appear in the inner aureole. Nebkas are accumulations of sand downwind from small objects, such as grain roots. Ripple-marks appear in exposed soil transverse to the direction of the wind; and accumulations of sand may be found upwind from vegetated field boundaries. The occurrence of these and other forms – variable from year to year – is governed by situation (village site, manured fields, or outer zone) and by the preceding season's rainfall. Mainguet et al., proposed that when deflation removes the upper soil horizon entirely, reversion to fallow allows a layer of moving sand to be reconstituted around the stems of the vegetation, creating conditions for a good crop of millet after three years. Such a cover of mobile sand is considered important by the farmers; it may conserve soil moisture. It may also protect the exposed B horizon from sheet erosion. Similar processes and forms occur in hundreds of Nigerian villages, including Dagaceri. On permanent fields, cultivation may not be inconsistent with homeostasis, even in conditions of almost constant wind action.[17]

On the drier soils of the Manga Grasslands, most cultivated land is limited to depressions, where moving sand dunes are absent. The necessity for fallows is recognised in agricultural practice. If cultivation should extend upslope, the virtual absence of a soil profile minimises the significance of deflation. All cultivated soils are low in nutrients (table 7.3), and yields are determined primarily by rainfall, whose annual fluctuations make it impossible to establish any trend of degradation in the longer term.[18]

The removal of trees for cultivation is likely to increase wind speeds at the surface. This is the justification for the creation, at considerable expense, of shelter belts. Rather than separating cultivation from tree plantations, however, the density of farm trees may be increased and fixed boundaries planted along a stabilised field pattern, which is already indigenous practice in the farmed parkland of such areas as the Kano Close-Settled Zone. The ecological stability of this area during two dry decades has been noteworthy.

If agriculture is to take place, some degree of degradation must be accepted as inevitable. Either the extent or the agronomic significance of aeolian action on arable land may be exaggerated at times; and there appear to be no empirical data on the changes brought about in semi-arid soils under cultivation cycles. The stabilisation of surface materials under intensive systems of cultivation offers a more practicable alternative to the extensive withdrawal of land from agricultural use.

Grazing pressure on pastures is most commonly attributed to *selective grazing* – eliminating preferred species, which tend to be replaced by less palatable ones; to *premature grazing* – eating plants before they reach the seed production stage; or to *trampling*. The result, according to the argument, is a qualitative degeneration of the pasture, or its physical destruction, as the herbaceous cover becomes discontinuous or excessively short and wind action proceeds to remove material from the surface. Soil properties undergo changes which lead to an alteration in species composition until a lower level equilibrium is reached (cf. Hare, 1983: 53–60).[19]

The case for *overgrazing* rests on the concept of carrying capacity. This capacity depends on the production, above the ground, of edible herbaceous matter, half of which can be ingested during the course of the year under optimal conditions (Boudet, 1975). The feed value of pastures varies with soil, vegetation, and season (Bremaud and Pagot, 1962), diminishing after November. It is possible, nevertheless, to estimate theoretical average carrying capacities for Sahelian pastures using climatic parameters, according to Le Houérou (1985).

Lacking usable data on livestock numbers that can be related to defined territories, the case for overgrazing is usually made indirectly on evidence of ecological degradation. This is a circular argument (Sandford, 1976; 1983: 11–17). The nature of rangeland management, and the possibility of alternative causes, may be ignored. Such evidence is certainly available in the Manga Grasslands, but there are grounds for scepticism about such an interpretation. Firstly the substitution of annual for perennial grasses cannot be assigned exclusively to selective grazing in view of rainfall trends in the last two decades and its common occurrence throughout the Sahel. Secondly, premature grazing has not inhibited the regenerative cycle of the grassland even under conditions of extreme stress, as demonstrated by its recovery in 1985. Thirdly, the appearance of rangeland dunes during the relatively wet years before 1969 suggests physical damage by grazing. But their subsequent enlargement and multiplication in the two dry decades following, as argued above, indicates a complementary, if not greater, role for aridity. Dunes developed after 1967 under conditions of strictly controlled grazing on the enclosed Ekrafane Ranch in Niger.[20] Finally, the growth of village perimeter dunes indicates without ambiguity a primary role for grazing by small livestock. But these formations are localised – as much of the evidence for rangeland deterioration turns out to be on inspection.

Recent research in two projects in Mali and Niger (the Sahel Primary Production Project and the Niger Range and Livestock Project), as reported by Stryker (1983: 183), has concluded that in Sahelian pastures receiving less than 500 mm of rainfall, 'there is little direct evidence of extensive rangeland degradation due to overgrazing in Mali and Niger; in some cases heavy grazing may have improved the rangeland'. Protein production per hectare compares very favourably with similarly dry areas of the USA and Australia. Furthermore, some measure of overgrazing is found to be economically optimal, and therefore some degradation of the rangeland is to be expected under present tenurial conditions. These conclusions call into question the idea of the Sahel as a fragile ecosystem under constant threat of extinction through overgrazing. Whether they apply equally to areas – such as the Inland Niger Delta (Krings, 1985) – where pastoral and agricultural uses compete for land is less certain.

What estimates of average carrying capacity must necessarily ignore is the fluctuations which occur from year to year as the result of rainfall variability. Yet these are of far greater moment to those who use the pastures than their average productivity. For land to be stocked at its *average* capacity means accepting losses of potential production in many years; to stock it at a *reduced* rate in order to avoid losses in all but, say, one year in twenty would be to waste forage on a massive scale, even assuming that others could be kept from using it. Sandford (1982; 1983: 38–40) has shown that from an economic standpoint, the costs of such a conservative stocking policy in terms of forgone production are not only quite high but, of course, rise with the variability coefficient of the rainfall: 'in areas with very unreliable rainfall (high coefficients of variation) even moderately conservative strategies are very costly'. Opportunistic strategies – increasing herd sizes and risking loss – make better *economic* sense.

Given a preference for opportunistic stocking strategies, the scope for loss is minimised by transhumance on an annual cycle, and by flexibility in the selection of pastures from year to year. Open rights to the use of grazing land are essential for such management, as is also freedom of movement across international borders.[21] While failures of management do occur, and both economic and ecological losses are sustained, the evidence provides adequate ground for scepticism about sweeping claims that overgrazing is the main cause of ecological degradation in West African rangelands. There is also reason to question the value of a concept dependent on theoretical average productivity in an environment subject to extreme variability.

The last two specific allegations of misuse – bush burning and 'bad' irrigation practices – need not detain us here. The first is rather rare in the drier semi-arid zone owing to the sparse vegetation cover, and is not practised much in the areas in which the present study was carried out. The second is generally confined to public-sector irrigation works where salinisation has resulted from design failures and poor drainage, such as the schemes along the Yobe River in northern Borno. There is no evidence of such problems yet arising on private, small-scale irrigation works.

Representative of the environmental misuse hypothesis is the following statement: 'Desertification is caused almost entirely by human misuse of the environment... This misuse, which is not necessarily the result of ignorance, takes the form of felling trees to provide fuel, overgrazing by domestic animals, and harmful agricultural practices...' (Cloudsley-Thompson, 1984; cf. 1978). The evidence reviewed here suggests, on the contrary, that such a statement underestimates the significance of the climatic factor. The term 'misuse' caricatures the relationship between ecology and indigenous land-use systems, exposing the conflict between science and ethno-science which was reviewed in chapter 1. It diverts attention away from the empirical study of land-use systems, whose basic rationality emerges unfailingly from almost every field investigation, and towards the application of conservationist dogma. But the rationale of indigenous land-use systems must become the basis for conservationary resource management for the simple reason that the land belongs to, and must continue to be occupied by, its present populations. The hypothesis of environmental misuse leads straight into the cul-de-sac of social coercion: 'Although it is theoretically possible to halt or even reverse desert expansion, the pressure of the human population does not usually permit recovery to take place... the necessary measures... include *population control, enforced emigration*, and *education*, as well as *control over the use of land and water*. Few Third World governments would dare to tackle such inflammatory issues...' (Cloudsley-Thompson, 1984; italics added).[22]

Thus we arrive at the third hypothesis, population pressure.

Population pressure

The hypothesis of population pressure, as commonly stated, may concede some degree of rationality to indigenous land-use systems but argues that people overcut, overcultivate or overgraze because the growth of population and livestock numbers forces them to do so. Individual or family rationality is in conflict with the common good. Population growth is regarded as an independent variable, and results from a historically recent reduction in the rate of mortality. A finite amount of land cannot support an increasing population under present technological conditions. So, if growth came about through external factors (such as improved health services), it will have to be slowed in the same way: by policies of population and/or livestock control.

This hypothesis rests on a weak data basis. Censuses are few, chronologically dispersed or incompatible between countries, even unreliable, and long delays occur in their publication. National vital rates are rarely representative of arid and semi-arid regions of the country. Rural and urban populations cannot be easily distinguished, still less rural groups (e.g. nomadic and sedentary) whose vital rates may differ significantly. For livestock populations, the position is even worse: FAO estimates are poor substitutes for surveys or censuses.[23]

Setting aside such problems, population pressure arguments are commonly based on some, if not all, of four questionable assumptions.

The first assumption – now less commonly found – is that human fertility behaviour lacks a rationale. Population growth – as in the theory of the demographic transition – results from a fall in mortality caused by factors external to the rural community, unaccompanied by a corresponding fall in fertility: a lag effect unrelated to the simple need for reproduction of the population.[24]

There is a coincidence between high fertility and the labour requirements of small-scale agriculture. Given a child mortality rate of up to 50 per cent, an ideal family size of five or more is consistent with an objective of securing male family labour sufficient to guarantee continuity of farm production, a modest long-term expansion in the scale of operations, and some security for old age.[25] Recurrent drought reinforces this rationale in two ways. Firstly, the best security against crop losses is to increase the size of one's cultivated holding, but when impoverished from the loss of crops and the selling of assets, recourse can no longer be had to hiring labour. Secondly, to have several men in the family is to open up the structure of alternative opportunities, of complementary specialisation: one may make mats, another sell his labour. Desertification adds its own quota of additional labour demands. Fuel needs to be brought from further afield, water needs lifting from greater depths, livestock have to be fed with laboriously cut browse.

The second assumption is that the productivity of land is fixed. The concept of the human carrying capacity of land was introduced into the literature on African rural systems by Allan (1965), who argued that when a threshold called the critical population density is exceeded, continued application of a given land-use system will cause ecological degradation. Such an event is implicit in most, if not all, formulations of the hypothesis of population pressure: 'Each piece of land has what we call its carrying capacity' (for humans and animals)... 'When that number is exceeded, the whole piece of land will quickly degenerate from overgrazing or overuse by human beings. Therefore, population pressure is definitely one of the major causes of desertification and the degradation of land' (Tolba, 1986;[26] cf. EMASAR, 1974; Le Houérou, 1976).

Since the concept of the human carrying capacity of the land was applied in the original case to native reserves in the formerly British Central Africa, where its operational assumptions were a fixed land area (defined by racial tenure policy), fixed agricultural technology, and exclusive dependence on farming for livelihood, it is not surprising that it has little practical value elsewhere (see Zelinsky, Kosinski and Prothero, 1970). However, this has not stopped its widespread use to underwrite general statements about desertification.

An alternative thesis (Boserup, 1965) proposed that population growth is a condition for agricultural intensification, although at the cost of diminishing marginal returns to labour. Amply supported by empirical studies in West Africa (Grove, 1961; Netting, 1968; Gleave and White, 1969; Mortimore, 1970, 1971), which affirm a general correlation between population density and the labour intensity of farming systems, this thesis has the advantage of making sense rather than nonsense of prevailing high fertility rates. Prothero (1962), in an early discussion of desertification in north-west Nigeria, suggested that an *increase* in population would facilitate a transition to more conservationary land use. Such a transition, as already noted, has occurred in the Kano Close-Settled Zone.

The limits to such intensification appear at present to be defined by the supply of manure or fertilisers. Organic manure, which is preferred by farmers, is to a large extent dependent on the population of domestic animals. In the Kano Close-Settled Zone, the density of small livestock rises towards the centre with the density of the human population (Hendy, 1977).[27] Nevertheless, manure applications are sub-optimal for desirable yields on the great majority of farms.

Conservationary land-use methods are labour-intensive. Erosion control (including terracing in hilly areas), tree planting, tending and harvesting, planting and maintenance of field boundaries, transporting and distributing manure or fertilisers, tending livestock in intensively farmed areas, not to mention increased labour inputs on normal agricultural operations, are not consistent with a low or declining population density, especially when introduced for the first time. There is, therefore, an inherent contradiction in calling simultaneously for intensified, conservationary land use and for family planning (Eckholm, 1984; Milas, 1984).

With regard to livestock breeding systems, in the northern Sahel at least, some empirical studies have shown that labour and capital are more limiting than land (Stryker, 1983). In the semi-arid zone, the reduction in grazing areas is more significant, since rangeland occupies the interstices in a cellular pattern of agricultural exploitation.[28] However, crop residues provide valuable fodder on agricultural land. Intensification in these systems depends on the possibilities for improving the productivity of rangeland, in turn dependent on questions of tenure and funds for investment. It seems unreasonable to exclude such a possibility.

The concept of carrying capacity rests implicitly on the third assumption, which is that of fixed technology (Ware, 1977). For example: one implication of population increase in the rural areas is 'a *vast increase in subsistence agriculture*, which, considering the limited availability of arable land, implies intensified overcultivation, greatly increased cultivation of marginal and submarginal lands and consequent accelerated land degradation, particularly in dryland and mountain regions' (Milas, 1984; italics added). It appears that the term 'subsistence agriculture' is used here in a technological sense, and that the possibility of technological change is excluded, as it was from the assumptions of Allan's original formulation.

The extent and possibilities of technical change in indigenous farming systems have been systematically overlooked by scientific research organisations, preoccupied with 'transformation' models of agricultural development. Examples of experimentation and innovation in one village were reviewed in chapter 4, and there is no reason to suppose that Dagaceri is exceptional in this respect. In northern Nigeria as a whole, positive evaluations of long-term trends towards intensification contrast quite markedly with the gloom that pervades much literature on francophone 'Sahelian' countries (cf. Franke and Chasin, 1980; Bonte, 1986). And contemporary integrated rural development programmes, based on intensification packages, render the assumption of fixed technology entirely obsolete.

The fourth and last questionable assumption – a 'single-sector' assumption – is that the continuing existence of a subsistence priority necessarily implies the total dependence of the family on its land for livelihood. This cannot have been a reasonable assumption in the native reserves of modern Zambia even in the 1940s, when Allan's ecological surveys inevitably ignored the place of copper mines in the family budgets of remote rural areas. But, in Africa, 'to a greater extent than in any other region, the fertility of the people is outstripping the fertility of their land in an area where the level of development is low and nearly all economic production is directly or indirectly related to the land' (Milas, 1984). The history of the Sahel suggests otherwise (Lovejoy and Baier, 1976). The desert-side economy depended on its trading networks based on the complementarity of different ecological zones. The diminishing reliability of rainfall, during the past two decades, in the semi-arid zone has reinforced the time-honoured strategies of diversification and mobility, and the exploitation of niches in the more secure economy of the south (as shown in chapter 4). The use of the option of movement may not mean permanent out-migration,[29] but rather, circulation – allowing interests to be maintained in two or more locations. The possibility of a progressive diminution of the subsistence priority, in favour of increased dependence on other sources of income, tends to be ignored in statements about population pressure. Yet this has been the effect of successive subsistence crises in Dagaceri and other villages like it.

If it is conceded that high fertility is rational within the constraints of certain land-use systems, that the productivity of land is not intrinsic but capable of improvement with additional inputs of labour (or capital), that technology has not been fixed in the past and should not be expected to be so in future, and that livelihoods may be supported increasingly from sources other than land, the population pressure hypothesis is revealed as a gross oversimplification.[30] Such a conclusion is borne out by the considerable variations that may be observed within the African semi-arid zone in the density of population, and the ability of some land-use systems to absorb considerable population increase with very little damage to the ecosystem (Kates, Johnson and Haring, 1977: 20). Comparative studies of human ecology in homogeneous ecological areas would further illumine this question, as these authors point out (p. 22). It does not, however, follow from generally relatively low densities in the arid and semi-arid zones that population pressure is not a factor in degradation (Timberlake, 1986: 28).

While it is a truism that without people there would be no exploitation, fewer people would not necessarily reduce degradation since some degradation is intrinsic in exploitation. More people, however (with the exhaustion of the supply of free land), may lead land-use systems along one of two pathways. Existing methods, applied increasingly often to the same land, may cause its degradation; or intensive methods may be introduced resulting in its conservation. The hypothesis that the first of these pathways is inevitable is not proven. The possibility of the second pathway has been demonstrated in some systems. Since not much can be done about the numbers of people already there, attention should be focused on strengthening and extending such systems.

This conclusion does not prejudice arguments for population stabilisation at national level, which are based on macro-economic considerations such as high urban unemployment, dependency ratios and costs of social welfare. There has been a noticeable turnabout in the attitudes of several African governments, including that of Nigeria, towards population stabilisation in the last decade. But as a specific solution to subnational problems of desertification, such a policy cannot be adequately supported by the population pressure hypothesis.

Structural hypotheses

Colonial rule disrupted precolonial social, political and economic relations in West Africa. Slavery was abolished, political hegemonies removed or inverted, bureaucracies interposed between rulers and people, Western education introduced, taxation in currency enforced, levies of labour, animals or money imposed; agricultural exports, labour selling, markets and trade were promoted; infrastructures were developed, and increased consumption encouraged. For some societies, the droughts of the seventies terminated a long process of disintegration. Tuareg grazing systems in the arid zone, formerly based on a class system of resource allocation, were severely disrupted, culminating in sedentarisation and even urbanisation (Toupet, 1975). Fulani systems, which were classless and thereby less prone to disintegration, were more resilient (Gallais *et al.*, 1977). The implications of twentieth-century changes for farming societies in the semi-arid zone were hardly less profound. Their significance for the present discussion lies in the extent to which they may be seen as contributing directly to ecological degradation.

According to Bonte (1986; 152), 'the fragility and vulnerability of the ecosystem are directly determined by systems of production.' Both herders and farmers were forced to produce more by the declining value of pastoral and agricultural labour – in terms of the commodities it could produce for the market – in relation to its social value. This they achieved by increasing the sizes of their herds and by cultivating more land. Excess grazing occurred on rangeland. Fallows were shortened or abandoned, soil fertility declined, and new land cultivated, reducing that available for grazing.

The 'groundnut hypothesis' occupies a central place in structural interpretations of ecological degradation in semi-arid West Africa (Comité d'Information Sahel, 1974; Copans, 1975, 1979; Franke and Chasin, 1980). Soil nutrients were exported along with the groundnuts and not adequately replenished by fertilisers. Average yields are supposed to have declined in the long term. Bare fields were exposed to erosion after the harvest. Deteriorating terms of trade caused further extensions to the cultivated area. Food production was neglected. 'Thus colonial profit-making in peanuts and African poverty at the producer level combined to set in motion a spreading wave of environmental degradation' (Franke and Chasin, 1980: 70). In Niger, where the area under groundnuts was estimated to have increased from 73,000 ha in 1934 to 432,000 ha in 1968,[31] an additional effect was arable encroachment into areas north of the limit for rain-fed agriculture (the 350 mm isohyet). Such encroachment, and the shortening of fallow cycles, reduced the grazing areas available for the pastoralists. 'Hunger is the main symptom... of a creeping agrarian and distributive crisis' (Brune, 1985).

Two empirical difficulties confront the 'groundnut hypothesis'. The first is the difficulty of weighing the effects on the soil of groundnut cultivation apart from those of the grains and grain legumes, in view of the widespread practice of intercropping. 'Cereals may benefit from association with a legume... legumes are probably a necessary rotational component under most conditions' (Jones and Wild, 1975: 140). Groundnuts support nitrogen-fixing root nodules and a harvested crop of 3 tons/ha of dry matter is estimated to leave behind 30–40 kg/ha of nitrogen in the soil (Bromfield, 1973, quoted in Kowal and

Kassam, 1978: 134). Nutrient balance sheets for three areas of Senegal under groundnut rotations could not distinguish between the effects of the different crops in the four-year rotations, but with no fertiliser at all the losses were calculated to be very considerable: 10–15 kg/ha of phosphorus, 42–63 kg/ha of potassium and 26–44 kg/ha of calcium (Tourte, reported in Jones and Wild, 1975: 211). Watson (1964) calculated that the amount of phosphorus in crops sold off the farm was 2 kg/ha for a 500 kg crop groundnuts, 1 kg/ha for a 250 kg crop of seed cotton, and 3 kg/ha for a 1,000 kg crop of grain; and that about 25,000 tons of single superphosphate were needed annually to replace the phosphate removed in northern Nigeria's export crop of groundnuts in the sixties (quoted in Jones and Wild, 1975: 143–4). If these estimates were sound, the shift from selling groundnuts to grain which has followed the depredations of rosette disease in Nigeria has potentially *more* significance for the long-term degradation of soils. But a satisfactory 'nutrient audit' for the groundnut economy is not yet available.

The second difficulty is that of establishing long-term downward trends in yields independent from the effects of rainfall and of short-term cycles of fallowing or fertilisation. FAO estimates of groundnut yields in 1969–70 – just before the impact of drought was felt – indicate just under 1 ton/ha for Nigeria and Senegal and considerably lower yields for all other West African countries (FAO, 1971). As the oldest and largest producers, these two countries might have been expected to show lower figures. Few data exist in northern Nigeria on which reliable estimates of long-term trends in crop yields under indigenous farming conditions can be based.

Writing of drought and famine, Copans (1979) states 'I cannot distinguish in the abstract or even statistically the effects of imperial domination and those of changes in the natural environment. They thoroughly dovetail.' Such a distinction is certainly difficult. But like any other, the structural hypothesis requires verification, and empirical support for it, I suggest, is still inadequate. Long-term degradation of farmland and rangeland has acquired the status of orthodoxy in the literature on the western Sahel, especially Senegal. In Nigeria, other than with respect to the last two decades, few have been prepared to make such claims. Are there fundamental differences? And if such degradation is proven, can the linkages proposed be demonstrated?

Resilience

It is necessary at this point to return to the distinction, emphasised throughout this study, between drought and desertification. It is more than a difference in time-scale. A meteorological drought, as an event, can meaningfully be understood in terms of cause and effect. The same is true of other events in the ecological realm – an outbreak of grasshoppers for instance. But a cause-and-effect paradigm is inadequate for understanding desertification, as the preceding review of the climatic, environmental misuse, and population pressure hypotheses has suggested. The last group, which I have inadequately termed structural hypotheses, alone comes to grips with the fact that ecological degradation is a process, linking the warp and woof of human societies with their natural environment, and expressing historical changes in them both. Faced with such complexity, the simple cause-and-effect paradigm has outlived its usefulness (cf. Copans, 1983; Garcia, 1981: 1).

So, also, has the technological evolutionary hypothesis advanced by the natural hazards school (Burton and Kates, 1964; White, 1974: 4–5) and discussed in relation to desertification by Heathcote (1980: 130–1). As Heathcote points out, this framework is simplistic when confronted with the diversity of patterns of response in empirical situations. But its primary reference is to 'natural hazard' events and, as argued earlier, desertification cannot in any sense be considered as an exogenous process, still less an event, in relation to human systems. And human behaviour goes far beyond the 'response' envisaged in what amounts to a 'covert environmental determinism' (Hewitt, 1983).

Another alternative is a crisis-recovery model. As illustrated by Warren and Maizels (1977: 81–99) this approach envisages a point in time when conservation policies are applied to the downward curve of desertification. Various pay-off curves can be projected, depending on the policy adopted (e.g. destocking rangelands, investing in terraces on cultivated slopes, introducing crop rotations), from the initial point of low returns to a steady (and higher) level. Implicit in this concept seems to be the idea that an equilibrium has been disturbed and can be regained. Garcia (1981) proposes a more formal application of systems theory in which the passing of a critical threshold (as in a major drought, or famine) causes the breakdown of the system – whether ecological or social. The question is whether a steady state or equilibrium has ever existed; whether sustainable yields as usually understood represent a realistic target; whether 'recovery', or continual adaptation, should be expected in an environment whose fundamental characteristic is uncertainty: 'uncertainty is the way of life for everything that lives in a dry environment' (Warren and Maizels, 1977: 86).

A way of incorporating uncertainty-as-norm into an understanding of desertification, as opposed to uncertainty-as-aberration, is suggested in the concept of ecosystem resilience as proposed by Holling (1973: 17; see Clark University, 1975; Lundholm, 1976). Resilience is contrasted to stability:

Resilience determines the persistence of relationships within a system and is a measure of the ability of these systems to absorb changes ... [it] is the property of the system and the persistence or probability of extinction is the result. Stability, on the other hand, is the ability of a system to return to an equilibrium state after a temporary disturbance ... [it] is the property of the system and the degree of fluctuation around specific states the result. [In] areas subjected to extreme climatic conditions the [faunal] populations fluctuate widely but have a high capability of absorbing periodic extremes of fluctuation. They are, therefore, unstable using the restricted definition above, but highly resilient.

Ecological degradation

I propose to extend the concept of resilience to human systems.

Uncertainty in the ecosystem arises fundamentally from the impact of rainfall variations on primary productivity. At Fété-olé in Senegal (Bille, 1974) – which is near Saint-Louis where annual rainfall varied in the range 144–691 mm during the century 1871–1970 – the production of dry matter above the ground varies 'normally' in the range 590–1,300 kg/ha. In 1972 it was practically nil. The principal determinant is the length of time when groundwater is available to the plants (50–110 days/yr). In pastoral areas, the effects of such fluctuations have been summarised as follows (Le Houérou, 1985: 177):

In general, variability in primary production is much greater than variability in rainfall. Variability in stock numbers is somewhat smaller than variability in rainfall, but variability in secondary production is very poorly documented; it would seem to be of the same order of magnitude as the former. And the ratio of livestock to human beings is inverse to rainfall. Thus overall, variability in climate is amplified in primary production, but dampened in the production of livestock and in the sustenance of human beings. This is primarily due to various adaptive strategies.

Among pastoral societies, such strategies include mobility in space, management of the species composition of livestock holdings, management of the habitat through such practices as burning, and managing the disposition of herds to maximise the use of available grazing. In these ways, according to Western (1982), nomadic pastoralism may achieve a higher level of 'food chain efficiency' than either wild animals or commercial ranching.[32] It is now widely conceded that few can compete with nomadic pastoralists in the efficiency of their adaptation to the spatio-temporal variability of the arid habitat.

'Resilient instability' was characteristic of the Tuareg whose economic interests extended beyond pastoralism to trade and socio-political domination of slaves and tributary farming communities. Thus Lovejoy and Baier (1976: 164–5; see also Baier, 1980) write of the desert-side economy that:

When drought occurred, the desert economy contracted, with nomadic Tuareg and farmers alike leaving for the extreme southern end of the trading network... Just as the Tuareg social structure telescoped during hard times, it expanded again when the weather improved or political conditions returned to normal... Nobles collected their personal followers and headed north... This spatial mobility was the basis of the social system, which adapted to the cyclical nature of the desert-edge climate.

The process was painful. According to Bonte (1986: 166–7)

Droughts can be seen as a means of resolving periodic crises of pastoral overproduction. They enforce a drastic reduction of production capacities and of the productivity of pastoral labour. Droughts also require the deployment of all the community internal and external relationships in order to reconstitute the herds and the decimated social groups. They therefore constitute a 'cyclical' dimension to the functioning of these pastoral systems of production. Historically they have provoked political restructuring, mass movement of human groups, and the establishment of new local hegemonies.

Turning to the agricultural semi-arid zone, Warren and Maizels (1977: 38) have likewise emphasised that shifting cultivation with its associated techniques makes intelligent use of a variable ecosystem. Light tilling and mulching preserve as much organic matter in the soil as possible, woody fallows bring up nutrients from depth and add litter to the soil, crop mixtures protect the surface from erosion and maximise the uptake of scarce nutrients, and drought-resistant varieties make the best use of available moisture. Spatial mobility is intrinsic, though on a time-scale of years.

However, farmers too are involved in complex political and economic structures, many of which necessitate spatial mobility, in patterns linking their semi-arid zone with more humid bioclimatic regions. As I have attempted to show, these patterns of adaptive behaviour account for the resilience of the human system, confronted as it is with periodic failures of production. Berg (1976) recognised that, notwithstanding heavy losses, 'strong elements of resilience and flexibility' and an ability to capitalise on regional interdependence characterised Sahelian economies.

The analogy with a resilient ecosystem raises the question whether stability, in the sense of equilibrium, is possible. Practically all proposals for the conservation of arid and semi-arid areas assume that it is. Warren and Maizels (1977: 82), for example, take it as axiomatic that 'long-term sustained yield is the only sensible aim of management in dry ecosystems even if considerable fluctuations from the mean are inevitable'. In order to achieve this, the carrying capacity of rangeland must be assessed and 'tracked' from year to year by suitable monitoring, and stocking policies enforced. Yet the difficulties of such assessment are admitted, as is also the alien character of this concept to the perception of those who use the range (*ibid.*: 71). Likewise, land capability analysis is considered essential for rain-fed agricultural use, together with coercive measures to protect land that is vulnerable to erosion. These suggestions are derived from the principles of recognising the integrity of, and collaborating with, the ecosystem. Unfortunately such 'top-down' land-use planning rarely recognises the integrity of, or collaborates with, the human system.

The stability view emphasizes the equilibrium, the maintenance of a predictable world, and the harvesting of nature's excess production with as little fluctuation as possible... A management approach based on resilience, on the other hand, would emphasize the need to keep options open, the need to view events in a regional rather than a local context, and the need to emphasize heterogeneity, [resulting in] the recognition of our ignorance; not the assumption that future events are expected, but that they will be unexpected. The resilience framework... does not require a precise capacity to predict the future, but only a qualitative capacity to define systems that can absorb and accommodate future events.

(Holling, 1973: 21).

A resilient ecosystem is far from being 'fragile' – a term that is conceptually misleading as well as incapable of definition. 'This fragility perspective

Ecological degradation

underlies much of the literature ... and has attained the status of a scientifically accepted truism. Yet there is contradictory evidence on the topic which suggests that dryland eco-systems may be far more resilient than is generally supposed' (Johnson, 1979: 26). Similarly the concept of 'vulnerability' (Garcia and Escudero, 1982) has not served the debate on desertification well. The extension of the idea of resilience from ecosystems to human systems directs emphasis away from a futile search for equilibrium to the strengthening of social adaptive behaviour.[33]

The year 1983 was better than 1973. He considers himself better off now than when we first met him [1974] because he now realises that there are many ways of making money – especially in trade – whereas then, all he did was make mats and grow groundnuts.

(Interview with elderly informant, Malam Z, in Dagaceri, 28 Oct. 1986)

9

POLICY DIRECTIONS

The 28 recommendations of the United Nations Conference on Desertification (UNEP, 1977c; 1978) ranged far and wide amongst both technical and social aspects of global ecological degradation and drought. The heart of the Plan of Action was the introduction or extension of land-use planning on 'ecologically sound' principles (recommendation 2), but account was also taken of the need to recognise the 'needs, wisdom and aspirations of the people' (recommendation 3). Technical proposals were principally directed to the major ecological subsystems – water resources, rangelands, rain-fed agricultural areas, irrigated areas, and woodland vegetation (recommendations 5–9), and to the development of alternative energy sources (19). There was also a group of proposals aiming to monitor and improve human welfare, including the strengthening of systems of insurance against drought (recommendations 12–17). The remaining recommendations covered survey, evaluation and monitoring (1, 11), science and education (18, 20), and national and international administrative, planning and financial aspects (21–8). Discussion of these global recommendations is beyond the scope of this study, based as it is on a small part of one African bioclimatic region. However, it may be noted that considerable criticism has attended the efforts of the United Nations Environment Programme whose Desertification Branch was given responsibility for co-ordinating the implementation of the recommendations (Timberlake, 1986: 92–7). Financial resources have been inadequate and, so far, progress has been slow: apparently 'the war against desertification is being lost' (Walls, 1984).[1]

Meanwhile, drought returned to semi-arid Africa in the eighties, and was countered, as before, by *ad hoc* activity on the part of governments, donors and voluntary agencies. The pressures of the emergency forced the priority to be placed on food aid, public employment programmes and water supply projects. If the danger has since receded, it may be doubted if the world is any better placed to confront the next African famine than it was in 1970.

In this study I have interpreted the linked problems of drought, famine and desertification by means of an adaptive framework and concluded that human systems in the semi-arid zone have demonstrated a capability for survival best described by the term resilience. As opposed to some alternative theoretical frameworks, this approach directs attention to past and present adaptive behaviour, which may provide a basis for policy priorities, consistent with a pragmatic view of the status quo. A conception of the human system as intrinsically resilient (and unstable) in a variable environment carries the discussion of policy response to somewhat broader ground than it normally occupies. Popular conceptions of official response are dominated by questions of drought prediction and relief; measures of land-use control; and technical anti-desertification projects. All these are 'top-down' concepts. By contrast, if strengthening the resilience of human systems is accepted as a primary objective, 'bottom-up' or grass-roots approaches become obligatory. The objective is now to match the intrinsic resilience of the ecosystems with a measure of resilience in human systems that can cope with the uncertainties caused by periodic failures of production. In the past, system resilience was achieved at the price of periodic and intense human suffering. The elimination of such suffering should be well within the reach of modern technical and administrative capabilities, in an age in which the sanctity of human life is protested as often as it is desecrated. The orientation is consistent with calls for increased autonomy and self-reliance (e.g. Raynaut, 1977; Van Arcadie, 1978; Bradley, 1980; UNEP, 1981; Van Apeldoorn, 1981; Timberlake, 1986).

Opportunistic land-use strategies are rational under conditions of recurrent agricultural or ecological drought. They carry the implication of heavy losses – periodically and unpredictably. Such losses can be dealt with in three ways, through insurance, diversification, and spatial mobility. All three have been almost ignored by official research and government policy until very recently. The discussion which follows is illustrative rather than comprehensive and, although set in a Nigerian context, has broader significance.

INSURANCE

Insurance is an attempt to even out the time stream of income by netting or smoothing periodic losses (Russell, 1970). Given a subsistence priority, storage of the means of subsistence (crops and animals) is the main insurance strategy available.[2] The groups of granaries to be seen on the periphery of every village or household in the semi-arid zone testify to the importance of this objective. So does the value attached to the possession of livestock. While the provision of veterinary services has helped to reduce the losses of livestock (especially cattle) from causes unrelated to drought, grain storage has benefited little from government's attentions. These have been directed almost entirely to technical research on reducing storage losses (which are small, for grain) by pest control or new structures (which are expensive).

The principal factors constraining grain storage in this bioclimatic zone are economic and not technical. The need for cash impels the sale of surplus grain which might otherwise form a reserve, and the tentacles of the market system ensure that this grain is exported out of the producing area. It has to be bought back at a higher price when the hungry season comes.

A case has been made elsewhere for grain reserves at the village level in drought-prone areas of Nigeria (Mortimore, 1978b; see Van Apeldoorn, 1981: 145–69). 70 per cent of farmers stated that their main preparation against drought in the future would be producing and storing more food. Storage costs are lower in dried earth granaries than for bag storage in depots, or silos, and losses are acceptably small (as little as 4 per cent: Hays, 1975; Giles, 1965). Central storage is subject to major limitations. Not only are storage costs higher but, because of the central location, assembling and distribution costs are high, and distribution may even become impossible when it is most needed, in the rainy season. The scale of operations of a central storage system, such as the Nigerian Strategic Grain Reserve, can never match the scale of need owing to limitations of government finance.[3] Central reserves may also suffer from a conflict of objectives between famine relief and price stabilisation (which are not always compatible), and are prone to inefficiencies and mismanagement.[4]

On the other hand, village level storage is cheap, and incurs no transport costs (and no disruption in the rainy season). The roles of producers, managers and beneficiaries are merged minimising conflict of interest. The scale of operations, and decisions about the disposal of stored grain, can be made in the context of the perceived needs of the community. And dependence on government's financial resources can be minimised.

Three problems confront the development of village grain reserves. The first is the low productivity of agriculture and the small size of surpluses. Grain might have to be purchased through the market as well as from local producers. However, the activities of village grain traders continue in all years and for most months of the year. Merely to retain this marketed fraction of output in the village would be a worthwhile achievement. The second problem is to ensure equitable distribution of grain from the reserve. The best way to counter inequity would be to manage the reserve at the level of the community, rather than the farm enterprise; new forms of co-operative management would need to be devised. The third problem is one of finance. The suggestion to finance village reserves from tax relief was overtaken by the abolition of community tax; Van Apeldoorn (*ibid.*) suggests that funds should be diverted from central storage projects to financing village grain reserves, since they would be more cost-effective.

Cereal banks have been tried in Burkina Faso; village communities are provided with revolving loans, and use them to buy grain at harvest time for release in the hungry season at controlled prices. An evaluation of their performance (Roche, 1984) has indicated only qualified success. The schemes may run into financial difficulties or fall foul of the incipient class structure of the village. This experience shows the difficulties confronting such schemes but does not invalidate the case for their existence. The underlying imperative is to strengthen local institutions and resource management, since responsible use of the ecosystem is the best form of insurance in the long run.[5]

The same principle is applicable to the reserve function of livestock. Storage of feedstock is more problematic than of grain, but is unavoidable in the long run; even now, there is extensive storage at village level of grass fodder, when its price is expected to rise. The main barrier to be overcome would be the reluctance of livestock owners to invest in the purchase of feedstocks when their prices were low, and pasture abundant. Pastoral co-operatives, which have been the subject of experimentation in the arid zone of Niger (Swift and Maliki, 1984), provide an institutional framework for herd reconstitution after a disaster, and also for cereal banks. They might be linked with stabilised grazing rights in the rangeland interstices of the semi-arid zone.[6]

The concept of insurance extends beyond grain or livestock reserves.[7] The maintenance of social networks of reciprocity, allowing adaptive response to crisis along some of the lines described in chapter 3, is a form of insurance, but lies beyond the scope of policy as usually viewed. Maintenance of a structure of alternative opportunities for earning income will be discussed in the next section. Mention should be made of a final category, 'ecological insurance'. Certain features of agricultural and pastoral practice – intercropping, spatial fragmentation of holdings, diversification of livestock, and grazing mobility – give some protection against hazards which are specific to crops, animals or places. The protection of trees on farm holdings provides not only browse for animals, but a range of edible materials, and assets which can be liquidated against contingencies (Chambers and Leach, 1987). Women's expert knowledge of famine foods available in the bush provides a minimal level of food insurance; the expertise represented in table 3.12 should be built on, not discarded (Lucas and Wickens, 1986). But crop (and animal) specialisation is now a characteristic of many major initiatives in the agricultural sector; irrigation projects and mechanised farms uproot protected farm trees; and the bush disappears rapidly as farm holdings expand, fuelwood cutters move out from the towns, and grazing pressure increases.

DIVERSIFICATION AND MOBILITY

Diversification is a fundamental objective of adaptive response to drought, to poverty, and to hunger, as risk spreading has always been an objective in manipulating the ecosystem. Economic diversification strengthens trade and interregional linkages, supports the extended family and provides a rationale for high fertility. It re-emphasises the importance of the informal manufacturing and services sector, and of a diversified ecology. This underlines the obvious truth that the household economy is multisectoral, although the full significance of the other sectors may be revealed only when the primary farming or livestock production sectors break down. Not only is diversity intrinsic in adaptive behaviour, but the effects of such a breakdown are also correspondingly complex.

Baker (1974a:18) has commented that 'separating society and economy was an essential failing in early social research amongst herding communities in Africa' and, in similar vein (1975:2), 'the division of responsibility into watertight sectoral compartments means that the administration is neither able to perceive the nature of an ecological problem, nor is it able to do anything meaningful about it'. Diversity forms the basis of the resilience of human systems in the semi-arid zone. Recognition of this fact is now growing,[8] but it remains to be seen whether policy responses can be as flexible, in view of the innate specialisation of governmental administrations. This specialisation characterises the upper levels where most matters concerning production are dealt with by technical departments. It is, however, absent at the lowest levels, where decisions on resource use are as natural to a village head as are matters concerning the payment of taxes. The intrinsic unity of rural systems – the human ecology of place or community – provides therefore another powerful reason for recovering the autonomy of the local community. One example may serve. A new borehole is installed by the government agency to whom water matters are delegated: a centralised technical department. But only at the level of the village (the consumer community) can its systemic ramifications be evaluated – changing mobility patterns for people and livestock, crop damage risks, inter-ethnic relations, technical oversight and maintenance, wastage and so on.

The multisectoral implications of drought and desertification carry the problem far beyond the scope of technical solutions and into the field of development planning as a whole. Van Apeldoorn (1981:145–69) argued that contingency planning must form an integral part of economic planning (which in Nigeria – until recently – used the vehicle of five-year plans). Spooner (1982:23) writes: 'it is unlikely that a frontal attack on the problem [of desertification], which attempts to rearrange human activities in direct management solutions, will have much impact. Desertification is intimately related to development and represents one aspect of the inadequacy of the development effort so far.'[9]

A strengthening of the structure of alternative opportunities will help to take pressure off the ecosystem. Le Houérou and Lundholm (1976) have argued the need for alternative investment opportunities to livestock as a way of constraining the growth of herds. Employment opportunities in industry, mineral exploitation (National Academy of Sciences, 1975), or in public works, tourism, solar energy, development and transport (Gilardi, 1978) might be created. Investment in secondary urban centres (World Bank, 1979; Mensching, 1985) might further this end, but the labour-absorptive capacity of the urban formal sector is likely to remain modest. More significant is the possibility of an enhanced role for the urban informal sector. There is no doubt that towns such as Nguru (in Nigeria) exercise a role in generating part-time employment and income-earning opportunities over a wide area. The multiplier effects of investments in formal sector projects are amplified by the small average value of most market transactions, and the rapid turnover in informal participation. 'The ease of entry, small capital requirements, and the low level of skills in the largely unregulated sector have attracted a large number of migrants unable to enter the formal sector. Policymakers should therefore remove the various discriminatory practices against this sector' (Adepoju, 1979).

Recognising diversity as a policy objective for the semi-arid zone therefore has major implications for migration policy. Conventionally, governments in Nigeria view internal population mobility as unidirectional (rural–urban), long-term (rather than short-term) and inefficient (contributing to urban unemployment, insecurity, and congestion). Most African governments (surveyed in 1976) wanted to stop or decelerate it (Clarke and Kosinski, 1982). Pastoral nomadism tends to be regarded as anachronistic, unconducive to good administration or education, and is expected to be superseded in time by 'resettlement' programmes. International mobility is viewed even more negatively as a threat to national security, employment and welfare, to the extent that Nigeria's borders were closed to other members of the West African Community from 1984 to 1986.

First of all, internal mobility is not restricted to rural–urban movement; rural–rural migration has by no means died out. During the seventies, substantial numbers of farming families migrated south from areas such as northern Sokoto into lightly populated woodland in the subhumid zone, there to establish new hamlets within the territories of existing settlements. Such spontaneous movements (cf. Wood, 1982) rarely find recognition in official policy (Adepoju, 1982); indeed this would be unnecessary since they have taken place for decades and will doubtless continue to do so. Access to land is accomplished through existing administrative institutions. Some such moves have crossed ethnic boundaries with little difficulty. Migration of this type usually represents a transfer of labour from land-scarce to land-free situations, and its encouragement furthers agricultural production as well as the welfare of those concerned. The same may apply to international rural–rural migration.

Secondly, governments rarely make a distinction between migration and circulation. A most important matter is the attitude of urban authorities to short-term circulation. For example, shortly after taking over the country in 1984, the Federal Military Government of Nigeria embarked at state level on a repatriation exercise direct at the street economy of the urban informal sector, demolishing unauthorised retail and service premises, traders' tables and kiosks, and banning itinerant traders from pavements and intersections. But the justifying slogan, 'War on Indiscipline', paid scant regard to the impact of such action on rural areas scarred by another drought. Colvin (1981:342–3) writes: 'migration is the major safety valve that Senegambian societies have for coping with disaster... Together with organised fasting within households, it is probably the major reason why so few died in the recent Sahelian drought.'

From the urban standpoint, circulation is less costly than migration, for the *masu cin rani* make few demands on urban services and do not put up houses or shanties. Their stake in the urban economy is limited while they retain farming interests at home. Circulation is less disruptive of family life, redistributes wealth from urban to rural areas, and disseminates useful knowledge (Nelson, 1976; Hugo, 1979). Tolerance of a high level of circulation in the economy puts a brake on permanent urbanisation, with the heavy social costs that implies. If governments wish to contain 'rural–urban drift' and prevent the emergence of shanty towns on the Latin American model, they need to accept a high level of short-term circulation within the foreseeable future.

With regard to the second category of internal mobility – pastoral nomadism – recent studies of mobile livestock-producing systems strongly suggest a counsel of caution on 'resettlement' policies. In areas subject to unreliable rainfall, labour-intensive mobility between pasture locations is efficient, probably more so than alternative resource management systems. Transhumance on a north–south axis is the only way whereby some short-lived pastures of annual grasses can be effectively used. Although in the Nigerian sector of the semi-arid zone these arguments are less powerful than further north, owing to somewhat greater and more reliable rainfall, the rationale of pastoral mobility should not be lightly cast aside.[10] Two conditions for the settlement of mobile pastoral groups will be guarantees of exclusive access to grazing and of alternative sources of fodder when the rain fails. This is likely to be applicable to all areas north of the subhumid (Guinea Savanna) zone.

International mobility, the third category, is highly relevant to the ECOWAS treaty which promises freedom of international movement to citizens of its member states for periods of up to 90 days. North–south mobility across Nigeria's borders is historically consistent, and on its continuation depends, at times, the effectiveness of diversification strategies in the arid and semi-arid zones of Niger.

The colonial governments turned a blind eye to most types of personal movement across the northern frontier of Nigeria. As Baier (1976) pointed out, the maintenance of a permeable northern border is healthy for the Niger livestock industry, which supplies a sizeable proportion of Nigeria's beef. In the years 1972–84, imports provided from 21 to 43 per cent of Nigeria's slaughter cattle according to official statistics (Federal Livestock Department, 1984: 25–9), as well as small livestock. Differential pricing policies on either side of the border encourage a smuggling economy (see, for example, Collins, 1976). The major border markets are on the Nigerian side, their tributary areas extending into Niger. It would harm Nigeria's interests as well as the spirit of ECOWAS if a policy of 'benign neglect' (Van Apeldoorn, 1981:162) were to be replaced by an attempt to reinforce colonial boundaries that are incongruent with the realities of ecology, society and economy.[11]

A policy orientation to support the resilience of human systems in the semi-arid zone calls, therefore, for intensified interregional dependence, and implies an evolution towards farming (if not livestock production) as an increasingly part-time occupation, either for individuals on a seasonal basis, or within the structure of the family. Part-time farming as a planning objective is both historically consistent and realistic in contemporary terms. What should be done for farming and livestock production systems in the light of these circumstances?

INTENSIFICATION

It would be misleading to suggest that resilience is the only characteristic of semi-arid human systems which has policy implications. Their other salient characteristic is demographic growth which, in conjunction with the finite supply of land for woodland grazing and cultivation, presents a choice between depopulation (if ecological degradation is an inevitable concomitant of demographic growth, as the population pressure hypothesis implies) and intensification (if the alternative model is correct). Depopulation by coercion may be discarded as a viable policy option.[12] That leaves intensification as an unavoidable policy objective for the semi-arid zone.[13] It has been advocated widely by ecologists (Wickens and White, 1979), anthropologists (Swift, 1982), geographers (Mensching, 1985; Achtnich, 1985) and others for the Sahel, and also for other countries of Africa (Darkoh, 1982). It is already the basis of agricultural development policy in the integrated rural development projects funded jointly by the World Bank and Nigerian state and federal governments, several of those extending into the semi-arid zone. Only three aspects will be discussed here, selected because they have intersectoral importance.

Stabilisation of land tenure is a precondition for intensified land use, and the need for it in the semi-arid zone is sharpened by the fact that the duality of farming and livestock production takes an ethnic expression to a considerable degree (Bernus, 1974b). As the supply of free land disappears, the division of the land resources between grazing and farming becomes the subject of an uneasy status quo or erupts into conflict. In the Maradi area (Bellot and Bellot-Couderc, 1979) the area available for pasture diminished from about 50 to 15 per cent between 1957 and 1975. In Dagaceri (see chapter 4) such a diminution was constrained only by administrative action.

For farmers, customary tenure (see chapter 2) no longer provides adequate protection of use rights. Shifting cultivators in the Manga Grasslands, for example, cannot protect long fallows from appropriation by others, and since Ful'be are also involved in farming, this may result in ethnic dispute. Neither can ownership of economic trees be sustained, except for the date palm. On irrigation schemes elsewhere, or near to towns, smallholders may be put under political pressure to sell their land to the wealthy. Customary tenure is not subject to registration and depends on witnesses, among whom the village head (who may be a political appointee) is powerful. Farming tenure does not necessarily imply the right of exclusion – of marauding cattle, for example.

For livestock producers, grazing rights depend on customary recognition, and automatically lapse when arable land encroaches on rangeland. Such encroachments eventually reduce the stocking capacity of pastures. Administrative recognition of grazing rights is given in some areas. Although complaints about encroachment may be contained in this way, such demarcations do not confer title on individuals or groups and may not, therefore, be defended in the courts. Nor do they prevent the invasion of the grazing areas by visiting herds, no matter what the implications may be for the residents. An increasing commitment to farming on the part of transhumant herders also poses a long-term threat to the survival of their pastures. In riverine areas, dry-season grazings on which many circuits of transhumance depend are threatened by increased farming activity, irrigation, flooding of dams, hydrological changes and other civil engineering projects.

A number of observers have therefore advocated stabilising pasture rights (Adams, 1975), to enable pastoralists to be masters of their own territories (Bernus, 1975), which is seen as a precondition for responsible herd management: 'pastoral territorialization, that is, the management by coherent social groups of established pasture lands and their careful exploitation, seems at present to be the necessary basis for the continuation and probably the survival of the pastoral groups' (Gallais, 1979). Such territorialisation would have to be consistent, however, with the primary need for mobility which remains fundamental to pastoral land use and allows both herds and pastures to survive drought (Swift, 1975). It would facilitate investments in pasture improvement, by means of re-seeding, for example, which seems to be a practical possibility in the semi-arid zone (De Leeuw, 1974; EMASAR II, 1978) if not in the arid zone further north (see Swift, 1977a).

As the third major party, government's own interests are no longer served by continuing fluidity in land tenure. Entrepreneur–farmers are busy acquiring title to extensive areas of unallocated woodland which thereby ceases to be available for grazing, for fuel harvesting, for future generations of farmers or merely as a reserve stock of public land. Considerations of equity are at stake, and the Land Use Act of 1978 made access to land for all Nigerians one of its cardinal principles. And while the law maintains the fiction of state ownership of all land, the state governments may soon find that they can gain access to none, save by paying compensation to those who got there first.

The stabilisation of land tenure impinges on ecological conservation in several ways. With regard to farmland, private tenure is conventionally regarded as more conducive to good management than is communal tenure (e.g. Lofchie, 1975). McCown *et al.* (1979:326) contrast the Kano Close-Settled Zone with the degraded environs of Nyala in the Sudan, where a failure to modify tenure from communal principles towards more private control resulted in deforestation and overgrazing. The introduction and registration of individual or family title to farmland should be a policy priority.

For grazing the position is less clear. Where herd management is on a community basis, individual tenure is inappropriate. Group ranches and related schemes have been tried or proposed in several parts of Africa (Oxby, 1981; see

Hjort, 1976). But the 'establishment of communal tenure systems that accommodate growth, conservation and equity objectives presents formidable challenges' (Lawry et al., 1983:257). Recognition of group rights to a particular piece of rangeland would run into difficulties when others had to be excluded; and to secure rights to one area, only to lose them in another, would undermine the basis of transhumance. Yet given the present trend in the land:population ratio, exclusive grazing tenure will be unavoidable if pastures are not to be destroyed completely.[14]

Government conservation projects, such as shelter belts and dune fixation schemes, obviously depend for their success on unambiguous definition of tenure. Furthermore, if compensation has to be paid for the land, they are usually uneconomic.[15] The protection and management of natural woodland is also dependent on tenure, and has frequently failed owing to contrary perceptions of forestry officials and the local people; policing is costly and ineffective. In Senegal, rural communities have been given back responsibility for managing natural woodland in their areas (Jackson et al., 1983: 53–4). Such an approach is consistent both with a concept of community tenure and an objective of strengthening local autonomy in resource management.

The relations between farming and herding constitute a second policy area relevant to the intensification of land use. The extent of mixing in the past has generally been underestimated because of the popular stereotypes of the Fulani as pastoralists and of other ethnic groups as farmers. Livestock are owned by persons at all points along a continuum from specialised animal husbandry to specialised farming (Fricke, 1979). But mixed enterprises may represent a step downwards for pastoralists, forced to take up farming because their herds are insufficient to support their families, whereas successful farmers invest their surplus in livestock (McCown et al., 1979).

Conventional wisdom has advocated a closer integration of livestock production with farming, as attempted by the mixed farming ideal in Nigeria (Van Raay, 1975: 165); and such an agropastoral economy (Gallais, 1972) would allow livestock production to be intensified by means, for example, of rain-fed or irrigated fodder-crop production (Club du Sahel, 1977). On the other hand, Delgado (1979) concluded from a study of a mixed farming and livestock-producing system in Burkina Faso that more, not less, specialisation is in the best interests of both the Fulani and the Mossi farmers. This conclusion was based on the conflicting labour demands of herd management and farming during the rainy season: 'the effort required to maintain the animals severely limits the ability of the herdsman to grow crops' (*ibid*.: 125). Resolution of this question is clearly important for the issues of land tenure just discussed, and for the future allocation of land use in the semi-arid zone.

The intensification of farming systems calls for increased supplies of manure, unless fertilisers are purchased. If contractual manuring arrangements are made with visiting herds, the additional livestock need not be owned by the farmers;

but such arrangements may now be on the decline, at least in Nigeria. Alternatively cattle, if owned by farmers, must be managed by mobile pastoralists during the greater part of the year. Cattle ownership, however, has been priced out of the reach of most farmers. Small livestock, therefore, must be a major component in an intensified farming system. They allow a shorter recovery cycle after drought – Dahl and Hjort (1976) estimated the annual growth rate of sheep and goats (in stock units) to be 11 per cent, compared with 3.4 per cent for cattle – and Wilson (1983) computed an index of animal production in Mali on which goats scored 565, sheep 588, cattle only 173 and camels 125.

Farm forestry (as exemplified in the Kano Close-Settled Zone) forms, with farming and livestock, a third pillar of intensification. Some trees (especially *Acacia albida* with its valuable pods) provide fodder for livestock, and contribute nitrogen to the soil (Dancette and Poulain, 1969; MAB, 1977; Van Voorthuizen, 1978). They provide construction timber, fuel, and a range of economic products of commercial and subsistence value. They represent a form of savings or insurance, as pointed out above. Farm forestry was neglected until recently by government forestry departments preoccupied with establishing plantations of fast-growing exotics. According to a recent report, in the francophone Sahelian countries these have performed below expectations, while costs have soared, so that it appears to be impracticable in the foreseeable future to meet the demand for fuelwood and other forest products (Jackson *et al.*, 1983:8). But while advocating improved management for natural forests, the report makes no reference to the productivity of indigenous species on farmland.[16] Official statements now advocate farm forestry as a strategy for controlling desertification (Catterson, *et al.*, 1987; cf. Von Maydell, 1977; but see Mann, 1978).

Farmers, as the Kano Close-Settled Zone shows, know the value of trees, and plant and protect them on permanent farmland under individual tenure. Tree nurseries and extension services are cheaper than planting shelter belts on land that has been the subject of compensation. A moderately dense population of farm trees appears to be an equally effective way of preventing soil erosion. But a reorientation towards the propagation of indigenous economic species (Verinumbe, 1987) must be accomplished within forestry departments. Some of the indigenous economic species, in particular *Balanites aegyptiaca*, are very hardy in dry conditions. Over-emphasis on economically neutral or even harmful exotics (such as *Eucalyptus* spp.) has probably restricted the impact of government forestry services in the semi-arid zone, notwithstanding the popularity of exotic shade trees in settlements. The compelling arguments in favour of farm forestry, at least in the semi-arid zone, are, firstly, the absorption of almost all the labour costs by the beneficiaries; and secondly the elimination of the need for public acquisition of scarce land.[17] It represents a realistic alternative to plantation forestry.

It is not my purpose to discuss the need for ecological monitoring in the semi-arid zone, which I take as axiomatic under conditions of intensification. A strong case for monitoring, especially of rangeland conditions, has been made (e.g. Rapp, 1976; Watson and Hemming, 1983), and the use of earth satellite data for such purposes both tested (Hellden, 1978) and reviewed (Zonneveld, 1978; Hock, 1984). In conjunction with ground radiometer measurements, Landsat data have been used to monitor albedo in Botswana (Ringrose and Matheson, 1986). The possibility of combining Landsat data with air photography in order to measure sequential change has also been confirmed (Mainguet *et al.*, 1976 FAO/UNDP, 1978), but whether remote sensing techniques can be extended from ecological monitoring to socio-economic parameters such as carrying capacity is much more doubtful (Reining, 1980).

CONCLUSION

Most scholars, whatever their ideological persuasion, will agree that colonialism reduced the autonomy of West African rural communities, and that national independence has done little to restore their control over economic forces affecting their land and livelihood. At this point, opinion divides. Calls for measures to strengthen autonomy, which are not confined to radical scholars, are as common as expert advocacy of greater control, 'education', and discipline. The technology for controlling desertification is available (Garduno, 1977; Mageed, 1986); Dregne (1983: 179) writes: 'Solutions to desertification problems in Africa are known and – in general – can be implemented readily if resources are available to do so.' Similarly, famine (or food crisis) management can be represented as a challenge well capable of being overcome given management techniques based on applied social science (Currey, 1984). But technology- and management-based solutions are essentially 'top-down' concepts and the record of such approaches in the Third World is not impressive. The priorities indicated in this chapter – insurance, diversification, and intensification – are consistent with autonomy rather than dependency. They suggest a redirection away from *projects* towards *policies*, from administrative *direction* towards *enabling*.

A strong scent of determinism pervades the literature about arid and semi-arid West Africa, in particular with respect to the linked problems of drought, famine and desertification, and predicted outcomes for the societies who inhabit the area tend to be pessimistic. What is interesting is that pessimistic determinism – if I may call it that – seems to lurk in studies representing a wide range of disciplines: climatology, ecology, economics, demography, economic history.[18] One may speculate about its source. Confidence in projections as a guide to the future tends to be widespread in science in the later twentieth century, whereas prediction is not a major objective of ethno-science. West Africans have survived droughts, famines and ecological degradation before by calling upon indigenous adaptive ingenuity. Indigenous knowledge therefore possesses survival value. Response has also been accomplished to changing opportunities and stresses as economies and societies undergo rapid transformation. The question is whether we are equipped to deny such a possibility in the future. Have conditions changed irrevocably? Are the forces of the modern world irresistible? Or are the pessimism and the determinism artifacts of an unequal cultural confrontation?

The empirical studies recounted in this book encourage, on the contrary, some hope. An adaptive framework is intrinsically optimistic in that it focusses on human capabilities. Such capabilities have more than academic interest. They provide a basis for evaluating policy directions more in line with peoples' basic needs than coercive measures ever can be. Coercion – whose consequences are difficult to anticipate – is beyond the resources of many governments. As these resources decline or stagnate, it makes better sense to rediscover and strengthen indigenous adaptive capability.

Arid and semi-arid environments call for special adaptive skills on the part of those who live there. The twentieth century has unwittingly transformed the socio-economic milieu of the inhabitants of these environments. The entitlement of such marginalised peoples to a fair share of resources is conceded as a matter of principle in debates about development. Whether such a principle is incorporated into policy making at the national level, where these peoples tend to form political minorities, is another matter. But the consequence of neglect will be the intensification of regional inequality, of peripheral dependency, and of periodic food crises. Therefore, ways must be found of supporting productive communities in such high-risk environments. This is more than a local problem, and calls for the commitment of national and international resources, going beyond the mere funding of localised projects to broad policies designed to support community resilience, and further the intensification of systems of production. The solutions of the past are not good enough. Slavery, pauperisation and emigration are bastard sons of drought.

NOTES

1 INTRODUCTION

1 In a market so well stocked with competing prescriptions, purchasers – from governments to students – may find a choice bewildering, especially when the ideological and power-bloc implications of policies are taken into account. Such perplexity is made worse by the growing scarcity of recent literature. A form of dependency rarely acknowledged, the 'book famine' raises the question: for whose benefit is the development debate in Western universities being conducted? There is risk of incongruence between the debates within African countries and without.
2 Both the terms 'fragile' and 'vulnerable' suffer from imprecision, and are never defined. The discussion of the concept of resilience will be taken up in chapter 8.
3 Dams and irrigation works were advertised by politicians and soldier–administrators as answers to drought. When they fell victim to water shortage, everyone was surprised. The South Chad Irrigation Project in Borno, Nigeria, stood high and dry in 1984 because Lake Chad could not be relied on to provide the quantities of water expected (Kolawole, 1987).
4 A battle which was won, conclusively, by the bush burners (in Nigeria). However, the issue is not dead. The governments of Gongola and Kaduna states, possibly unaware of history, banned bush burning in 1985 in an attempt to contain desertification, and raised the stakes by threatening imprisonment (*The New Nigerian*, 29 Nov. 1985).
5 In a letter to the Assistant Commissioner at Gaborone, Chief Lentswe of Mochudi wrote; 'For your information I would state that I am unaware of "any wholesale or indiscriminate cutting down and destruction of trees" in my portion of the Protectorate.' The anonymous comment on this letter, which is on display in the Mochudi Museum, states that both men knew that many trees had been cut down and sold for pit props in the South African mines.
6 This is considered to be the greatest acceptable degree of risk, but farming was going on north of this line in the 1960s. The WMO possibilities are based on rainfall *before* the seventies. Work published in 1982 estimated that at a station near Niamey (well south of the limit), serious yield reduction in 'traditional' millet cultivation can be

expected in two years out of ten (Agnew, 1982).
7 An alternative regionalisation using the Budyko ratio gives an aridity index of > 1.5 for an area approximately corresponding to the semi-arid zone (UNEP, 1977b). A different division between arid and semi-arid is provided by MAB (1979). These regionalisations have greater practical value than older attempts to define a savanna climate (Swami, 1973).
8 It should be noted that these were calculated on the basis of rainfall *before* the seventies. (See note 6 above.)
9 Hubert argued that the fixed dunes in the Sahel, previously understood to mean that the desert had retreated northwards, were still subject to sand movement (ripple marks) and therefore were *forming* under vegetation.
10 The rainfall data available at the time were extremely scanty, and Hubert's interpretation controversial (Chudeau, quoted in Renner, 1926).
11 He misplaced the 'outer borders' of the Sahara at a number of locations, including the sand bluffs of the Manga Grasslands north of Geidam – which puts the whole study area of chapter 7 into the Sahara.
12 Failure to provide the basis of such estimates for critical evaluation had filled the desertification literature with unsupported statements.
13 Apart from ignoring grazing and other uses of land, this definition seems inoperable in a context of subsistence production.
14 The estimates were based on standard definitions of the terms 'moderate', 'severe', and 'very severe' provided by the organisers of the questionnaire survey. There is scope for considerable variability of interpretation, as well as accuracy, in such subjective evaluations. On the other hand, little reliance can be placed on statistical data such as those cited without source by Dregne (1983: 183–4) purporting to show that of 5.5 m ha of rain-fed cropland in the 'arid lands' of Nigeria, 5.2 m ha are affected by desertification; and of 30m ha of rangeland, 28m ha are so affected. Continued reliance on such 'guestimates' ten years after the establishment of the Desertification Branch of UNEP demonstrates the failure of UNEP to fulfil its monitoring responsibility, a charge made by the Independent Commission on International Humanitarian Issues (Timberlake, 1986: 92).
15 For example, the statement: 'trends in the price of charcoal, which can be ascertained in markets, is [*sic*] an excellent indicator of scarcity of firewood, pressure on firewood reserves, deforestation, and thus desertification. Because charcoal is primarily a market commodity and firewood primarily a subsistence commodity, charcoal price is more reliable as an indicator' (Reining, 1978: 63).
16 Nevertheless, the computer analysis and results of the survey were not completed and presented to government until July 1975. Research based in universities is vulnerable to conflicting priorities, however urgent in social terms.
17 Acceptable translation of this term has proved elusive (chapter 3).

Notes to pages 20–34

18 A major reason for this is the straightjacket imposed by the requirements of the Ph.D. system in foreign universities. Follow-up studies, which might verify the conclusions of such analyses, have rarely been considered: a Ph.D. candidate must break new ground!
19 The complexity of individual transactions, transfers and sharing is such that an attempt to do so would be pointless in a cross-sectoral study. Rather the family is viewed as a nexus of economic (and other) decisions relating to the welfare of its members.
20 It may be questioned whether some of the funds spent on high-cost research in the rural sector might have been more effectively distributed amongst a larger number of smaller projects. (The direct costs of the research reported in this book from 1973 to 1986 were less than ₦5,000 – then about S£4,000.)

2 FROM FEAST TO FAMINE?

1 *The African Mail*, 6 Mar. 1914, quoted in J.S. Hogendorn (1978:113). Excepting here, all references to tons are to metric tons.
2 A crop of 3 tons/ha of dry matter fixes about 150 kg of nitrogen (Kowal and Kassam, 1978: 133–6).
3 Data for 1957–58 indicate an increase in the cropped area of guinea corn, millet, cowpeas and groundnuts of 59, 48, 48 and 120 per cent (respectively), and in production of 67, 22, 25 and 70 per cent. These estimates were for one season only and were prone to under-recording (Federal Office of Statistics, n.d.).
4 Contrary to alarming predictions, unsupported by evidence, which were issued from time to time (for example Trevallion, 1966:103).
5 In Niger and other francophone countries, following association with the EEC in 1965, French guarantees were removed and the groundnut price fell.
6 The studies were located in the Malumfashi area of Kaduna State, where the data were collected from a population of 42,493. The results are broadly consistent with those obtained in the Garki district in Kano State (a WHO Malaria Control Project) and by Faulkingham and Thorbahn (1975) in a village near Madawa in Niger.
7 The law was expressed in the *Land Tenure Law* (1962) of Northern Nigeria, itself based on colonial enactments, and in turn followed (but not fundamentally altered) by the *Land Use Act* of 1978. See, amongst others: Meek (1946), Rowling (1952), McDowell (1964), Starns (1974), Ega (1979); in a broader Nigerian context, Famoriyo (1979).
8 Such developments might be for additional private building plots on city perimeters; schools and hospitals; industrial projects; government farms; dams; or redistribution of landholdings prior to irrigation development.
9 The British found it convenient to invoke this theory in 1903, following the precedent set by the founders of the Sokoto Caliphate, who claimed the right by conquest in 1803–4.

10 The colourful and oft-quoted account of Hastings (1925: 111; see Grove, 1973: 135; Hill, 1972: 285; Watts, 1983a: 288) was validated by an elderly eye witness in Danbatta District in 1973, who could remember seeing refugees dying on the road in their flight southwards; the villagers had no food to give them.
11 Ungogo District, for example, with over 70,000 people, was allocated 100 tons of rice and 100 tons of sorghum and millet, approximately one bag for 35 people, in four months.
12 In Niger, in the nine months following the harvest of 1972, 110,000 tons of groundnuts were exported while the government appealed for food aid of 100,000 tons of cereals (Derrienic, 1976: 37).

3 DROUGHT IN THE 1970s

1 See, amongst many: Dalby and Harrison Church, 1973; Bouquet, 1974; Sheets and Morris, 1974; Pitte, 1975; Caldwell, 1975; Copans 1975; Glantz, 1976a; Derrick, 1977; Dalby, Harrison Church and Bezzaz, 1977.
2 Such a conceptualisation has limitations. The pattern of distribution in a single year is unlikely to conform in detail with that arrived at after averaging many years' rainfall. But the latitudinal framework is a necessary starting point for an analysis of West African rainfall; see chapter 6.
3 The definition of the end of the growing season (see note c under table 3.1) was relaxed by one ten-day period.
4 In a rare record of such variations, three stations at Kano situated 15, 13 and 8 km apart recorded monthly totals in May–September 1985 differing by 0–83 per cent ($\bar{x}$ = per cent/month).
5 Well deepening is a normal maintenance operation and would always be attempted before a well was abandoned.
6 The interviews were carried out in the Hausa language by four interviewers working under the author's supervision. Shoe-string finance and restricted transport were added to limitations of time.
7 Sample surveys in some other states, carried out by students of the University of Ibadan from July to September 1974, have been reported by Oguntoyinbo and Richards (1977, 1978).
8 Some comment should be made on the trustworthiness of these estimates, since village heads had nothing to lose by exaggerating the extent of losses and may even have anticipated more help from the State Drought Relief Committee if they thought the information would be used by government. Against the possibility of exaggeration three points may be made: (1) village heads and district authorities were always informed that the information was being gathered on behalf of the university and not on behalf of either local government or drought relief administrations; (2) analysis of the data has not produced evidence of falsification; (3) the concentration of the most serious crop failures in northern districts, and the lower estimates for early millet (the hardier crop), conform with expectations.

9 The *takardar jumla* or tax assessment sheets maintained at district level contained an impressive amount of statistical information in respect of each village area, including demographic, livestock, agricultural and craft production data. But inspection usually revealed that livestock figures in particular were guesses at best, and might not change for several consecutive years.

10 In some areas of Kano State, where markets and road access were available, starving cattle were sold before they died to middlemen who went to the villages and purchased them, then transported them on the hoof (when able to walk) or in handcarts and trucks, eventually to reach Kano for slaughter. At Malamaduri dying cattle were leaving for Kano at the rate of 400–500 per week, worth a mere ₦1,600–2,000 (normally, ₦16,000–20,000), in mid-July 1973. The trade had been going on thus for four months.

11 Some village heads administered up to thirty small hamlets or wards, and comprehensive reporting, particularly for small livestock, seems improbable. And they were not always on confidential terms with nomads in transit.

12 To minimise the effects of error in the reports, Van Apeldoorn summarised the data in classes (₦50–99, 100–149, 150–199, 200 or more per ton). The midpoints of these classes have been used for calculating the mean (Van Apeldoorn, 1978a; vol. 1, 144–6).

13 A divergence between prices in major markets and in outlying villages, which formed a majority in the survey, seems quite probable. The internal grain trade, whose improved performance in 1974 may explain the slightly lower peak of that year (according to the CWRs), would have less impact in such places. Relief grain was being released in substantial quantities in some areas, and would have depressed market prices somewhat. But the higher peak in 1974 that is suggested by the village heads' reports is consistent with the progression of hardship known to follow the catastrophe of 1973, the social broadening of impoverishment and inflation in the economy.

14 Cf. Watts (1983a: 382–3), who gives price data for Katsina, and figure 4.4. Price data in the crop and weather reports is so variable spatially (as Watts rightly emphasises) – and incomplete – that the choice of stations has much influence on the pattern. The inclusion of southern Kano State, as well as Metropolitan Kano, in the series charted in figure 4.4 accounts for some differences with figure 3.5.

15 Using different grain prices, Watts (1983a: 387) obtains a different conclusion.

16 No claim is made for great accuracy or depth of this data; but all subsequent checks have tended to confirm its reliability.

17 Such equivocation is defended on the ground that the object of the survey was not to establish production relations, but to determine patterns of decision making in response to drought. In practice, large families were rare in these isolated communities; residential, kinship, farm-producing and food-consuming units overlapped to a high degree.

18 In 1974, planting took place in all five villages after heavy rain early in May, but this was followed by 5–7 weeks without rain. Seedlings withered and died extensively. Eighty-seven per cent said they had no seed for a further planting. Yet plant they did, and 1974 brought a fair harvest.
19 The number of fields – a surrogate for farm size given the impossibility of measuring all holdings – correlated significantly with output of millet, guinea corn and groundnuts.
20 On the one hand an informant might be expected to understate his yields in relation to expectations if he thought it might influence relief distribution; on the other, if better off than average, he might fear the information would leak to other villagers in greater need. However, every effort was made to dissociate the enquiry from government, and the desperate plight of almost everybody militated against guile.
21 That is, assuming that the 1972 harvest produced a third of a year's requirements and that of 1973 less than a tenth (table 3.5) on average.
22 The interviews were carried out before the villages had benefited from relief food distributed by the Kano State Drought Relief Committee. This grain arrived in time to support most families relatively adequately during the ensuing farming season. It should be remembered that a description of nutritional status at the family level does not take account of variations between members when, according to custom, the family head eats first, and women or some children last.
23 As shown by Norman et al. (1982: 144–6) in three villages near Zaria, one of which had a large Fulani element.
24 Of the large number of adaptations available in theory (e.g. new crop varieties, irrigation, new tools, new cultivation practices), few were actually available owing to limitations on the supply of information, inputs and capital.
25 Seed for the grain crops was scarce.

Principal sources of seed for grain plantings, 1973 and 1974

	Five villages: farmers (%) 1974	Two districts: villages (%) 1973	1974
Reserves	17	27	7
Market	35	32	26
Gift	{35	38	55
Relief		3	11

26 The spreading tentacles of Kano's fuelwood trade have reached lightly populated woodlands 300 km away.

27 Variability in choice of strategy was not confined to individuals and families, but was also discernible between villages. The sale of animals was most common where the incidence of animal ownership was highest. Weaving was mentioned in only one village, where the craft was established. *Kaba* work was mentioned by over 80 per cent in one village, but not once in the weavers' village. Such patterns conform with the known variability of rural northern Nigeria.

28 It is possible that such a correlation was due to variable frankness, whereby a village head who admitted one was as likely to admit the others.

29 According to the *New Nigerian*, a farmer in Dikwa, his three wives and twelve children were living entirely on these termite granaries. The farm lay on the *firki* clay soils where late-maturing guinea corn (*Sorghum dura*) normally remains in the fields until February. The attacks on the underground termitaria could be interpreted as the recovery of stolen property, or alternatively as an approach to arthropod domestication! (See note 32.)

30 Like the nicknames given to personalities, they may be used without thought or even knowledge of their original meaning.

31 The balance was made up by answers which attempted to distinguish between the preceding famines. Most informants could not attempt such a distinction.

32 Among the optimistic minor reasons were the lack of necessity to raid termitaria (but see note 29) and to grind up one's calabashes to eat.

33 A full account of government relief has been given by Van Apeldoorn (1978a: vol. 2, 53–90).

34 Not 100 per cent Muslim, however, as sometimes claimed. Scattered groups of non-Muslim Hausa (Maguzawa) live in the study area, as well as some Muslim communities considered by the orthodox to be heretical.

35 Including a commentator in the *New Nigerian* (15 Aug. 1973) who alleged that in one town adopted as a refuge by the ostracised women, it immediately rained.

36 Bima Hill, north of Gombe, has Mahdist associations.

37 These quotations have been loosely translated from Hausa to English by the interviewers. The question was also met with agnosticism: 'No one can tell! It is left for anyone to give his suggestion. It certainly stands wholly on one's thought. If one thinks it is due to donkeys, it is so', and outright rejection: 'I don't like this type of question and anyone who insists on asking [it] will be out with me! More especially someone I don't know!' It is obvious that interviews of this type are far from ideal for approaching interpretative issues.

38 One reason given was 'Allah mun tuba' (God, we repent); another, 'because God would not like us to suffer from more hardship'. However, it seems that the omnipotence rather than the mercy of God was dominant in people's minds.

4 THIRTEEN YEARS IN THE LIFE OF A VILLAGE

1 Only 18 household compounds were enumerated in detail (12 per cent).
2 Only eight families were enumerated in detail (15 per cent).
3 Census data in Nigeria are unreliable owing to under-counting, or inflation for political purposes, and the last published returns were for the census of 1963 (Mortimore, 1984). Local administration tax lists and cattle tax records are known to undercount, while cattle vaccination data include migrant as well as resident herds. Air photos can yield quite accurate measurements of the spatial growth of settlements, but coefficients for estimating population have not been worked out.
4 The account which follows is based almost entirely on information given by Manga; it should not all be assumed to apply to the Hausa.
5 The measure of expected yields is very approximate since it may not take account of changes in farm or family size.
6 The sample population in Dagaceri had more adults than the Zaria population and the mean annual temperature is 3 °C higher. These differences may be self-cancelling. Dayi is cooler than Dagaceri. Both areas are wealthier, and although dietary patterns are similar, Dagaceri may be more dependent on grain.
7 At 343 cal/gm of guinea-corn flour and 387 of millet (Simmons, 1976:114).
8 A loss of 20 per cent in weight occurs in grinding millet into flour, and of 37.5 per cent in threshing. For guinea corn, which was not measured in Dagaceri, the loss of weight in grinding is assumed to be 23 per cent (Smith, 1955: 237) and in threshing 30 per cent (Smith, 1955: 237, 239; Hays, 1975). Smith measured a 42 per cent loss in the weight of millet on threshing at Zaria.
9 Special Sahelian Office (1973). Faulkingham and Thorbahn (1975) surprisingly describe as a 'bumper crop' in southern Niger a yield of 123.5 kg per head, in 1969, when, according to them, rainfall was ideal.
10 The timing and duration of these visits were dictated by the author's institutional commitments. The visits took place at the following times: 1: May 1974; 2: July 1975; 3: October 1975; 4: March 1976; 5: May 1976; 6: December 1976; 7: June 1978; 8: July 1978; 9: July 1979; 10: July 1980; 11: July 1981; 12: June 1982; 13: July 1983; 14: February 1985; 15: January 1986; 16: October 1986.
11 The prices are collected by the local staff of the ministry in terms of market measures (the *al-husseini* and *tiya*) which are multiplied up to prices per ton. Error factors and gaps in the series are minimised by using the state average. The data are usually treated with caution by researchers (e.g. Van Apeldoorn, 1978a: vol. 1, 117, 134).
12 Called *karambo*, this insect was not identified.
13 The regional index (table 4.3) conceals remarkable divergences

Notes to pages 95–100

between the western stations, which received 52, 68, 65 and 87 per cent of the mean (Kano) and 31, 79, 43 and 44 per cent (Zinder) for the four months, June–September, and the eastern stations, which received 193, 109, 32 and 91 per cent (Nguru), and 128, 157, 95 and 37 (Maine). Dagaceri had the worst of both patterns: a bad start and an early finish.

14 The annual rainfall records for six stations in southern Niger (Birnin Konni, Tahoua, Maradi, Zinder, Maine Soroa and Nguigmi) give a composite index of 100 per cent of normal in 1980, falling to 71, 61, 54 and 50 in the four succeeding years to 1984. Errors, however, in respect of individual months have passed uncorrected into international records (*US Monthly Climatic Data*). The same stations recorded rainfall in August 1984, on average 32 per cent of normal.

15 That this adverse outcome was local in extent is suggested by good harvest reports from villages 50–100 km east of Dagaceri.

16 Identification by Martin Fisher, Department of Biological Sciences, Bayero University, Kano. The reason for the huge multiplication and southward migration of *Jaculus jaculus* (probable subspecies: *favonicus*) is not known, but it may be speculated that it was due to the abundant growth of grass in the previous season. The population dynamics of animals specialising in granivory in desert ecosystems are closely related to the supply of seed (Brown, Reichman and Davidson, 1979).

17 Identification by J. Ayertey, Department of Biological Sciences, Bayero University, Kano. Other species found were *O. nigeriensis* and *Kraussaria angulifera*. J. Davies (personal communication) reports that a similar outbreak of *O. senegalensis* took the inhabitants of Gumel by surprise in the harvest season of 1974. See Hergert (1975); Cheke, Fishpool and Forrest (1980); Fishpool and Popov (1981).

18 Analysis of the 1957 and 1964 classes at Birniwa Primary School (opened in 1938) showed that 60 per cent of the students had left the area by 1978, either by proceeding further in the educational system, or to seek employment (K.M. Brown, in Mortimore, 1976: 42).

19 I hope that the inadequacy of this account demonstrates the need for longitudinal monitoring in which economic data can be correlated with rainfall and prices collected *in the village*.

20 Termed the Birniwa Association. Pullan (1962) recorded 10 per cent clay and silt, 0.16 per cent carbon, an infiltration rate of 2.7 and a coarse sand: fine sand ratio of 1:3.2 in the top 24 inches of the profile.

21 Latrine manure is not collected because there are no latrines. It is, however, applied directly, with no charge for transport.

22 This rate is lower than in the Kano Close-Settled Zone where applications of 3.7 tons/ha were reported in the sixties (Mortimore, 1967).

23 Until a supply was established at Birniwa in 1982, chemical fertilisers were not used.

24 Abalu *et al.* (1983: 49) suggest that damage to seedlings may occur within four weeks of sowing, and emphasise the need to know the

maximum delay that is permissible before negative effects on yields may be expected.
25 As far as the Manga are concerned Abalu et al. (1983: 40) are incorrect in claiming that only old women and unmarried girls are employed on the farm, and then only for planting and harvesting.
26 A producing family may differ from a co-resident family in that married sons living separately may combine with their father (or, after their father's death, with each other).
27 Abalu et al. (1983: 39) quote an erroneous claim that average farm size in the area is less than a hectare. See note a, table 4.7.
28 The sale of land was introduced recently by the Hausa. Manga practice did not formerly include sale or pledging. Near towns such as Birniwa renting (H *haya*) of land by strangers may occur.
29 Abalu et al. (1983: 35–7) observed or reported the following pests on millet: headworm (*Raghuva* spp.), stem borer (*Acigona ignefusalis*), black ants (*Messor galla*), grasshoppers; on cowpea: green leafhoppers (*Empoasoxa* spp.); and the following diseases: *Striga nermontheca* on millet and *S. gesneroides* on cowpea; downy mildew, smut and ergot on millet; *Septovia* leaf spot on cowpea; on guinea corn, smut, oval leaf spot and anthracnose.
30 A typical reply to the question: 'Did farmers grow enough food this year?' is: 'They did if they planted on a sufficient scale.'
31 The attitude of the Manga to shifting cultivation, as opposed to rotational bush fallowing, is informed by both custom and necessity. The movement of whole communities to fresh land was common in the past, but in Dagaceri, it is now averred, 'In Allah bai ba ka ba, ba za ka samu ba': 'if God doesn't bless you with it, you won't find it [by moving]'. In any case, new land is scarce. But farther east (see chapter 7), the movement of villages or groups of families to new sites, or to join other villages, still occurs.
32 The name *'dan arba'in* is not however new to northern Nigeria. It was used for early-maturing varieties more than fifty years ago (Bargery, 1934).
33 There are several varieties. The possible significance of *guna* as a replacement for the groundnut in the village economy is supported by an interesting parallel in the marketing arrangements. Davies (1979a) reports that urban traders in 1974 visited farmers in Gumel Emirate to promote its production – as the groundnut had been promoted in Kano sixty years earlier (Hogendorn, 1978: 86–8). The railway continued to play a part: the areas visited all had railway stations.
34 These were known by the following names (in order of maturation): 'Yar Dangwanki, 'Yar Washa, 'Yar Kumbale, Makaho da Wayo, and 'Yar Labe. The first two matured with the early millet. Palatability, lodging and other characteristics were variable (J. Davies, personal communication).
35 One 'early adopter' said that he learnt of its use on overland pilgrimage through the Sudan, and had been using it since 1965.
36 It should be noted that livestock prices vary by 100 per cent or more

according to sex, age and condition. Generalisation should be treated with caution.
37 The production and consumption of fuelwood in the Kano region are the subject of a forthcoming report of the Rural Energy Research Project, Department of Geography, Bayero University (United Nations University).
38 The Dagace was withholding allocating fallow land to newly arrived, destitute farming families in 1985 for fear that clearance and burning would further reduce available grazing. It was reported that even among the resident Manga those without access to fodder were having to sell their animals.
39 Other opportunities in the commercial sector require smaller investments but offer a low turnover (butchering, leatherworking, retailing).
40 Reserved seed is selected from the previous harvest. Purchased seed, also selected, costs more than food grain.
41 Labour is recruited by personal invitation. M's labour force fluctuates from six to twenty persons at a time.
42 A was interviewed before the harvest failure of 1986. Others in his position had only a month's food assured in October of that year. Nevertheless, A's circumstances do not suggest a downward cycle of impoverishment, dependency and exploitation as proposed by Wisner (1977a) on the basis of a family case study in Meru district, Kenya.
43 The larger of two resident grain traders estimated that he sold (outside) 6 tons of grain bought in the village during the dry season of 1985–86, admitting that the poorest often sold at glut prices.
44 In 1976–77, an attempt had been made to open a weekly market at Dagaceri, which might have intensified its integration into the money economy, but the new market was set up at a neighbouring village, Kukangiwa, instead. It was not successful.

5 WIDER HORIZONS

1 Such a general practice need not be confused with the *zawiyya* or lodging house maintained by an Islamic brotherhood for its members (Paden, 1975: 141–3).
2 The survey was constrained by the prior claims of the rural survey which was being carried out at the same time. It consisted of unsupervised questionnaire interviews conducted in Hausa with individual adults on a search-and-interview basis.
3 Ninety-four per cent had travelled by road, reflecting the new ubiquity of the minibus in rural areas.
4 The Hausa term (singular, *Buzu*) for Tuareg of mixed descent, or for any persons wearing Tuareg clothing.
5 Many Buzaye interviewed were primarily farmers. But pastoralists who had lost all their livestock were commonly encountered in Kano.
6 Buzaye did not settle permanently. On securing employment, a guard

(H *gadi*) would arrange for his own replacement before departing to attend to his interests at home. Whole families lived on the streets.
7 Including begging, which, as a recognised occupation, provides the believer with an opportunity for self-improvement (Mensah, 1977).
8 Evidence from interviews obtained in 1985 is supported by rainfall data; total rainfall in 1973, as a percentage for the 1931–60 mean, was as follows: Tahoua: 62, Maradi: 55, Zinder: 55 (in Niger); Kano: 48, Nguru: 46 (in Nigeria).
9 However, 8 per cent reported uncompleted journeys, and said that the men had not returned for the usual planting season. They expected them when the rains improved, or even after the harvest.
10 These patterns were little different from those reported before Kakaduma.
11 Including, in times of drought, prayers for rain.
12 These are not shown in figure 5.7.
13 One informant worked for four seasons, 1972–75, doing harvest piece work at ₦14 per portion taking 20 man-days.
14 'One who lodges Barebari traders' (Bargery, 1934): i.e., Kanuri or Manga.
15 Deaths 'from rain' (exposure) in three consignments in 1979–80 were one out of 20, one out of 22 and five out of 15. Charges for transport, which are paid through the *fatoma* at the destination, are waived if an animal arrives dead.
16 One trader, regretting neglecting his farms in 1979, admitted that 'Arziki yana cikin kasa' (good fortune is in the soil).
17 This *fatoma* would rent a house for a year in any part of greater Lagos, and accommodate lodgers who, when known, could obtain loans. Later, K used a Hausa *maigida*.
18 A fugitive occupation, for cutting trees is forbidden in the vicinity of Dagaceri.
19 Throughout the period in question (April 1984–March 1986), Nigeria's borders were closed to her ECOWAS neighbours.

6 TWO DRY DECADES

1 This may be compared with 48 per cent in 1973. It is difficult to account for the claim that 'in contrast to the very dry conditions of 1968 and 1973, the Sahelian zone of Nigeria has been enjoying satisfactory rainfall in recent years' (World Climate Programme, 1983: 6). In the years 1980–84 Nguru received 60, 76, 72, 42 and 54 per cent of the 1931–60 mean. Nguru is closer still than Kano to the 'Sahelian Zone'.
2 Only twice before had August received less rainfall than both July and September, in years both falling close to normal (1951 and 1970).
3 The rainfall recorded at Kano Airport in 1985 was not representative of the area. Two stations situated within 20 km recorded 20–5 per cent more. But they confirmed the absence of the August peak.

4 Davy et al. (1976: 27) found 'slight indications that rainfalls in June and July are a little more reliable than those in October and September respectively'.
5 In any case the mean is distorted by skewed distributions and is usually higher than the median for rainy-season months in semi-arid West Africa (Davy et al., 1976: 25; Glantz and Katz, 1977).
6 Winstanley (1985) provides no details of the stations used, the geographical extent of his subSaharan and tropical zones, the methods used to check the instrumental records in order to produce 'homogenous data sets in which I can have confidence', the hydrological data (which are claimed to support his thesis) nor of the documentary sources consulted.
7 See also Palmer (1986). Bryson (1973, 1974) proposed the hypothesis that progressive cooling of Arctic relative to tropical sea surface temperatures would modify the temperature gradient from the Equator to the Pole and consequently the disposition of the Hadley circulation which governs precipitation in the tropics, thereby restricting the influence of monsoonal rainfall in West Africa.
8 Jasinski and Karnovitz (1985) claim that 'during the hot, dry African summer of 1984... satellite data were used to verify the seriousness of drought in Africa, providing the basis for an advisory to Agency for International Development missions in Africa that warned that Nigerian [sic: Nigérien?] farmers could lose as much as 50 per cent of their crops because of the drought'. It seems unlikely that the warning added to what was already known by farmers. Another proposal for forecasting that has been put forward is the use of geopotentials (Bunting et al., 1975; Elston and Dennett, 1977).
9 For example: 'at Niamey Ville, rainfall probabilities show that from Week 25 (18–24 June) to Week 37 (10–16 Sept.) there is in 8 out of 10 years a definite chance of receiving at least 5 mm per week' (Sivakumar et al., n.d.: 21).
10 Owen and Folland (1987) concluded that while the soil moisture mechanism amplifies rainfall differences in their model, sea surface temperatures are mainly responsible for initiating these differences. The albedo and soil moisture feedback theories are postulated on the assumption that deforestation and clearance by human or animal populations are responsible for the alteration of vegetation over large areas. Berkovsky (1984) calculated that 'albedo would have to be changed over a substantial region, approximately 250 km, in order to feel the effect of changed albedo'.
11 A second theory suggests that particles of decayed vegetation in the atmosphere provide condensation nucleii. This could conceivably be important in rain formations at the commencement of the rainy season, when high continental air mingles with humid monsoonal air along the ITD. Its effect on rainfall would be positive.
12 With regard to the southern hemisphere, Nicholson (1983) points out that statistically significant persistence of drought had not – till then – been observed in the Kalahari. See Lamb (1966).

13 A controversy occurred with the foresters who believed that forests conserve water (Barber and Dousse, 1964).
14 His further proposition – that the removal of the vegetation would increase ground and air temperatures, causing air currents which would divert the rain-bearing winds elsewhere – is less plausible! Prothero (1962) reviewed some evidence on well water supplies in north-western Nigeria, concluding that early colonial claims of deterioration were ill-founded.
15 The prevalence of the conception of groundwater as a free good to be raided at will is demonstrated by the Thousand Boreholes programme initiated by the Abubakar Rimi administration in Kano state in the eighties. Foreign consulting and contracting firms were told to find water in locations selected on political grounds and no attempt was made to estimate reserves, recharge or off-take.
16 Used in 1973; possibly not controlled for daily changes in pressure.
17 But the scale resolution of the 1979 mapping was inferior owing to a smaller number of observations.
18 An analysis of hydrological data on surface river flow cannot be attempted here, and the seasonal fluctuations in wells are rarely observed.

7 SHIFTING SANDS

1 The term Manga Grasslands was given to the area by the Anglo-French Forestry Commission (1937) and Collier and Dundas (1937). The name Manga is also given to an extensive area north-west of Lake Chad, now largely unsettled.
2 The northern portion of the Grasslands is covered by the excellent maps published by the Institut Géographique National (Paris) at the scale 1:200,000. These maps are detailed and reliable, incorporating a classification of vegetation types and a dense network of benchmarks. Sadly, the maps published at 1:250,000 for areas south of the border, by the Federal Surveys of Nigeria, are almost devoid of settlement, vegetation, or height data, and misplace the international border by up to 1.5 km.
3 How vestigial is the Yobe may be gauged from the fact that *before* the effects of the droughts of 1972 and 1973 were felt, the estimated annual average flow fell from $6,167 \times 10^6$ m^3 upstream to 1,850 at Gashua and, notwithstanding an addition from the Komadugu Gana, to 986.8 at Damasak and a mere 444 at Lake Chad. These huge losses are due to evaporation and seepage (Land Resources Division, 1972: vol. 1).
4 Also, the acquisition of these data in the quantities and at the dates required presents almost insuperable problems in Nigeria.
5 The value of the air photo archives of Nigeria is not sufficiently recognised by either federal or state governments. Irreplaceable boxes of prints lie forgotten and without inventory under thick layers of dust and leaking ceilings.

6 A third set which should have been taken in 1976–77 at the scale 1:25,000 stopped short of the study area for reasons known only to the contractor and the Director of Federal Surveys.

The air photography of 1950 and 1969 was consulted at the Directorate of Overseas Surveys, London, by permission of the Director of Federal Surveys, Lagos.

7 On each set of photographs, a primary sample was identified by overlaying a grid of squares on the non-overlapping portions of the photographs, from which a secondary sample of squares of 2.25 ha (in 1950) and 4 ha (in 1969) was selected systematically. The relevant details of the air photo sampling are given below.

Air photo sampling

		1950	1969
Scale (approx.)		1:30,000	1:40,000
Study area (km^2)		1,480	1,840
Primary sample:	km^2	638	1,392
	percentage of study area	43	76
Secondary sample:	size of blocks (ha)	2.25	4
	number of blocks	3,300	2,175
	km^2	74.2	87.0
	percentage of primary sample	11.6	6.2
	percentage of study area	5.0	4.7

The frequency pattern of land cover types is skewed in favour of the major classes, and several of the smaller cannot be considered to have been reliably measured with such a sparse sampling design. The area sampled falls below the coverage of 17 per cent recommended for detecting sequential land cover changes in Nigerian savanna conditions (Field and Collins, 1986).

8 This conclusion is based on field surveys conducted up to 1986, which indicate that the *kwari* banks have been stable since 1950 except for a few localities where moving dunes may be observed (cf. the case studies discussed below).

9 Defence of such a miscellaneous data base can be offered on the ground that the relative objectivity of air photographs, accuracy of 'ground truth' measurements, and depth or insight of interview data are mutually compensatory.

10 Such as Kumagunnam, Garunguna, Maimalari (figure 7.1).

11 Government nurseries set up in 1979 at Kaska and Bulatura, with funds from the Federal Arid Zone Afforestation Committee, were closed down five years later for want of funds.

12 Such trees must tolerate high salt concentration in the soil.

13 It has become commonplace to point out that no reliable data on livestock distribution are available in Nigeria.

14 According to IEMVPT (1979), the Grasslands are dominated by *Aristida mutabilis* and are classified as part of the Sahelo-Saharian Sector of the Sahelian Domain. This work does not distinguish between the treeless plains shown in figure 7.1 and the wooded steppe to the north (*Acacia raddiana, A. senegal*).

15 Side-looking airborne radar imagery has limited value in savanna conditions (see Parry and Trevitt, 1979), and the maps produced by the Federal Department of Forestry contributed little to what was already known of The Manga Grasslands.

16 The experts do not agree on yields. The Land Resources Division study (1972: vol. 4:81; *but* see map 10) gives 1,100–2,230 kg of dry matter/ha for the *Andropogon–Aristida* community, and less than 1,100 for *Aristida* and *Cenchrus*. Rippstein *et al.* estimate an average of 1,200 kg/ha for *Aristida/Cenchrus* grassland, which they say may surpass 2,000 (1972: 172–4). This is more in line with the favourable evaluation held by users in the Manga Grasslands.

17 Expert assessment is uniformly pessimistic. They have low moisture-holding capacity, low inherent fertility, a fair to poor capacity to benefit from fertilisers. They are subject to a moderate or severe erosion hazard, should never be used for mechanised cultivation and, even under hand cultivation, only perennial crops are recommended, along with stringent erosion control (Land Resources Division, 1972: map 12).

18 However, samples from 40 cm show that the clay/silt fraction rises to 16 per cent on a stable dune in the Grasslands, whereas at 50 km further west, the increase (from 11.4 at 15 cm to 12.6 at 40 cm) is very small. This demonstrates the sorting effect of recent aeolian action near to the surface in the Grasslands (data by courtesy H. Mensching, University of Hamburg).

19 So does the total area of the Manga villages, a proxy for population which remained unchanged at 270 ha. The Ful'be also cultivate, but their settlements are insufficiently distinct on air photographs to be reliably identified or measured.

20 The invisibility of the second type to the coarse-grained sampling method used in table 7.1 accounts for the appearance of stability in sand formations conveyed in that table.

21 Air photo interpretation detects ungrassed surfaces accurately, but may miss small shrubs considered, on ground traverses, to be significant evidence of stabilisation. Resident observers are most struck by the vertical development of dunes. Thus there is scope for error.

22 The western end of the transect was tied to a neem tree in the village market, and the eastern end to a datum established in 1982. There is an error factor of 1 m. The bearing of 70° is that of the dominant wind in the dry season.

23 See note 21. The surveys conducted in 1986 strictly delimited the area of mobile sand and excluded areas colonised by *Leptadenia pyrotechnica* which may appear unvegetated on a 1:40,000 photograph. In 1986 the

dune formations extended more than half a kilometre along the banks of the *kwari* to north and south, but only the dunes identified in 1969 were measured.
24 Farmers from Ngelsandi marketed their cassava (grown in the deep *kwari* farms) in Nigeria throughout the period (1984–86) when the border was officially closed.
25 It should be noted that while the dominant sand-moving winds blow from the north-east, during the short rainy season south-westerlies remodel the micro-topography of the dunes (as shown by the reverse slip-slopes behind the dune crests in figures 7.3 and 7.6A and B).
26 Indirectly, the prevailing system of shifting cultivation contributes to the cyclical changes in the settlement pattern which, as the case histories show, are closely associated with the growth and decline of the village perimeter dunes. The Manga here do not operate a bush fallowing system such as that observed in Dagaceri. The land available for rain-fed cultivation in the shallow *kwari* is not considered to be scarce, and when a family or group of families – for whatever reason, agricultural or other – decides to move, no attempt is made to retain tenurial rights to farm land. Occupation of vacant *kwari* land can be negotiated anywhere with the leaders of the community having territorial jurisdiction; and settlement may be accomplished either by joining such a village or by founding a new one. In the deep *kwari*, on the other hand, date palms are considered always to belong to their first owners; and rights to farm or collect natron are protected.
27 It is easy to conclude, when passing along a motor route, that there are fewer settlements than there really are.

8 INTERPRETATION

1 Bonte (1986: 156) is misled in thinking that the proponents of the desertification theory in the 1930s claimed that climate was its only cause, as he is in believing that it was 'initially elaborated' by 'English geographers'; see chapter 1, where a review of the literature indicates that the origins of the idea were as impeccably French as the word itself.
2 In view of the complexity of famines, and the mobility they may engender, it seems hopeless to search for demographic patterns of general applicability (Hugo, 1984).
3 Adaptive response has been the subject of several attempts to discern an inherent structure or temporal ordering of alternatives (e.g., Mortimore, 1973; Wood, 1976; Watts, 1983a: 436). But the variable circumstances of individuals within the community and of communities in different places are not consistent with the application of a general model.
4 Blaikie (1981) writes of a household 'access profile' – or assemblage of resources (land, labour, capital, livestock, implements, non-agricultural income, market access, etc.).

5 It was disrupted technologically in that the time-honoured balance between annually cultivated fields and fallow was no longer observed.
6 Complaints of the breakdown of social relations are as old as the recorded history of famine itself: 'Each man has become a thief to his neighbour' (inscription on the Nile dated 4247 BC, quoted from Graves by Keys *et al.* (1950: 5); Dando, 1980: 73–4).
7 Even in a region as remote as northern Darfur, increased mobility, trading and commercialisation of livestock and of millet, occurred in 1968–74 (Abdull-Jalil, 1980).
8 Though whether it was in any sense in an equilibrium state is more controversial (cf. Dresch, 1959, 1975; Swift, 1977a, b).
9 Drawing on East African experience, O'Keefe and Wisner (1975) likened the drought–famine process to a 'game' in which classes of antagonistic players – e.g. landless and rich peasants – were forced into competitive roles, old functional strategies were out-moded by, for example, changes in land tenure and the rules of the game altered as market forces penetrated everywhere.
10 The village was Rijiyar Tsamiya (Ahmed, 1976, reported in Van Apeldoorn, 1981: 57–61). This case became the basis of some sweeping arguments in Kates *et al.* (1981).
11 Richards (1986) has reported, from a very different ecological zone, an equally vigorous system of options for coping with the annual hungry season. On-farm adaptations are complemented by several alternative sources of income and by hunting and collecting; help may be obtained from kin, patrons, through loans or pledges. Famine is, of course, an extension in time and social impact of the common experience of the hungry season.
12 Watts (1983b) makes a number of criticisms of adaptation as a frame of reference. I cannot share his opinion that it necessarily assumes a 'neo-Darwinian' view of societies as biological populations, that it cannot handle societies in transition, and that it displaces the notion of mind from man to the ecosystem, depriving man of objectives. No such mechanistic transfer of the biological concept is intended here (see chapter 1).

As for Marxism, it is unexceptionable to state that 'labour is the active and effective relation between society and nature', but to restrict such relations to the material is to ignore the significance of a religious framework in influencing social behaviour, a topic I have most inadequately touched on in chapter 4.
13 Conceding the inadequacies of all general migration theory, Zelinsky (1979) upholds the need to search for one. Goldschieder (1979), however, explicitly rejects any single transition model of urbanisation and migration. Chapman and Prothero (1983) review the available explanatory models and conclude that there is need for a more integrated explanation.
14 These conclusions do not support the hypothesis advanced by Kates, Johnson and Haring (1977: 35–6) which links migration directly with desertification. According to them, migration is a

response to the long-term decrease in productivity associated with desertification... Initially, migration represents a short-term adjustment to a temporarily limited period of adverse conditions... only when drought continues for several consecutive years, often exposing underlying desertification or raising doubts in refugees' minds about future resource base productivity, do migrants plan to remain... or to seek alternative livelihood options... Longer-term adaptations generally involve the establishment of more or less permanent ties between two economic systems, one rural and one urban, and the periodic movement of labour between them... Yet the very success of urban migrants often undercuts the viability of the traditional sector... the productivity of the rural resource base decreases as the labour available is no longer sufficient to maintain yields. Migrants in this situation become both the cause and consequence of desertification... those left behind are increasingly marginalised and a debilitating retrogressive spiral can be instituted. What initially appeared to be a stable dual economy becomes a prescription for decline at the margins.

Firstly, no trend towards the substitution of circulation by permanent urban migration is discernible; secondly, a worsening shortage of farm labour has not yet been demonstrated; thirdly, no 'debilitating retrogressive spiral' is obviously noticeable. On the contrary, if labour scarcity removes land from cultivation, the ecosystem should be expected to benefit.

15 On the other hand, domestic animals are a part of the ecosystem, and Bernus (1977a: 81) found that fresh tree growth was most common where ruminants were available to ingest the seeds.

16 B. Peyre de Fabregues (personal communication), who reports that *C. biflorus* has replaced *Andropogon* spp. on the Ekrafane Ranch in Niger, under controlled grazing conditions.

17 No mention has been made of water erosion on cultivated land, although it does occur on finer soils and has even been observed on the slopes of fixed dunes in central Niger (Delwaulle, 1973; Talbot and Williams, 1978).

18 Declining yields of some major crops in Kordofan (Sudan) between 1961 and 1973 (DECARP, 1976) appeared to be inversely related to the extension of the cultivated area (Ibrahim, 1978); but in a multiple correlation analysis of yield data for the period 1952–80, 85 per cent of the variation was explained by two rainfall parameters (Olsson, 1983). Therefore, the claim that millet cultivation is a major agent of desertification in the Sudan (see Mensching and Ibrahim, 1977) should be treated with reserve.

19 'The degradation of rangelands is the most widespread form of desertification associated with animal-based livelihood systems, and, *even in the absence of climatic change*, has resulted in impoverishment of, and in physical and social hardship among, many dryland pastoral communities' (UNEP, 1977a: 20; italics added). Such a view has been considered so orthodox during the past decade that it is hardly ever considered necessary to support it with facts.

20 B. Peyre de Fabregues personal communication. Moving sand covered 4.5 per cent of an area of 3 million ha in Sud-Tamesna in 1985, according to interpretation of Landsat MSS data (De Wispelaere and Peyre de Fabregues, 1986:53).
21 Thus, when herds from Gouré entered the Dagaceri area in 1984–85, although their owners had never visited the area before, they were allowed access to the grazing reserves occupied by local Ful'be. The grazings were expected to last for about three weeks, after which time the herds proceeded elsewhere.
22 This point of view is quite common amongst ecologists. Since imported food is not the solution 'the only alternative must be a very strict control of the human and livestock populations' (Wickens and White, 1979:240).
23 Sandford (1976, 1983) is sceptical of the conventional orthodoxy that livestock numbers have grown at a substantial rate in the longer term.
24 In fact, the improvements in health-care provision that are conventionally supposed to account for falling mortality may be hard to find in remote semi-arid communities.
25 Double this number was suggested for the Sahel in 1970 (Caldwell, quoted in Eckholm, 1984).
26 Dr M.K. Tolba is (1987) Executive Director, United Nations Environment Programme, and was Secretary-General of the United Nations Conference on Desertification, 1977.
27 Cattle densities decline, as they are more dependent on free grazing than on browse and residues.
28 It should be remembered that farming pastoralists are themselves responsible for some of this reduction.
29 As declared by Eckholm (1984), who bases his claim on 'heavy out-migration' of Mossi people from Burkina Faso.
30 Recognition of this fact by specialists unfortunately may not forestall extreme statements of the 'Malthusian–Red Cross' view (Garcia, 1981), such as the following recently given space in a meteorological journal: 'The present famine is a manifestation of Nature's way of ridding a *superfluous* population from areas unable to support it due to the harsh climate... the present rush to save the dying thousands in Ethiopa and adjoining countries will only act as a temporary *palliative*. The saved children will grow up, the population will not *fall as Nature intends*, and the crisis will be renewed, inevitably, in the near future... [as Nature] does with wild life populations' (Spink, 1985; italics added).
31 To fall again to 165,000 ha in 1977–78 (Grainger, 1982:16).
32 Food chain efficiency is defined as consumption by trophic level over production by trophic level.
33 I suggest that the formal apparatus of systems theory is inappropriate (cf. Garcia and Escudero, 1982). People, not systems as such, should be the primary focus.

9 POLICY DIRECTIONS

1. Of $456 million per year required to finance the Plan of Action to Combat Desertification (at 1977 prices) only $43,500 had been received by 1983 in a special account created for countries needing financial assistance (from five developing countries!).
Meanwhile the global costs of stopping desertification were revised in 1980 to the figure of $4.5 billion per year until the end of the century (UNEP, 1983). By the end of 1983, UNEP had spent about $20 m (only 8 per cent of its spending) on 43 projects in various countries (UNEP, 1984). However, APPER has allocated $3.41 bn of its proposed budget of $128.1 bn to drought and desertification control.
2. The oft-used likeness with Joseph's famous strategy in ancient Egypt is only partly correct. Joseph's unusual early warning system told him how *many* kine were thin and fat.
3. To supply 15 million affected persons with a year's supply of grain in 1973–74 would have required nine times more than the entire national planned capacity in 1980 – and that drought lasted for *two* years.
4. Price stabilisation calls for controlled sale of part of stocks; famine relief for free distribution of the entire stock. Centralised food relief systems took a year to get going effectively in 1973–74 and were scandalised by misappropriation of funds (e.g. the Kano State Drought Relief Committee). In December 1985, following a bumper harvest, the Nigerian Grains Board was seen as the guarantor of the farmer's right to a market for his surplus.
5. Autonomy is the obvious answer to the process of marginalisation that, according to many authors, has beset rural communities in the semi-arid zone (see chapter 8). But it is also obligatory if adaptive systems are to be supported and extended.
6. Van Apeldoorn (1981) has suggested that simply to improve market access for owners of livestock in remote areas would strengthen the reserve function of livestock.
7. United Nations Conference on Desertification (recommendation 17) listed the following: financial insurance schemes for crops and livestock, methods of supporting the purchasing power of small farmers and livestock owners, diversion of tax revenues into insurance food reserves, special livestock risk insurance including traditional institutions, relief employment, emergency water supplies and transport (UNEP, 1977d).
8. But the point is by no means generally conceded by specialists. For example, in a recent volume on farming systems research (Moock, 1986), the Foreword (by Pius Okigbo) recognises the multisectoral basis of decisions by farming households and the need for interdisciplinary research, but Norman and Baker (in the same volume) assert that 'FSR teams cannot afford to *get lost in academic studies* of social complexity and dynamics' (p. 54; italics added).

9 But the technocentric approach is alive and well; a recent restatement says that the solution lies in 'understanding the causes of desertification and adopting measures to counteract it, in undertaking engineering work to bring water where it is needed, and in altering land-use practices' (White, 1986).

10 Nomadic Fulani are aware that they occupy a niche that no one else desires: 'Only our people are willing to suffer' said a woman while loading water onto a donkey for the long journey from a village well to her camp in the waterless forest.

11 Such realities are recognised in the proposals to reorganise the West African livestock sector into zones – raising zones, finishing zones etc. – arranged in a north–south pattern based on ecological rather than political boundaries (see Van Dyne, 1975). But this radical proposal has gained little support.

12 The Ethiopian resettlement programme is the exception that proves the rule. Carried out by a centralised authoritarian regime, involving a docile population and no international frontiers and spurred by a quite exceptional subsistence crisis, it is unlikely to be replicated in West African conditions. Spontaneous redistribution of the population, however, should be encouraged.

13 It is doubtful whether it is an option for the arid pastoral zone so the restricted zonal application of the ensuing comments should be stressed. A major study in 1974 (MIT, 1974; Matlock and Cockrim, 1976) argued that the benefits of intensification are too low to justify higher levels of economic activity in the Sahel. This argument has been abandoned in the semi-arid zone, as the popularity of intensification strategies shows.

14 In the extensively grazed arid zone, grazing tenure may be best effected through group ownership of wells (Haaland, 1977: 191); free access to government boreholes caused localised overgrazing in Niger in the sixties (Bernus, 1974a). But in the semi-arid zone pastoralists make free use of village wells.

15 See Ojo *et al.* (1987); Otegbeye and Ogigirigi (1987); Turabu (1987). Even in the USA shelter belts cost from five to ten times the value of the land they are planted on (Worster, 1979: 222).

16 Notwithstanding a 'balance sheet' for Niger which estimates annual wood requirements to be a million tons in excess of available increment from natural forest and plantations. How is the deficit met?

17 An urgent priority is to find out why farmed parkland has *not* developed around many rural and urban settlements in Senegal and the Sudan (for example) and what constraints need to be overcome – an absence of individual land tenure, an unfavourable water budget, an unfavourable attitude to tree management? – in order to realise the intrinsic advantages of growing trees.

18 Hill (1986: 171–4) offers a fierce critique of 'doomsday economics'.

BIBLIOGRAPHY

Abalu, G.O.I., 1975 'Supply response to producer prices: a case study of groundnut supply to the Northern States Marketing Board'. *Savanna*, 4/1; 33-40.

Abalu, G.O.I., Abdullahi, Y.A., Ajayi, O., Fisher, N.M., Manzo, S.K., Musa, H.L., Ogunbile, A.O. and Voh, J.P., 1983 *Exploratory survey of the farming systems of north-eastern Kano state, Nigeria*. Working Paper 4, Department of Agricultural Economics and Rural Sociology, Institute for Agricultural Research, Ahmadu Bello University, Zaria.

Abdu, Peter Shehu, 1976 *Drought-caused migration around Illela. The case of Tozai, Sonnani, and Amarawa, Gwadabawa district, Sokoto division in Sokoto state, Nigeria*. B.A. Dissertation, Department of Geography, Ahmadu Bello University, Zaria.

Abdull-Jalil, Musa Adam, 1980 'Some comments on the effect of drought on socio-economic formations in northern Darfur'. In: Carl Duisberg Ges., 1980: 97-100.

Achtnich, W., 1985 'Agrarproduktion im Sahel: Möglichkeiten und Grenzen'. *Die Erde*, 116: 159-68.

Adams, M.E., 1975 'A development plan for semi-arid areas in Western Sudan'. *Experimental Agriculture*, 2: 277-87.

Adams, W.M. and Grove, A.T. (eds.), 1985 *Irrigation in tropical Africa. Problems and problem solving*. Cambridge African Monograph 3, African Studies Centre, University of Cambridge.

Adebayo, S.I. and Mohammed, I., 1987 'Rainfall, crop failure and drought. An appraisal for 1983-84 experience in relation to normal in Kano state'. In: Sagua, V.O. et al. (eds.), 1987: 93-104.

Adefolalu, D.O., 1983 'Desertification of the Sahel'. In: Ooi Jin Bee (ed.), 1983: 402-38.

Adepoju, Aderanti, 1979 'New conceptual approaches to migration in the context of urbanization: the case of Africa south of the Sahara'. In: IUSSP, 1979: 30-3.

1982 'Population redistribution: a review of governmental policies'. In: Clarke, J.I. and Kosinski, L.A. (eds.), 1982: 58-67.

Agboola, S.A., 1979 *An agricultural atlas of Nigeria*. Oxford University Press.

Agnew, C.T., 1982 'Water availability and the development of rainfed agriculture in south-west Niger, West Africa'. *Transactions, Institute* of British Geographers, NS 7: 419–57.

Ahmed, M.L., 1976   Socio-cultural organisation and the sharing of the impact of a natural hazard, in Rijiyar-Tsamiya, Daura administrative area, Kaduna state. B.A. Dissertation, Department of Geography, Ahmadu Bello University, Zaria.

Aliyu, Mohammed, 1976   Modifications of agricultural techniques to the drought situation in Illela village, Gwadabawa district, Sokoto state 1972/73. B.A. Dissertation, Department of Geography, Ahmadu Bello University, Zaria.

Allan, William, 1975   The African husbandman. Edinburgh: Oliver and Boyd (reprinted 1977 at Westport, Connecticut: Greenwood Press).

Amin, Samir (ed.), 1974   Modern migrations in West Africa. Oxford University Press, for the International African Institute.

Anglo-French Forestry Commission, 1937   Report of the Anglo-French Forestry Commission, December 1936–February 1937. Nigeria Sessional Paper 37 of 1937. Lagos: Government Printer.

Anyadike, R.N.C., 1982   'Natural vegetation in relation to the moisture situation in West Africa'. Bulletin, Institut Fondamentale d'Afrique Noire, Sér. A, 44: 221–33.

Armstrong, R., 1967   'The nightwatchmen of Kano'. Middle Eastern Studies, 3: 269–82.

Aubréville, A., 1949   Climats, forêts et désertification de l'Afrique tropicale. Paris: Société d'Editions Géographiques, Maritimes et Coloniales.

Ayoade, J.O., 1977   'Perspectives on the recent drought in the Sudano–Sahelian region of West Africa with particular reference to Nigeria'. Archiv für Meteorologie, Geophysik und Bioklimatologie, Ser. B, 25: 67–77.

Baba, J.M., 1975   Induced agricultural change in a densely populated district, a study of the existing agricultural system in Kura district and the projected impact of the Kano River Irrigation Project. Ph.D. Thesis, Ahmadu Bello University, Zaria.

in press   'The problem of rural inequality on the Kano River Project, Nigeria'. In: Swindell, K. et al. (eds.), in press.

Baier, Stephen, 1976   'Economic history and development: drought and the Sahelian economies of Niger'. African Economic History, 1: 1–16.

1980   An economic history of Central Niger. Oxford: The Clarendon Press.

Baker, Randall, 1974(a)   Perceptions of pastoralism. Development Studies Discussion Paper 3, University of East Anglia.

1974(b)   'Famine: the cost of development?' The Ecologist, 4: 170–5.

1975   The administrative trap. Development Studies Discussion Paper 5, University of East Anglia.

Baldwin, K., 1957   The Niger Agricultural Project. Oxford: Blackwell.

Ball, Nicole, 1976   'Understanding the causes of the African famine'. Journal of Modern African Studies, 14: 517–22.

1978   'Drought and dependence in the Sahel'. International Journal of Health Services, 8: 271–98.

Barber, W. and Dousse, B., 1964   'Rise in the water table in parts of

Potiskum division, Bornu – further observations'. *Geological Survey of Nigeria Reports*, 1523.

Barbour, K.M. and Prothero, R. Mansell (eds.), 1961 *Essays on African population*. London: Routledge and Kegan Paul.

Bargery, G.P., 1934 *A Hausa–English dictionary and English–Hausa vocabulary*. Oxford University Press.

Barral, H. and Benoit, M. 1977 'Nature et genre de vie au Sahel. L'année 1973 dans le nord de la Haute-Volta'. In: Gallais, Jean *et al*., 1977: 91–112.

Bein, F.L., 1980 'Response to drought in the Sahel'. *Journal of Soil and Water Conservation*, 35: 121–4.

Bellot, J.M. and Bellot-Couderc, B., 1979 'Sécheresse et élevage au Sahel'. *Cultures et Développement*, 11: 47–67.

Bennett, J.W., 1976 'Anticipation, adaptation, and the concept of culture in anthropology'. *Science*, 192: 847–53.

1978 'A rational-choice model of agricultural resource utilization and conservation'. In: Gonzalez, N.L. (ed.), 1978: 151–85.

Berg, Elliott, 1965 'The economics of the migrant labour system'. In: Kuper, Hilda (ed.), 1965: 160–81.

1976 'The Sahelian economies – an overview'. In: Paylore, Patricia and Haney, Richard, A. Jnr. (eds.), 1976: 44–60.

Berkovsky, L., 1984 'Rainfall patterns in the desert'. In: World Climate Programme 1984b: 71–2.

Bernus, Edmond, 1974(a) 'Possibilités et limites de la politique d'hydraulique pastorale dans le Sahel Nigérien'. *Cahiers d'ORSTOM*, Sér. Sci. Hum., 11: 119–26.

1974(b) 'L'évolution récente des relations entre éleveurs et agriculteurs en Afrique tropicale'. *Cahiers d'ORSTOM*, Sér. Sci. Hum., 11: 137–43.

1975 'Human geography in the Sahelian zone'. In: MAB, 1975: 61–74.

1977(a) *Case study on desertification. The Eghazer and Azawak region, Niger*. United Nations Conference on Desertification, 1977. Nairobi: United Nations Environment Programme.

1977(b) 'Les éleveurs face à la sécheresse en Afrique Sahelienne: exemples Nigériens'. In: Dalby, David *et al*. (eds.), 1977: 140–7.

1979 'Exploitation de l'espace et désertification en zone Sahélienne'. *Travaux de l'Institut de Géographie de Reims*, 39/40: 49–59.

1980(a) 'The Eghazer and Azawak region, Niger'. In: Biswas M.R. and A.K. (eds.), 1980.

1980(b) 'Famines et sécheresse chez les Touaregs Sahéliens: les nourritures de substitution'. *Africa*, 50: 1–7.

Bernus, Edmond and Savonnet, G., 1973 'Les problèmes de la sécheresse dans l'Afrique de l'Ouest'. *Présence Africaine*, 88: 113–38.

Berry, Leonard, 1984(a) (With the United Nations Sudano–Sahelian Office) *Assessment of desertification in the Sudano-Sahelian region 1978–1984*. UNEP/GC. 12/Background Paper 1. Nairobi: United Nations Environment Programme.

1984(b) 'Desertification in the Sudano-Sahelian region 1977–1984'. *Desertification Control Bulletin* (UNEP), 10: 23–8.

Berry, L., Hankins, T., Kates, R.W., Maki, L. and Porter, R.W., 1971 *Human adjustments to agricultural drought in Tanzania. Pilot investigations.* Natural Hazard Research Working Paper 19/Research Paper 13, Dar es Salaam: Bureau of Resource Assessment and Land Use Planning.

Bille, J.C., 1974 'Recherches écologiques sur une savane sahélienne du Ferlo septentrional, Sénégal: 1972, année sèche au sahel'. *La Terre et la Vie,* 28: 5–20.

Birks, J.S., 1978 *Across the savannas to Mecca. The overland pilgrimage route from West Africa.* London: C. Hurst.

Biswas, M.R. and A.K. (eds.), 1980 *Desertification associated case studies prepared for the United Nations Conference on Desertification,* 2 vols. Oxford: Pergamon.

Blaikie, Piers, 1981 *Capitalism, the peasantry and environmental degradation: some issues of method.* Development Studies Discussion Paper 81, University of East Anglia.

Bonfiglioni, A.M., 1985 'Evolution de la propriété animale chez les Wodaabe du Niger'. *Journal des Africanistes,* 55: 29–37.

Bonte, Pierre, 1986 'The Sahel: transformation and drought'. In: Garcia, Rolando V. and Spitz, Pierre (eds.), 1986: 150–81.

Boserup, Ester, 1965 *The conditions of agricultural growth.* London: Allen and Unwin.

Boudet, G., 1972 'Désertification de l'Afrique tropicale sèche'. *Adansonia,* Sér. 2, 12: 505–24.

1975 'Pastures and livestock in the Sahel'. In: MAB, 1975: 29–34.

1976 'Mali:regional study and proposals for development'. In: Rapp, Anders *et al.* (eds.), 1976: 137–53.

1977 'Désertification ou remontée biologique au Sahel'. *Cahiers d'ORSTOM,* Sér. Bio., 12: 293–300.

1979 'Quelques observations sur les fluctuations du couvert végétal Sahélien au Gourma Malien et leurs conséquences pour une stratégie de gestion sylvo-pastorale'. *Bois et Forêts des Tropiques,* no. 184: 31–44.

Bougères, Jacques, 1979 'L'état de dégradation des formations sableuses du Sahel Voltaïque ou l'urgence d'une intervention'. *Travaux de l'Institut de Géographie de Reims,* 39/40: 91–101.

Bouquet, Christian, 1974 'Le déficit pluviométrique au Tchad et ses principales conséquences'. *Les Cahiers d'Outre-mer,* 77: 245–70.

Boutrais, Jean, 1977 'Une conséquence de la sécheresse: les migrations d'éleveurs vers les plateaux Camerounais'. In: Dalby, David *et al.* (eds.), 1977: 127–39.

Bovill, E.W., 1921 'The encroachment of the Sahara on the Sudan'. *Journal of the African Society,* 20: 174–85; 259–69.

Bovin, M., 1985 'Nomades "sauvages" et paysans "civilisés": Wo'daa'be et Kanuri au Borno'. *Journal des Africanistes,* 55: 53–73.

Bowden, M.J. *et al.*, 1981 'The effect of climatic fluctuations on human populations: two hypotheses'. In: Wigley, T.M. *et al.* (eds.), 1981: 479–513.

Bradley, A.K., Macfarlane, S.B.J., Moody, J.B., Gilles, H.M., Blacker, J.G.C. and Musa, B.D., 1982(a) 'Malumfashi Endemic Diseases Research Project, XIX. Demographic findings: population structure and fertility'. *Annals of Tropical Medicine and Parasitology*, 76: 381–91.

1982(b) 'Malumfashi Endemic Diseases Research Project, XX. Demographic findings: mortality'. *Annals of Tropical Medicine and Parasitology*, 76: 393–404.

Bradley, P., 1980 'Agriculture in the Mauritanian Guidimaka: environmental or social problems?. In: De Brandt, J. *et al.* (eds.), 1980: 269–80.

Bradley, P.N., 1973 'The delayed start of the 1973 rains at Samaru, Nigeria'. *Savanna*, 2/1: 78–81.

1977 'Vegetation and environmental change in the West African Sahel'. In: O'Keefe, Phil and Wisner, Ben (eds.), 1977: 34–54.

Bray, T.M., 1977 'Universal Primary Education in Kano state: the first year'. *Savanna*, 6/1: 3–14.

1981 *Universal Primary Education in Nigeria. A study of Kano state* London: Routledge and Kegan Paul.

Breman, H. and Cisse, A.M., 1977 'Dynamics of Sahelian pastures in relation to drought and grazing'. *Oecologia*, 28: 301–15.

Bremaud, O. and Pagot, J., 1962 'Grazing lands, nomadism and transhumance in the Sahel'. *Arid Zone Research* (UNESCO), 18: 311–24.

Bromfield, A.R., 1973 'Uptake of sulphur and other nutrients by groundnuts in northern Nigeria'. *Experimental Agriculture*, 9: 55–8.

Brooke, Clarke, 1967 'Types of food shortages in Tanzania'. *The Geographical Review*, 57: 333–57.

Brown, J.H., Reichman, O.J. and Davidson, D.W., 1979 'Granivory in desert ecosystems'. *Annual Review of Ecology and Systematics*, 10: 201–27.

Brune, S., 1985 'Hungerkrise im Sahel: Natur- oder Sozialkatastrophe?. *Die Erde*, 116: 185–95.

Bryson, R.A., 1973 'Drought in Sahelia: who or what is to blame?'. *The Ecologist*, 3: 366–71.

1974 'A perspective on climatic change'. *Science*, 184: 753–60.

Bunting, A.H., Dennett, M.D., Elston, J. and Milford, J., 1974 'Weather and climate in the Sahel. An initial study'. *Conference on International Development Problems in the Sahel, Bellagio, 1974*. Washington, DC: National Academy of Sciences.

1975 'Seasonal rainfall forecasting in West Africa'. *Nature*, 253: 622–3.

1976 'Rainfall trends in the West African Sahel'. *Quarterly Journal of the Royal Meteorological Society*, 102: 59–64.

Burkill, H.M., 1985 *The useful plants of West tropical Africa. Vol. 1, families A-D.* Kew: Royal Botanic Gardens.

Burton, I. and Kates, R.W., 1964 'The perception of natural hazards in resource management'. *Natural Resources Journal*, 3: 412–41.

Burton, Ian, Kates, Robert W. and White, Gilbert F., 1978 *The environment as hazard*. New York: Oxford University Press.

Caldwell, J.C., 1975 *The African drought and its demographic implications*. New York: The Population Council.

—— 1977 'Demographic aspects of drought: an examination of the African drought of 1970–74'. In: Dalby, David *et al.* (eds.), 1977: 93–100.

Caldwell, J.C., Addo, N.O., Gaisie, S.K., Igun, A. and Olusanya, P.O. (eds.), 1975 *Population growth and socio-economic change in West Africa*. New York: Columbia University Press, for the Population Council.

Caldwell, J.C. and Okonjo, C. (eds.), 1968 *The population of tropical Africa*. London: Longman.

Campbell, I., 1974 'Human mismanagement as a major factor in the Sahelian drought tragedy'. *The Ecologist*, 4: 164–9.

Carl Duisberg Ges., 1980 *Seminar on desertification problems, Khartoum*. Cologne: Carl Duisberg Gesellschaft.

Carter, J.W. and Barber, W., 1958 'The rise in the water table in parts of Potiskum division, Bornu'. *Geological Survey of Nigeria Records*, 1958: 5–13.

Catterson, T.M., Gulick, F.A. and Resch, T., 1987 'Rethinking forestry strategy in Africa: experience drawn from USAID activities'. *Desertification Control Bulletin* (UNEP), no. 14: 33–40.

Central Bank of Nigeria, 1973, 1974 *Economic and Financial Review*, October.

Chamard, C. and Courel, M.F., 1975 'Contribution à l'étude géomorphologique du Sahel. Les forme dunaires du Niger occidental et de la Haute Volta septentrionale'. *Bulletin ASEQUA*, no. 44–45: 55–66.

—— 1979 'Contribution à l'étude du Sahel Voltaïque. Causes et conséquences de la dégradation du couvert végétal des dunes. Secteur de Menegu-Bidi (Département du Sahel – Sous-Préfecture de l'Oudalan)'. *Travaux de l'Institut de Géographie de Reims*, 39/40: 75–90.

Chambers, Robert and Leach, Melissa, 1987 *Trees to meet contingencies: savings and security for the rural poor*. Discussion Paper 228, Institute of Development Studies, University of Sussex.

Chambers, Robert, Longhurst, Richard and Pacey, Arnold (eds.), 1981 *Seasonal dimensions to rural poverty*. London: Frances Pinter.

Chapman, Murray and Prothero, R. Mansell, 1983 'Themes on circulation in the Third World'. *International Migration Review*, 17: 597–632.

—— (eds.),1985 *Circulation in population movement. Substance and concepts from the Melanesian case*. London: Routledge and Kegan Paul.

Charney, J.G., 1975 'Dynamics of desert and drought in the Sahel'. *Quarterly Journal of the Royal Meteorological Society*, 101: 193–200.

Charney, J.G., Quirk, W.J., Chow, S.-H. and Kornfield, J., 1977 'A comparative study of the effects of albedo change on drought in

semi-arid regions'. *Journal of the Atmospheric Sciences*, 34: 1366–85.
Charney, J.G. and Stone, P.H., 1975 'Drought in the Sahara: a biogeophysical feedback mechanism'. *Science*, 187: 434–5.
1976 'Reply'. *Science*, 191: 100–2.
Cheke, R.A., Fishpool, L.D.C. and Forrest, G.A., 1980 '*Oedaleus senegalensis* (Krauss.) (Orthoptera: Acrididae: Oedipodinae): an account of the 1977 outbreak in West Africa and notes on eclosion under laboratory conditions'. *Acrida*, 9: 107–32.
Chevalier, A., 1950(a) 'La progression de l'aridité, du desséchement et de l'ensablement et la décadence des sols en Afrique Occidentale française'. *Comptes Rendus Hebdomadaires des Séances de l'Académie des Sciences*, 230: 1550–3.
1950(b) 'Mesures urgentes à prendre pour entraver le desséchement, l'ensablement et la décadence des sols et de la végétation en Afrique et spécialement au Soudan français'. *Comptes Rendus Hebdomadaires des Séances de l'Académie des Sciences*, 230: 1720–3.
1950(c) 'Programme de reboisement de lutte contre la sécheresse et d'aménagement agraire en Afrique Occidentale française'. *Comptes Rendus Hebdomadaires des Séances de l'Académie des Sciences*, 230:1991–4.
1950(d) 'Régéneration des sols et de la végétation en Afrique Occidentale française'. *Comptes Rendus Hebdomadaires des Séances de l'Académie des Sciences*, 230: 2064–6.
1950(e) 'La protection de la nature et les parcs-réserves de l'Afrique Occidentale française'. *Comptes Rendus Hebdomadaires des Séances de l'Académie des Sciences*, 230: 2140–2.
1952 'La décadence des sols et de la végétation en Afrique Occidentale française et la protection de la nature'. *Bois et Forêts des Tropiques*, no. 16: 335–53.
Cissoko, S.-M., 1968 'Famines et épidemies à Toumbouctou et dans la Boucle du Niger du XVI au XVIII siècle'. *Bulletin, Institut Fondamentale d'Afrique Noire*, Sér. B, 30: 806–21.
Clark University, 1975 'Distinctive environmental problems of the West-African Sahelian-Sudanic least-developed countries. A note'. Background paper for the Bellagio Conference. In: National Academy of Sciences, 1975.
Clarke, J.I. and Kosinski, L.A. (eds.), 1982 *Redistribution of population in Africa*. London. Heinemann.
Cline-Cole, R.A., Falola, J.A., Main, H.A.C., Mortimore, M.J., Nichol, J.E. and O'Reilly, F.D., 1987 *Wood fuel in Kano. Final report of the Rural Energy Research Project, Bayero University, Kano*. Department of Geography, Bayero University for the United Nations University.
Cloudsley-Thompson, John L., 1978 'Human activities and desert expansion'. *The Geographical Journal*, 144: 416–23.
1984 'Open letter: the footprint in the sand'. *Environmental Conservation*, 11/2: 95–7.
Club du Sahel, 1977 *Proposals for a strategy for drought control and development in the Sahel*. Paris: Organisation for Economic Co-operation and Development.

Cochème, J. and Franquin, P., 1967 *A study of the agroclimatology of the semi-arid areas south of the Sahara in West Africa.* Rome: Food and Agriculture Organisation of the United Nations.

Cohen, Abner, 1969 *Custom and politics in urban Africa: a study of Hausa migrants in Yoruba towns.* University of California Press.

Cohen, Ronald, 1967 *The Kanuri of Bornu.* New York: Holt, Rinehard and Winston.

Collier, F.S. and Dundas, J., 1937 'The arid regions of Northern Nigeria and the French Niger Colony'. *Empire Forestry Journal,* 16: 184–94.

Collier, P., 1983 'Oil and inequality in rural Nigeria'. In: Ghai, D. and Radwan, S. (eds.), 1983: 191–217.

Collins, John Davison, 1976 'The clandestine movement of groundnuts across the Niger–Nigeria boundary'. *Canadian Journal of African Studies,* 10: 259–78.

Colvin, L.G. et al., 1981 *The uprooted of the western Sahel. Migrants' quest for cash in the Senegambia.* New York: Praeger.

Comité d'Information Sahel, 1974 *Qui se nourrit de la famine en Afrique? Le dossier politique de la faim au Sahel.* Paris: Maspéro.

Commins, Stephen, K., Lofchie, Michael F. and Payne, Rhys (eds.), 1986 *Africa's agrarian crisis. The roots of famine.* Boulder, Colorado: Lynne Rienner.

Commonwealth Development Corporation, 1971 *Nigeria: review of livestock industry. Outline of a national cattle development programme and identification of projects suitable for IBRD financing.* London: Commonwealth Development Corporation.

Copans, Jean (ed.), 1975 *Sécheresses et famines du Sahel. 1, Ecologie, dénutrition, assistance. 2, Paysans et nomades.* Paris: Maspéro.

1979 'Droughts, famines and the evolution of Senegal (1966–1978)'. *Mass Emergencies,* 4: 87–93.

1983 'The Sahelian drought: social sciences and the political economy of underdevelopment'. In: Hewitt, K. (ed.), 1983: 83–97.

Coppock, J.T. (ed.), 1979 *Agriculture and food supply in developing countries. Papers presented at the Meeting of the Commission on World Food Problems and Agricultural Productivity of the International Geographical Union, Nigeria, 1978.* Department of Geography, University of Edinburgh.

Corkill, N.L., 1949 'Dietary change in a Sudan village following locust invasion'. *Africa,* 19: 1–11.

C T F T, 1975 'The role of the forester in land use planning in the Sahel'. In: MAB, 1975.

Currey, B., 1984 'Coping with complexity in food crisis management'. In: Currey, B. and Hugo, G. (eds.), 1984: 183–202.

Currey, B. and Hugo, G. (eds.), 1984 *Famine as a geographical phenomenon.* Dordrecht: D. Reidel.

Dahl, Gudrun and Hjort, Anders, 1976 *Having herds. Pastoral herd growth and household economy.* Studies in Social Anthropology 2, Department of Social Anthropology, University of Stockholm.

Dalby, David and Harrison Church, R.J. (eds.), 1973 *Drought in Africa.*

Proceedings of the 1973 Symposium. Centre for African Studies, School of Oriental and African Studies, University of London.

Dalby, David, Harrison Church, R.J. and Bezzaz, Fatima (eds.), 1977 *Drought in Africa 2*. African Environment Special Report 6. London: International African Institute.

Dalziel, J., 1937 *The useful plants of West tropical Africa. An appendix to 'the flora of West tropical Africa'*. London: Crown Agents.

Damen, M.C.J., Sicco Smit, G. and Verstappen, H. Th. (eds.), 1986 *Remote sensing for resources development and environmental management. Proceedings of the 7th International Symposium, Enschede, 25–29 August, 1986*. Rotterdam: Balkema.

Dancette, C. and Poulain, J.F., 1969 'Influence of *Acacia albida* on pedoclimatic factors and crop yields'. *African Soils*, 14: 143–84.

Dando, William A., 1980 *The geography of famine*. London: Edward Arnold.

Darkoh, Michael B., 1982 'Population expansion and desertification in Tanzania'. *Desertification Control Bulletin* (UNEP), no. 6: 26–33.

Daudu, Ismaila, 1976 *Modification of agricultural techniques to the drought situation in Mungurun village, Hadejia administrative area, Kano state, 1972/73*. B.A. Dissertation, Department of Geography, Ahmadu Bello University, Zaria.

Davies, James, 1979(a) 'Water melon comeback in North'. *Noma* (Institute for Agricultural Research, Ahmadu Bello University, Zaria), 2/2: 17–20.

1977(b) 'Benniseed in Hadejia and Gumel'. *Noma* (Institute for Agricultural Research, Ahmadu Bello University, Zaria), 2/3: 17–20.

Davy, E.G., 1974 'Drought in West Africa'. *WMO Bulletin*, 23: 18–23.

Davy, E.G., Mattei, F. and Solomon, S.I., 1976 *An evaluation of climate and water resources for development of agriculture in the Sudano-Sahelian zone of West Africa*. Special Environmental Report 9 (WMO no. 459). Geneva: World Meteorological Organisation.

De Brandt, J., Mandi, P. and Seers, D. (eds.), 1980 *European studies in development. New trends in European development studies*. London: Macmillan.

De Castro, Josué, 1952 *Geography of hunger*. London: Victor Gollancz.

De Leeuw, P.N., 1974 'Livestock development and drought in the northern states of Nigeria'. *Nigerian Journal of Animal Production*, 1/1: 61–73.

1976 'Fodder resources and livestock development in north-east Nigeria'. *Savanna*, 5/1: 61–74.

De Wispelaere, G. and Peyre de Fabregues B., 1986 *Action de recherche méthodologique sur l'évaluation des ressources fourragères par télédétection dans la région du Sud-Tamesna (Niger). Rapport de seconde année*. Maisons-Alfort: Institut d'Elevage et de Médecine Vétérinaire des Pays Tropicaux.

DECARP, 1976 *Sudan Desert Encroachment and Rehabilitation Programme*. Khartoum: Ministry of Agriculture, Food and Natural Resources, Agriculture Research Council, National Council for Research.

Delgado, Christopher F., 1979 *The Southern Fulani farming system in Upper Volta: a model for the integration of crop and livestock production in the West African savannah.* African Rural Economy Paper 20, Department of Agricultural Economics, Michigan State University, Ann Arbor.

Delwaulle, J.C., 1973 'Résultats de six ans d'observations sur l'érosion au Niger'. *Bois et Forêts des Tropiques*, 150: 15–36.

Dennett, M.D., Elston, J., and Rodgers, J.A., 1985 'A reappraisal of rainfall trends in the Sahel'. *Journal of Climatology*, 5: 353–61.

Dennett, M.D., Rodgers, J.A. and Stern, R.D., 1983 'Independence of rainfalls through the rainy season and the implications for the estimation of rainfall probabilities'. *Journal of Climatology*, 3: 375–84.

Depierre, D. and Gillet, H., 1971 'Désertification de la zone sahélienne du Tchad. Bilan de 10 années de mise en défense'. *Bois et Forêts des Tropiques*, 139: 3–25.

Derrick, Jonathan, 1977 'The Great West African Drought, 1972–74'. *African Affairs*, 76: 537–86.

1984 'West Africa's worst year of famine'. *African Affairs*, 83: 281–300.

Derrienic, Hervé et al., 1976 *Famines et dominations en Afrique. Paysans et éleveurs du Sahel sous le joug.* Rennes: Université de Haute Bretagne.

Dregne, H.E., 1970 *Arid lands in transition.* Publication 90. Washington, DC: American Association for the Advancement of Science.

1983 *Desertification of arid lands.* Advances in Desert and Arid Land Technology and Development 3. Hardwood Academic Publishers.

Dresch, J., 1959 'Les transformations du Sahel Nigérien'. *Acta Geographica*, June: 3–12.

1975 'Reflections on the future of the semi-arid regions'. In: Richards, Paul (ed.), 1975: 1–9.

Du Preez, J.W. and Barber, W., 1965 *The distribution and chemical quality of groundwater in Northern Nigeria.* Bulletin 36. Kaduna: Geological Survey of Nigeria.

Du Preez, J.W. and Richards, H.J., 1958 'The hydrology of Gumel emirate, Kano province'. *Geological Survey of Nigeria Records*, 1955: 65–70.

Duffill, M.B., 1986 'Hausa poems as sources for social and economic history'. *History in Africa*, 13: 35–88.

Dupree, Herb and Roder, Wolf, 1974 'Coping with drought in a preindustrial, pre-literate farming community'. In: White, Gilbert F. (ed.), 1974: 115–19.

Durand, J.-H., 1977 'A propos de la sécheresse et de ses conséquences au Sahel'. *Les Cahiers d'Outre-mer*, 30: 383–403.

Dutt, Romesch, 1900 *Famines and land assessments in India.* London: Kegan Paul, Trench, Trübner.

Dyson-Hudson, N., 1972 'The study of nomads'. In: Irons, W. (ed.), 1972: 2–30.

Eckholm, Erik, 1976 *Losing ground. Environmental stress and world food prospects.* New York: Pergamon.

1977 'The other energy crisis'. In: Glantz, Michael H. (ed.), 1977a: 39–56.

1984 'Poverty, population growth and desertification'. *Desertification Control Bulletin* (UNEP), no. 10: 37–41.

Eckholm, Erik and Brown, Lester, 1977 *Spreading deserts: the hand of man.* Worldwatch Paper 13. Washington, DC: Worldwatch Institute.

Ega, L.A., 1979 'Security of tenure in a transitory farming system: the case of Zaria villages in Nigeria'. *Agricultural Administration*, 6: 287–98.

Elston, J. and Dennett, M.D., 1977 'A weather watch for semi-arid lands within the tropics'. *Philosophical Transactions of the Royal Society of London*, Ser. B, 278: 593–609.

EMASAR, 1974 *The ecological management of arid and semi-arid rangelands in Africa and the Near East, an international programme. Report of an expert consultation, May, 1974.* Rome: Food and Agriculture Organisation of the United Nations.

EMASAR, II, 1977 *EMASAR Phase II. Vol. 1, Les pays Sahéliens: développement et vulgarisation dans le domain pastoral. Eléments d'une stratégie.* Rome: Food and Agriculture Organisation of the United Nations.

1978 *EMASAR Phase II. Vol. 8, Sudan: proposals for grazing land development.* Rome: Food and Agriculture Organisation of the United Nations.

Etkin, Nina L., 1981 'A Hausa herbal pharmacopoeia: biomedical evaluation of commonly used plant medicines'. *Journal of Ethnopharmacology*, 4: 75–98.

Etkin, Nina L. and Ross, Paul J., 1982 'Food as medicine and medicine as food. An adaptive framework for the interpretation of plant utilization among the Hausa of Northern Nigeria'. *Social Science Medicine*, 16: 1559–73.

1983 'Malaria, medicine and meals: plant use among the Hausa and its impact on disease'. In: Romanucci-Ross, L. *et al.* (eds.), 1983: 231–59.

Famoriyo, Segun, 1979 *Land tenure and agricultural development in Nigeria.* Ibadan: Nigerian Institute for Social and Economic Research.

FAO, 1966 *Agricultural development in Nigeria, 1965–1980.* Rome: Food and Agriculture Organisation of the United Nations.

1971 *Production Yearbook.* Rome: Food and Agriculture Organisation of the United Nations.

1975 *Animal Health Yearbook.* Rome: Food and Agriculture Organisation of the United Nations.

FAO/UNDP, 1978 *Technical Report 2 to the Federal Government of Nigeria: Remote Sensing. Section 4, Land use change investigation.* Hunting Technical Services – FAO/UNDP Forest Development Project NIR/71/546. Rome: Food and Agriculture Organisation of the United Nations.

FAO/UNEP, n.d. (?1983) *Provisional methodology for assessment and mapping of desertification.* Rome: Food and Agriculture Organisation of the United Nations.

Farmer, G. and Wigley, T.M.L., 1985 *Climatic trends for tropical Africa. A research report for the Overseas Development Administration.* Climatic Research Unit, University of East Anglia.

Faulkingham, R.H., 1977 'Ecologic constraints and subsistence strategies: the impact of drought in a Hausa village, a case study from Niger'. In: Dalby, David *et al.* (eds.), 1977: 148–58.

Faulkingham, R.H. and Thorbahn, P.F., 1975 'Population dynamics and drought, a village in Niger'. *Population Studies*, 29: 463–77.

Faure, H. and Gac, J.-Y., 1981 'Will the Sahelian drought end in 1985?' *Nature*, 291: 475–8.

Federal Department of Forestry, 1977 [Nigeria], Vegetation and land use, scale 1:250,000 (Sheet ND 32–12) Ibadan.

Federal Livestock Department, 1984 *Nigerian Livestock Information Service. 1984 Annual Report/Bulletin.* Lagos: Federal Livestock Department.

Federal Ministry of Agriculture, 1974 *Agricultural development in Nigeria, 1973–1985.* Lagos: Joint Planning Committee, Federal Ministry of Agriculture and Natural Resources.

Federal Office of Statistics, n.d. (1961) *Agricultural Sample Survey, 1955–60. Bulletins 1–4.* Lagos: Federal Office of Statistics.

1972 *Consolidated results of crop estimation surveys, 1965–66 to 1967–68.* Rural Economic Survey. Lagos: Federal Office of Statistics.

Ferguson, D.S., 1967 *The Nigerian beef industry.* Cornell International Development Bulletin 9, Cornell University.

Field, N.J. and Collins W.G., 1986 'Land use from aerial photographs in the Nigerian savanna'. In: Damen, M.C.J. *et al.* (eds.) 1986: 435–40.

Fisher, H.H., 1927 *The famine in Soviet Russia. The operations of the American Relief Administration.* New York: Macmillan.

Fishpool, L.D.C. and Popov, G.B., 1981 'The grasshopper faunas of the savannas of Mali, Niger, Benin and Togo'. *Bulletin, Institut Fondamentale d'Afrique Noire*, Sér. A, 43: 275–429.

Fishwick, R.W., 1970 'Sahel and Sudan Zones of Northern Nigeria, North Camerouns and the Sudan'. In: Kaul, R.N. (ed.), 1970: 59–85.

Flohn, H. and Nicholson, Sharon E., 1980 'Climatic fluctuations in the arid belt of the Old World since the Last Glacial Maximum; possible causes and future implications'. *Paleoecology of Africa and the Surrounding Islands* (Balkema), 12: 3–22.

Folland, C.K., Palmer, T.N. and Parker, D.E., 1986 'Sahel rainfall and worldwide sea temperatures, 1901–85'. *Nature*, 320: 1–5.

Forrest, T.G., 1977 'Agricultural policies in Nigeria, 1900–78'. In: Heyer, J. *et al.* (eds.), 1977: 222–58.

Franke, R.W. and Chasin, B.H., 1980 *Seeds of famine: ecological destruction and the development dilemma in the West African Sahel.* Monclair New Jersey: Allanheld, Osmun.

Fricke, Werner, 1979 *Cattle husbandry in Nigeria. A study of its ecological conditions and socio-geographical differentiations.* Geographisches Institut der Universität Heidelberg (published in German, 1969).

Fuglestad, Finn, 1974 'La Grande Famine de 1931 dans l'ouest nigérien: réflexions autour d'une catastrophe naturelle'. *Revue Française*

d'Histoire d'Outre-mer, 61: 18–33.
1983   A history of Niger 1850–1960. Cambridge University Press.
Gallais, Jean, 1972   'Essai sur la situation actuelle des relations entre pasteurs et paysans dans le Sahel ouest-africain'. In: Etudes de géographie tropicale offertes à Pierre Gourou, pp. 300–13. Paris: Mouton.
1979   'La situation de l'élevage bovin et le problème des éleveurs en Afrique occidentale et centrale'. Les Cahiers d'Outre-mer, 32; 113–44.
Gallais, Jean et al., 1977   Stratégies pastorales et agricoles des Sahéliens durant la sécheresse 1969–1974. Travaux et Documents de Géographie Tropicale, 30. Bordeaux: Centre d'Etudes de Géographie Tropicale.
Gano, D.Z., 1983   Hawking in the city: needs, problems and physical planning solutions – a case study of Kano. M.Sc. Thesis, Department of Urban and Regional Planning, Ahmadu Bello University, Zaria.
Gapp, K.S., 1935   'The universal famine under Claudius'. Harvard Theological Review, 28: 258–65.
Garcia, Rolando, V. (ed.), 1981   Drought and man. The 1972 case history. Vol. 1, Nature pleads not guilty. Oxford and New York: Pergamon.
Garcia, Rolando V. and Escudero, J.C. (eds.), 1982   Drought and man. The 1972 case history. Vol.2, The constant catastrophe: malnutrition, famines and drought. Oxford and New York: Pergamon.
Garcia, Rolando V. and Spitz, Pierre (eds.), 1986   Drought and man. The 1972 case history. Vol.3, The roots of catastrophe. Oxford and New York: Pergamon.
Garduno, M. Anaya, 1977   Technology and desertification. United Nations Conference on Desertification, 1977. Nairobi: United Nations Environment Programme.
Ghai, D. and Radwan, S. (eds.), 1983   Agrarian policies and rural poverty in Africa. Geneva: International Labour Organisation.
Gilardi, J.-C., 1978   'Contribution au concept d'externalité: l'exemple de la désertification'. Mondes en Développement, no.24: 840–69.
Giles, P.H., 1975   The storage of cereals by farmers in Northern Nigeria. Samaru Research Bulletin 42, Institute for Agricultural Research, Ahmadu Bello University, Zaria.
Gillet, H., 1975   'Plant cover and pastures of the Sahel'. In: MAB, 1975: 21–7.
Glantz, Michael H. (ed.), 1976(a)   The politics of natural disaster: the case of the Sahel drought. New York: Praeger.
1976(b)   Value of a reliable long range climate forecast for the Sahel: a preliminary assessment. Boulder, Colorado: National Center for Atmospheric Research.
(ed.), 1977(a)   Desertification: environmental degradation in and around arid lands. Boulder, Colorado: Westview Press.
1977(b)   'The value of a long range weather forecast for the West African Sahel'. Bulletin of the American Meteorological Society, 58: 150–8.
1980   'Man, state and the environment: an enquiry into whether solutions to desertification are known but not applied'. Canadian Journal of Development Studies, 1/1: 75–97.

Glantz, Michael H. and Katz, R.W., 1977 'When is a drought a drought?' *Nature*, 267: 192–93.

Glantz, Michael H. and Orlovsky, Nicolai, 1983 'Desertification: a review of the concept'. *Desertification Control Bulletin* (UNEP), no.9: 15–22.

Gleave, M.B. and White, H.P., 1969 'Population density and agricultural systems in West Africa'. In: Thomas, M.F. and Whittington, G.W. (eds.), 1969: 273–300.

Goddard, A.D., Mortimore, M.J. and Norman, D.W., 1975 'Some social and economic implications of population growth in rural Hausaland'. In: Caldwell, J.C. et al. (eds.), 1975: 321–36.

Goldscheider, Calvin, 1979 'Modernization, migration and urbanization'. In: IUSSP 1979: 20–3.

Goldstein, Sidney, 1978 *Circulation in the context of total mobility in Southeast Asia*. Papers of the East–West Population Institute 53, Honolulu: East–West Centre.

Gonzalez, N.L., (ed.) 1978 *Social and technological management in dry lands. Past and present, indigenous and imposed*. American Association for the Advancement of Science Selected Symposium 10. Boulder, Colorado: Westview Press.

Gould, W.T.S. and Prothero, R. Mansell, 1975 'Space and time in African population mobility'. In: Kosinski, Leszek A. and Prothero, R. Mansell (eds.), 1975: 39–49.

Grainger, Alan, 1982 *Desertification. How people make deserts, how people can stop and why they don't*. London: Earthscan (International Institute for Environment and Development).

Green, Leslie, 1974 'Migration, urbanization and national development in Nigeria'. In: Amin, Samir (ed.), 1974: 281–304.

Greene, M.H., 1974 'Impact of the Sahelian drought in Mauritania, West Africa'. *The Lancet*, 1 June: 1093–7.

—— 1975 'Impact of the Sahelian drought in Mauritania'. *African Environment*, 1/2: 11–21.

Grégoire, E., 1983 'Un système de production agro-pastoral en crise: le terroir de Gourjaie (Niger)'. In: *Enjeux fonciers en Afrique noire*, pp. 202–11. Paris: Editions Karthala.

Gregory, S., 1982 'Spatial patterns of Sahelian annual rainfall, 1961–1980'. *Archiv für Meteorologie, Geophysik und Bioklimatologie*, Ser. B, 31: 273.

Grove, A.T., 1952 *Land and population in Katsina Province*. Kaduna: Government Printer.

—— 1961 'Population densities and agriculture in Northern Nigeria'. In: Barbour, K.M. and Prothero, R. Mansell (eds.), 1961: 115–36.

—— 1973 'A note on the remarkably low rainfall of the Sudan Zone in 1913'. *Savanna*, 2/2: 133–8.

—— 1977(a) 'Desertification in the African environment'. In: Dalby, David et al. (eds.), 1977: 54–64.

—— 1977(b) 'Desertification'. *Progress in Physical Geography*, 1: 296–310.

Grove, A.T. and Pullan, R.A., 1964 'Some aspects of the Pleistocene paleogeography of the Chad Basin'. In: Howell, F. Clark and Bourlière, François (eds.), 1964: 230–45.

Haaland, Gunnar, 1977 'Pastoral systems of production: the socio-cultural context and some economic and ecological implications'. In: O'Keefe, Phil and Wisner, Ben (eds.), 1977: 179–93.

Hall, A.E., Cannell, G.H. and Lawton, H.W. (eds.), 1979 *Agriculture in semi-arid environments*. Ecological Studies 34. Heidelberg: Springer Verlag.

Hammer, Turi (Digernes), 1977 *Wood for fuel – energy crisis implying desertification. The case of Bara, the Sudan*. Bergen: Geografisk Institutt.

Hankins, Thomas D., 1974 'Response to drought in Sukumaland, Tanzania'. In: White, Gilbert F.(ed.), 1974: 98–104.

Hare, F. Kenneth, 1983 *Climate and desertification: a revised analysis*. World Climate Programme 44. Geneva: World Meteorological Organisation.

    1984(a) 'Recent climatic experience in the arid and semi-arid lands'. *Desertification Control Bulletin* (UNEP), no. 10: 15–22.

    1984(b) 'Climate and desertification'. *WMO Bulletin*, 33: 288–95.

    1985 *Climate variations, drought and desertification*. WMO Report no. 653. Geneva: World Meteorological Organisation.

Harrison, M.N. and Jackson, J.K., 1958 *Ecological classification of the vegetation of the Sudan*. Forests Bulletin 2. Khartoum.

Hart, J.K., 1974 'Migration and the opportunity structure: a Ghanaian case study'. In: Amin, S (ed.), 1974: 321–42.

    1982 *The political economy of West African agriculture*. Cambridge University Press.

Hastings, A., 1925 *Nigeria days*. London: Bodley Head.

Hays, Henry M., 1975 'The storage of cereal grains in three villages of Zaria Province, northern Nigeria'. *Savanna*, 4/2: 117–24.

Heathcote, R.L. (ed.), 1980 *Perception of desertification*. Tokyo: United Nations University.

    1983 *The arid lands: their use and abuse*. London: Longman.

Heijnen, Joop and Kates, R.W., 1974 'North-east Tanzania: comparative observations along a moisture gradient'. In: White, Gilbert F.(ed.), 1974: 105–14.

Hellden, Ulf, 1978 *Evaluation of Landsat-2 imagery for desertification studies in northern Kordofan, Sudan*. Department of Physical Geography, University of Lund.

Helleiner, G.K., 1966 *Peasant agriculture, government and economic growth in Nigeria*. Yale University Press.

Hendy, C.R.C., 1977 *Animal production in Kano state and the requirements for further study in the Kano close-settled zone*. Land Resource Report 21. Tolworth: Land Resources Division, Overseas Development Administration.

Hergert, G.R., 1975 'An analysis of grasshopper problems in Kano state'. *Samaru Agricultural Newsletter* (Institute for Agricultural Research, Ahmadu Bello University, Zaria), 17: 91–4.

Hewitt, K. (ed.), 1983 *Interpretations of calamity*. Risks and Hazards Series

Bibliography

1. London: Allen and Unwin.

Hewitt, K. and Burton, I., 1971 *The hazardness of a place.* Toronto University Press.

Heyer, J., Roberts, P. and Williams, G. (eds.), 1977 *Rural development in tropical Africa.* London: Macmillan.

Hill, Alan G. (ed.), 1985 *Population, health and nutrition in the Sahel. Issues in the welfare of selected West African communities.* London: Routledge and Kegan Paul.

Hill, Polly, 1972 *Rural Hausa: a village and a setting.* Cambridge University Press.

 1977 *Population, prosperity and poverty. Rural Kano 1900 and 1970.* Cambridge University Press.

 1982 *Dry grain farming families. Hausaland (Nigeria) and Karnataka (India) compared.* Cambridge University Press.

 1986 *Development economics on trial.* Cambridge University Press.

Hills, E.S. (ed.), 1966 *Arid lands. A geographical appraisal.* London: Methuen.

Hinchley, M.T. (ed.), 1978 *Symposium on drought in Botswana.* Gaborone: The Botswana Society.

Hinds, Jeremy, 1978 'Famine in northern Nigeria. Muslim reaction in the press'. *Bulletin of Christian Institutes of Islamic Studies*, 1/1-2: 38-49.

Hjort, A., 1976 'Cattle insurance blocks – an alternative to group ranches'. In: Rapp, Anders et al. (eds.), 1976: 207-15.

Hock, Joan C., 1984 'Monitoring environmental resources through NOAA's polar orbiting satellites'. *ITC Journal*, 4: 263-8.

Hogendorn, Jan S., 1978 *Nigerian groundnut exports: origins and early development.* Zaria: Ahmadu Bello University Press.

Holling, C.S., 1973 'Resilience and stability of ecological systems'. *Annual Review of Ecology and Systematics*, 4: 1-23.

Holy, L., 1980 'Drought and change in a tribal economy: the Berti of northern Darfur'. *Disasters*, 4: 65-71.

Horowitz, Michael M., 1972 'Ethnic boundary maintenance among pastoralists and farmers in the Western Sudan (Niger)'. *Journal of Asian and African Studies*, 7: 105-14.

 1976 'Sahelian pastoral adaptive strategies before and after drought'. In: Paylore, Patricia and Haney, Richard A. Jnr. (eds.), 1976: 26-35.

Howell, F. Clark and Bourlière, François (eds.), 1964 *African ecology and human evolution.* London: Methuen.

Hoyle, B.S. (ed.), 1974 *Spatial aspects of development.* Chichester: Wiley.

Hubert, Henry, 1920 'Le desséchement progressif en Afrique Occidentale'. *Bulletin du Comité d'Etudes Historiques et Scientifiques d'AOF*: 401-67.

Hugo, Graeme J., 1979 'New conceptual approaches to migration in the context of urbanization: a discussion based on Indonesian experience'. In: IUSSP, 1979: 25-30.

 1984 'The demographic impact of famine: a review'. In: Currey, B. and Hugo, G. (eds.), 1984: 7-31.

Hulme, M., 1984 '1983: an exceptionally dry year in central Sudan'.

Weather, 39: 281–85.

Huntingdon, Ellsworth, 1915 *Climate and civilization.* Yale University Press.

Hussein, Abdul Mejid (ed.), 1976 *Rehab. Drought and famine in Ethiopia.* African Environment Special Report 2. London: International African Institute.

Hutchinson, J. and Dalziel, J.M., 1954–68 *The flora of West tropical Africa,* 2nd edn (rev. Keay, R.W.J. et al.). London: Crown Agents.

Hutchinson, P., 1985 'Rainfall analysis of the Sahelian drought in the Gambia'. *Journal of Climatology,* 5: 665–72.

Hyden, Goran, 1983 *No shortcuts to progress.* London: Heinemann.

Hyder, D.N. (ed.), 1978 *Proceedings of the First International Rangeland Congress.* Denver, Colorado: Society for Range Management.

Ibrahim, Fouad, 1978 'Anthropogenic causes of desertification in Western Sudan'. *GeoJournal,* 2–3: 243–54.

IEMVPT, 1979 *Bassin de Lac Tchad. Synthèse des études agropastorales,* 1:1,000,000. Maison–Alfort: Institut d'Elevage et de Médecine Vétérinaire des Pays Tropicaux.

Ilesanmi, O.O., 1973 'An empirical formulation of an ITD rainfall model for the tropics: a case study for Nigeria'. *Journal of Applied Meteorology,* 10: 882–91.

ILO, 1981 *First things first. Meeting the basic needs of the people of Nigeria. Report to the government of Nigeria by a JASPA Basic Needs Mission.* Addis Ababa: International Labour Organisation.

Ireland, A.W., 1962 'The little dry season in southern Nigeria'. *The Nigerian Geographical Journal,* 5: 7–20.

Irons, W. (ed.), 1972 *Perspective on nomadism.* Leiden: E.J. Brill.

IUSSP, 1979 *The territorial mobility of population: rethinking its forms and functions.* IUSSP Paper 13. Liege: International Union for the Scientific Study of Population.

Jackson, J.K., Taylor, G.F. and Condé–Wane, C., 1983 *Management of the natural forest of the Sahel region.* Club du Sahel Paper D(83) 232. Paris: Organisation for Economic Co-operation and Development–Permanent Interstate Committee for Drought Control in the Sahel (CILSS).

Jasinski, J. and Karnovitz, A., 1985 'Weather satellites: tracking the drought'. *Weatherwise,* 38: 79.

Jedrej, M.C., 1985 'Sudan: living with the drought'. *Bulletin of the University of Aberdeen African Studies Group,* no.21: 10–11.

Jiya, Mamman, 1974 'Report of the effect of drought conditions on livestock production in North Western State: 1972/73'. *Nigerian Journal of Animal Production,* 1/1: 17–23.

Johnson, D.L. (ed.), 1977 'The human dimensions of desertification'. *Economic Geography,* 53 (special issue).

1979 'Management strategies for drylands: available options and unanswered questions'. In: Mabbutt, J.A. (ed.), 1979: 26–35.

Jones, Brynmor, 1938 'Desiccation and the West African colonies'. *The Geographical Journal,* 91: 401–23.

Jones, D.G., 1960 'The rise in the water table in parts of Daura and Katsina emirates, Katsina province'. *Geological Survey of Nigeria Records*, 1957: 24–6.

Jones, G.R.J. and Eyre, S.R., 1967 *Geography as human ecology*. London: Edward Arnold.

Jones, M.J. and Wild, A., 1975 *Soils of the West African savanna. The maintenance and improvement of their fertility*. Harpenden: Commonwealth Agricultural Bureau.

Kano State Drought Relief Committee, 1974 *Second progress report on drought relief operations in Kano state*. Kano: Office of the Military Governor.

Kassas, Mohammed, 1985 'Deforestation, desertification, and soil loss'. *Desertification Control Bulletin* (UNEP), no. 12: 12–19.

Kates, R.W. et al., 1981 *Drought impact in the Sahelian-Sudanic zone of West Africa: a comparative analysis of 1910–15 and 1968–74*. Background Paper 2, Centre for Technology, Environment and Development, Clark University, Worcester, Massachusetts.

Kates, R.W., Ausubel, J.H. and Berberian, M. (eds.), 1985 *Climate impact assessment*. Chichester: Wiley, for the Scientific Committee on Problems of the Environment.

Kates, R.W., Johnson, D.L. and Haring, K.J., 1977 *Population, society and desertification*. Background Document, United Nations Conference on Desertification, 1977. Nairobi: United Nations Environment Programme.

Katz, R.W., 1978 'Persistence of subtropical African droughts'. *Monthly Weather Review*, 106: 1017–21.

Kaul, R.N. (ed.), 1970 *Afforestation in arid zones*. The Hague: Junk.

Keay, R.W.J., Onochie, C.F.A. and Stanfield, D.P., 1964 *Nigerian trees* (rev.), 2 vols. Ibadan: Federal Department of Forest Research.

Keys, A. et al., 1950 *The biology of human starvation*. Minneapolis: University of Minnesota Press.

Khan, M.A. 1984 'Hawking business at the major traffic lights in Kano'. Kano Management Forum, Bank of the North – Department of Management Sciences, Bayero University, Kano.

Kidson, J.W., 1977 'African rainfall and its relation to upper air circulation' *Quarterly Journal of the Royal Meteorological Society*, 103: 441–56.

King, J.G.M., 1939 'Mixed farming in Northern Nigeria'. *Empire Journal of Experimental Agriculture*, 7: 271–98.

King, Roger, 1975 'Experiences in the administration of co-operative credit and marketing societies in northern Nigeria'. *Agricultural Administration*, 2: 195–208.

1977 'Cooperative policy and village development in northern Nigeria'. In: Heyer, J. et al. (eds.) 1977: 259–80.

Kolawole, A., 1987 'Environmental change and the South Chad Irrigation Project (Nigeria)'. *Journal of Arid Environments*, 13: 169–76.

Kolawole, M.I., 1974 'Economic aspects of private tractor operations in the savanna zone of Western Nigeria'. *Savanna*, 3/2: 175–84.

Kosinski, Leszek A. and Prothero, R. Mansell (eds.), 1975 *People on the move*. London: Methuen.

Kowal, J.M. and Adeoye, K.B., 1973 'An assessment of aridity and the severity of the 1972 drought in northern Nigeria and neighbouring countries'. *Savanna*, 2/2: 145–58.

Kowal, J.M. and Kassam, A.H., 1973 'An appraisal of drought in 1973 affecting groundnut production in the Guinea and Sudan Savanna areas of Nigeria'. *Savanna*, 2/2: 159–64.

　　1975 'Rainfall pattern in the Sudan Savanna region of Nigeria'. *Weather*, 30: 24–28.

　　1978 *Agricultural ecology of savanna. A study of West Africa*. Oxford: The Clarendon Press.

Kowal, J.M. and Knabe, D.T., 1972 *An agroclimatological atlas of the northern states of Nigeria*. Institute for Agricultural Research, Ahmadu Bello University, Zaria.

Kraus, E.B., 1977 'Subtropical droughts and cross-equatorial energy transports'. *Monthly Weather Review*, 105: 1009–18.

Krings, T., 1985 'Viehhalter contra Ackerbauern: eine Fallstudie aus dem Nigerbinnendelta (Republik Mali)'. *Die Erde*, 116: 197–206.

Kuper, Hilda (ed.), 1965 *Urbanization and migration in West Africa*. University of California Press.

Kura, Auwalu Gwio, 1976 *Social and economic adaptations to the drought in Bedde division, Borno state*. B.Sc. Dissertation, Department of Sociology, Ahmadu Bello University, Zaria.

Labaran, Abubakar, 1986 *Correlates of agrarian capitalism: land appropriation, concentration and the emergent agrarian structure in the Sokoto region*. M.Sc. (Land Resources) Thesis, Department of Geography, Bayero University, Kano.

Labouret, H., 1941 *Paysans d'Afrique Occidentale*. Paris: Gallimard.

Lamb, H.H., 1966 'Climate in the 1960s'. *The Geographical Journal*, 132: 183–213.

　　1974 'Drifting towards drought'. *Geographical Magazine*, 46: 455.

　　1977 'Some comments on the drought in recent years in the Sahel-Ethiopian zone of North Africa'. In: Dalby, David *et al.* (eds.), 1977: 33–7.

Lamb, P.J., 1978 'Large-scale tropical Atlantic circulation patterns associated with sub-Saharan weather anomalies'. *Tellus*, 30: 240–51.

　　1982 'Persistence of sub-Saharan drought'. *Nature*, 299: 46–8.

　　1983 'Sub-Saharan rainfall update for 1982: continued drought'. *Journal of Climatology*, 3: 419–22.

Lamprey, H.F., 1975 *Report on the desert enroachment reconnaissance in northern Sudan: 21 October to 10 November, 1975*. Nairobi: United Nations Environment Programme – United Nations Educational, Scientific and Cultural Organisation (typescript).

Land Resources Division, 1972 *The land resources of North-east Nigeria*, 5 vols. Land Resources Study 9. Surbiton: Land Resources Division, Overseas Development Administration.

Landsberg, H.E., 1975 'Sahel drought; change of climate or part of climate?'. *Archiv für Meteorologie, Geophysik und Bioklimatologie*, Ser. B, 23: 193–200.

Laurent, C.K., 1968 'The use of bullocks for power on farms in Northern Nigeria'. *Bulletin of Rural Economics and Sociology* (University of Ibadan), 3: 235–61.

Lawry, Steven W., Riddell, James C., and Bennett, John W., 1983 'Land tenure policy in African livestock development'. In: Simpson, James R. and Evangelou, Phylo (eds.), 1983: 245–59.

Laya, D., 1975 'Interviews with farmers and livestock owners in the Sahel'. *African Environment*, 1/2: 49–93.

Le Houérou, H.N., 1976 'Rehabilitation of degraded lands'. In: Rapp, Anders *et al.* (eds.), 1976: 189–206.

1985 'Pastoralism'. In: Kates, R.W. *et al.* (eds.), 1985: 155–85.

Le Houérou, H.N. and Lundholm, B., 1976 'Complementary activities for the improvement of the economy and the environment in marginal drylands'. In: Rapp, Anders *et al* (eds.), 1976: 217–29.

Leftwich, Adrian and Harvie, Dominique, n.d. (1986) *The political economy of famine. A preliminary report on the literature, bibliographic resources, research activities and needs in the UK.* Discussion Paper 116, Insititute for Research in the Social Sciences, Institute of Social and Economic Research, Department of Politics, University of York.

Leng, Gunter, 1982 *Desertification. A bibliography with regional emphasis on Africa.* Schwerpunkt Geographie, Fachbereich 1, Universität Bremen.

Leow, K.S. and Ologe, K.O., 1981 'Desertification: a precipitation consideration'. 24th Annual Conference. Kano: The Nigerian Geographical Association.

Lewicki, T., 1974 *West African food in the Middle Ages.* Cambridge University Press.

Lofchie, Michael F., 1975 'The political and economic origins of African hunger'. *Journal of Modern African Studies*, 13: 551–67.

1986 'Africa's agricultural crisis: an overview'. In: Commins, Stephen K. *et al.* (eds.), 1986: 3–18.

Longhurst, Richard, 1984 *The energy trap. Work, nutrition and child malnutrition in northern Nigeria.* Cornell International Nutrition Monograph Series 13, Cornell University.

Lough, J.M., 1980 'West African rainfall variations and tropical Atlantic sea surface temperatures'. *Climate Monitor*, 9/5: 150–7.

1986 'Tropical Atlantic sea surface temperatures and rainfall variations in subSaharan Africa'. *Monthly Weather Review*, 114: 561–70.

Lovejoy, P.E., 1986 *Salt of the desert sun.* Cambridge University Press.

Lovejoy, P.E. and Baier, S., 1976 'The desert-side economy of the central Sudan'. In: Glantz, Michael H. (ed.), 1976a: 145–75.

Lucas, Grenville Ll. and Wickens, Gerald E., 1986 'Arid land plants – the data crisis'. Kew:Royal Botanic Gardens.

Lundholm, B., 1976 'Adaptations in arid ecosystems'. In: Rapp, Anders *et al.* (eds.), 1976: 19–28.

MAB, 1975 *The Sahel. Ecological approaches to land use.* MAB Technical Note 1. Paris: United Nations Educational, Scientific and Cultural

Organisation.
1977 *Development of arid and semi-arid lands: obstacles and prospects.*
   MAB Technical Note 6. Paris: United Nations Educational, Scientific and Cultural Organisation.
1979 *Map of the world distribution of arid regions.* MAB Technical Note 7. Paris: United Nations Educational, Scientific and Cultural Organisation.
Mabbutt, J.A. (ed.), 1979 *Proceedings of the Khartoum Workshop on Arid Lands Management, 1978.* Tokyo: The United Nations University.
1985 'Desertification of the world's rangelands'. *Desertification Control Bulletin* (UNEP), no. 12: 1–11.
Mabogunje, Akin L., 1970(a) 'Migration policy and regional development in Nigeria'. *Nigerian Journal of Economic and Social Studies,* 12: 243–62.
1970(b) 'Systems approach to a theory of rural – urban migration'. *Geographical Analysis,* 2: 1–18.
1972 *Regional mobility and resource development in West Africa.* Montreal: McGill-Queens University Press.
Mabogunje, Akin, L. and Gana, Jerry, 1981 *Rural development in Nigeria: case study of the Funtua Integrated Rural Development Project, Kaduna state, Nigeria.* United Nations Centre for Regional Development, University of Ibadan.
McCown, R.L., Haaland, G. and De Haan, C., 1979 'The interaction between cultivation and livestock production in semi-arid Africa'. In: Hall, A.E. *et al.* (eds.), 1979: 297–332.
McDonnell, G., 1964 'The dynamics of geographic change: the case of Kano'. *Annals, Association of American Geographers,* 54: 355–71.
McDowell, C.M., 1964 *An introduction to the problems of land ownership in Northern Nigeria.* Institute of Administration, Ahmadu Bello University, Zaria.
Macleod, N.H., 1976 'Dust in the Sahel: cause of drought?' In: Glantz, Michael H. (ed.), 1976a: 214–31.
McTainsh, G.H., 1980 'Harmattan dust deposition in northern Nigeria'. *Nature,* 286: 587–8.
1984 'The nature and origin of the aeolian mantles of central northern Nigeria'. *Geoderma,* 33: 13–37.
1985 'Desertification and dust monitoring in West Africa'. *Desertification Control Bulletin* (UNEP), no. 12: 26–33.
1986 'A dust monitoring programme for desertification control in West Africa'. *Environmental Conservation,* 13: 17–25.
McTainsh, G.H. and Walker, P.H., 1982 'Nature and distribution of Harmattan dust'. *Zeitschrift für Geomorphologie,* 26: 417–35.
Maddox, G.H., 1986 'Njaa: food shortages and famines in Tanzania between the wars'. *International Journal of African Historical Studies,* 19: 17–34.
Mageed, Y.A., 1986 *Anti-desertification technology and management. Assessment of water resources in arid and semi-arid regions.* Nairobi: United Nations Environment Programme.

Mainguet, Monique, 1980 'L'interdépendence des mécanismes éoliens dans les zone arides du Sahara et dans leurs marges sahéliennes; ses effets sur la propagation de la désertification'. In: Meckelein, W. (ed.), 1980: 107–23.

Mainguet, Monique, Canon, L. and Merrer, J.Y., 1976 'Recherche sur les photographies aériennes et les images-satellites, d'indices de désertification dans le Sahel, à l'est du Niger'. *Photo-Interprétation*, 3: 19–41.

Mainguet, Monique, Canon-Cossus, Lydie and Chemin, Marie Christine, 1979 'Dégradation dans les régions centrales de la République du Niger: degré de responsabilité de la nature du milieu, de la dynamique externe et de la mise en valeur par l'homme'. *Travaux de l'Institut de Géographie de Reims*, no. 39–40: 61–73.

Mainguet, Monique and Cossus, L., 1980 'Le Sahel, bordure méridionale du Sahara: étude de géographie'. *Historiens et Géographes*, June–July: 813–30.

Maley, J., 1981 *Etudes palynologiques dans le bassin du Tchad et paléoclimatologie de l'Afrique nord-tropicale de 30,000 ans à l'époque actuelle*. Travaux et Documents 129. Paris: Office de la Recherche Scientifique et Technique Outre-mer.

Mallory, W.H., 1926 *China: land of famine*. Special Publication 6. New York: American Geographical Society.

Mann, H.S. (ed.), 1978 *Arid Zone Research and Development. Proceedings and Selected Papers: International Symposium, Jodhpur, 1978*. Jodhpur: Central Arid Zone Research Institute.

Mason, B.J., 1976 'Towards the understanding and prediction of climatic variations'. *Quarterly Journal of the Royal Meteorological Society*, 102: 473–98.

Matlock, W.G. and Cockrim, E.L., 1976 'Agricultural production systems in the Sahel'. In: Glantz, Michael H. (ed.), 1976a: 232–55.

Mauny, Raymond, 1961 *Tableau géographique de l'Ouest Africain au Moyen Age d'après les sources écrites, la tradition et l'archéologie*. Dakar: Institut Français d'Afrique Noire.

Mbithi, P.M. and Wisner, B., 1972 *Drought and famine in Kenya: magnitude and attempted solutions*. Institute of Development Studies, University of Nairobi.

Meckelein, W. (ed.), 1980 *Desertification in extremely arid environments*. Stuttgarter Geographische Studien 95, Geographisches Institut der Universität Stuttgart.

Meek, C.K., 1946 *Land law and custom in the colonies*. Oxford University Press.

Meillassoux, C., 1974 'Development or exploitation: is the Sahel famine good business?' *Review of African Political Economy*, 1: 27–33.

Mensah, Ebow, 1977 'A note on the distribution of beggars in Zaria'. *Savanna*, 6/1: 73–6.

Mensching, Horst G., 1980 'The Sahelian zone and the problems of desertification. Climatic and anthropogenic causes of desert encroachment'. *Palaeoecology of Africa and the Surrounding Islands*

(Balkema), 12: 257–66.
1982 'Nomads and farmers in the West African Sahel – problems of competing land use'. *Applied Geography and Development*, 20: 7–19.
1983 'The development of small and middle-sized settlements as a measure to combat desertification, with special reference to the Sahelian Zone'. *ITCC Review* (Association of Engineers and Architects in Israel), 12.
1985 'Die Sahelzone – Probleme ohne Lösung?' *Die Erde*, 116: 99–108.
Mensching, Horst and Ibrahim, Fouad, 1977 'The problem of desertification in and around arid lands'. *Applied Sciences and Development*, 10: 7–43.
Michon, Paul, 1973 'Le Sahara avance-t-il vers le sud?'. *Bois et Forêts des Tropiques*, 150: 3–14.
Milas, Seifulaziz, 1984 'Desert spread and population boom'. *Desertication Control Bulletin* (UNEP), no.11: 7–16.
Miles, M.K. and Folland, C.K., 1974 'Changes in the latitude of the climate zones of the northern hemisphere'. *Nature*, 252: 616.
Ministry of Agriculture, n.d. 'The dying rivers of the Northern Region of Nigeria'. Kaduna: Northern Nigeria Ministry of Agriculture.
Mischlich, A., 1943 'Religiöse und weltliche Gesänge der Mohammedaner aus dem Sudan'. *Studien zur Auslandskunde: Afrika*, 2/3: 129–98.
MIT, 1974 *A framework for evaluating long-term strategies for the development of the Sahel-Sudan region*, 10 vols. Centre for Policy Alternatives, Massachussetts Institute of Technology, Boston.
Mkunduge, G.L., 1973 'The Ukaguru environment: traditional and recent responses to food shortages'. *Journal of the Geographical Association of Tanzania*, 8: 63–85.
Monod, T. (ed.), 1975 *Pastoralism in tropical Africa*. Oxford University Press, for the International African Institute.
Moock, J.L. (ed.), 1986 *Understanding Africa's rural households and farming systems*. Boulder, Colorado: Westview Press.
Morgan, W.B. and Pugh, J.C., 1969 *West Africa*. London: Methuen.
Mormoni, Zainab Ladi, 1976 *The effect of the 1972/73 drought on the economic role of women in Hurumi hamlet, Kausani, Kano state, Nigeria*. B.A. Dissertation, Department of Geography, Ahmadu Bello University, Zaria.
Mortimore, Michael, 1967 'Land and population pressure in the Kano Close-Settled Zone, Northern Nigeria'. *The Advancement of Science*, 23: 677–88. In: Prothero, R. Mansell (ed.), 1972: 60–70.
1968 'Population distribution, settlement and soils in Kano province, Northern Nigeria, 1931–62'. In: Caldwell, J.C. and Okonjo, C. (eds.), 1968: 298–306.
1970 'Population densities and rural economies in the Kano Close-Settled Zone, Nigeria'. In: Zelinsky, Wilbur *et al.* (eds.), 1970: 280–8.
1971 'Population densities and systems of agricultural land use in northern Nigeria'. *The Nigerian Geographical Journal*, 14: 3–15.
1972 'Some aspects of rural–urban relations in Kano, Nigeria'. In:

Vennetier, P. (ed.), 1972: 871–88.
1973 'Famine in Hausaland, 1973'. *Savanna*, 2/2: 103–8.
1974 'The demographic variable in regional planning in Kano state, Nigeria'. In: Hoyle, B.S. (ed.), 1974: 129–46.
1975(a) *A provisional report on the effects of the drought in Kano state, 1972–4.* Department of Geography, Ahmadu Bello University, Zaria.
1975(b) *A report on drought responses among the farmers of northern Kano state, 1972–4.* Department of Geography, Ahmadu Bello University, Zaria.
1976 *Changes in agrarian structure in Dagacheri village, Kano state, Nigeria.* Study commissioned by the Human Resources, Institutions and Agrarian Reform Division, FAO. Department of Geography, Ahmadu Bello University, Zaria.
1977 'Northern Nigeria'. *World Atlas of Agriculture, vol. 4, Africa.* Novara: Instituto Geografico de Agostini.
1978(a) 'Livestock production'. In: Oguntoyinbo, J.S. *et al.* (eds.), 1978: 240–60.
1978(b) 'Grain reserves at the village level in drought-prone areas of Nigeria'. In: Van Apeldoorn, G.J. (ed.), 1978b: 220–6.
1979 'The supply of urban foodstuffs in northern Nigeria'. In: Coppock, J.T. (ed.), 1979: 45–66.
1982 'Framework for population mobility: the perception of opportunities in Nigeria'. In: Clarke, J.I. and Kosinski, L.A. (eds.), 1982: 50–7.
1984 'Nigeria'. In: Nag, Prithvish (ed.), 1984: 146–61.
Mortimore, Michael and Wilson, J., 1965 *Land and people in the Kano close-settled zone.* Occasional Paper 1, Department of Geography, Ahmadu Bello University, Zaria.
Motha, R.P., Le Duc, S.K., Steyaert, L.T., Sakamoto, C.M. and Strommen, N.D., 1980 'Precipitation patterns in West Africa'. *Monthly Weather Review*, 108: 1567–78.
Mukhtar, Ibrahim, 1974 'Drought status in Kano state'. *Nigerian Journal of Animal Production*, 1/1: 28–31.
Mustapha, Shettima, 1977 'The encroachment of the desert in the central Bilad al-Sudan: but is it really encroaching?'. Third International Conference on the Bilad al-Sudan, University of Khartoum, 1977.
Nag, Prithvish (ed.), 1984 *Census mapping survey.* New Delhi: Concept Publishing, for the International Geographical Union Commission on Population Geography.
National Academy of Sciences, 1975 *Arid lands of subSaharan Africa.* Washington, DC: National Academy of Sciences.
National Committee on Arid Zone Afforestation, 1978 *A report by the National Committee on Arid Zone Afforestation prepared for the Federal Military Government of Nigeria.* Lagos: Federal Department of Forestry.
Ndaks, George David, 1976 *The effects of the 1972–73 drought on the*

agricultural community of Sandamu in Daura emirate, Kaduna state. B.A. Dissertation, Department of Geography, Ahmadu Bello University, Zaria.

Nelson, J.M., 1976 'Sojourners and new urbanites: causes and consequences of temporary versus permanent cityward migration in developing countries'. *Economic Development and Cultural Change*, 24: 721–57.

Netting, Robert, McC., 1968 *Hill farmers of Nigeria. Cultural ecology of the Kofyar of the Jos Plateau.* Seattle: University of Washington Press.

Nicholson, Sharon E., 1978 'Climatic variations in the Sahel and other African regions during the past five centuries'. *Journal of Arid Environments*, 1: 3–24.

 1979 'Revised rainfall series for the West African subtropics'. *Monthly Weather Review*, 107: 620–3.

 1980 'The nature of rainfall fluctuations in subtropical West Africa'. *Monthly Weather Review*, 108: 473–87.

 1982 *The Sahel: a climatic perspective.* Club du Sahel Paper D(82)187. Paris: Organisation for Economic Co-operation and Development – Permanent Interstate Committee for Drought Control in the Sahel (CILSS).

 1983 'Sub-Saharan rainfall in the years 1976–80: evidence of continued drought'. *Monthly Weather Review*, 111: 1646–54.

Norman, D.W., 1972 *An economic survey of three villages in Zaria Province. 2, Input–output study. Vol. 1, Text.* Samaru Miscellaneous Paper 37, Institute for Agricultural Research, Ahmadu Bello University, Zaria.

Norman, D.W. and Baker, D.C., 1986 'Components of farming systems research, FSR capability, and experiences in Botswana'. In: Moock, J.L. (ed.), 1986: 36–57.

Norman, D.W., Simmons, Emmy B. and Hays, Henry M., 1982 *Farming systems in the Nigerian savanna. Research and strategies for development.* Boulder, Colorado: Westview Press.

Northern Nigeria, 1966 *Northern Nigeria Statistical Yearbook.* Kaduna: Government Printer.

Norton, C.C., Mosher, F.R. and Hunton, B., 1979 'An investigation of surface albedo variations during the most recent Sahel drought'. *Journal of Applied Meteorology*, 18: 1252–62.

Offodile, M.E., 1971 'Ground-water level fluctuations in the East Chad Basin of Nigeria'. *Geological Survey of Nigeria Reports*, 1494.

Ogallo, L., 1979 'Rainfall variability in Africa'. *Monthly Weather Review*, 107: 1133–9.

Ogundana, Babafemi, 1978 'Ports and external trade'. In: Oguntoyinbo, J.S. *et al.* (eds.), 1978: 338–50.

Oguntoyinbo, J.S., 1974 'Land use and reflection coefficient (albedo) map for southern parts of Nigeria'. *Agricultural Meteorology*, 13: 227–37.

 1981 'Climatic variability and food crop production in West Africa'. *GeoJournal*, 5: 139–49.

Oguntoyinbo, J.S., Areola, O.O. and Filani, M. (eds.), 1978 *A geography of Nigerian development*. Ibadan: Heinemann.

Oguntoyinbo, J.S. and Richards, Paul, 1977 'The extent of intensity of the 1969-73 drought in Nigeria: a provisional analysis',. In: Dalby, David *et al.* (eds.), 1977: 114-26.

—— 1978 'Drought and the Nigerian farmer'. *Journal of Arid Environments*, 1: 165-94.

Ojo, G.O.A., Onyewotu, L.O.Z. and Ujah, J.E., 1987 'Use and management of shelterbelts'. In: Sagua, V.O., *et al.* (eds.), 1987: 251-8.

O'Keefe, Phil and Wisner, Ben, 1975 'African drought – the state of the game'. In: Richards, Paul (ed.), 1975: 31-9.

—— (eds.), 1977 *Land use and development*. African Environment Special Report 5. London: International African Institute.

Olatunbosun, Dupe, 1975 *Nigeria's neglected rural majority*. Ibadan. Nigerian Institute for Social and Economic Research.

Olofin, E.A., 1982 'Some effects of the Tiga Dam and Reservoir on the downstream environment in the Kano River Basin, Kano state/Nigeria'. *Beiträge zur Hydrologie*, 3: 11-28.

—— 1985 'Climatic constraints to water resource development in the Sudano-Sahelian Zone of Nigeria'. *Water International*, 10: 29-37.

Olowu, J.A.I. and Uzoma, J.U., 1965 *Availability of groundwater in the Chad Basin of Bornu and Dikwa emirate, Northern Nigeria. A description for the general reader*. Washington, DC: United States Geological Survey.

Olsson, L., 1983 'Desertification or climate? Investigation regarding the relationship between land degradation and climate in the central Sudan'. *Lund Studies in Geography*, Ser. A, Physical Geography, 60: 1-36.

Omotosho, J.B., 1985 'The separate contributions of line squalls, thunderstorms and the monsoon to the total rainfall in Nigeria'. *Journal of Climatology*, 5: 543-52.

Onyemelukwe, J.O.C., 1974 'Some factors in the growth of West African market towns: the example of pre-Civil War Onitsha, Nigeria'. *Urban Studies*, 11: 47-59.

Ooi Jin Bee (ed.), 1983 *Natural resources in tropical countries*. Singapore University Press, for the Commonwealth Geographical Bureau.

Orogun, Ewenosore, 1986 *Land administration on the Kano River Project*. M.Sc. (Land Resources) Thesis, Department of Geography, Bayero University, Kano.

Otegbeye, G.O. and Ogigirigi, M.A., 1987 'Improvement of drought resistant tree species for plantation development in the semi-arid zones of Nigeria'. In Sagua, V.O. *et al.* (eds.), 1987: 259-69.

Otterman, J., 1981 'Satellite and field studies of man's impact on the surface in arid regions'. *Tellus*, 33: 68-77.

Owen, J.A. and Folland, C.K., 1987 'Modelling the influence of sea surface temperatures on tropical rainfall'. *Proceedings of the International Geographical Union Study Group on Recent Climatic Change*, Sheffield.

Oxby, Clare, 1981 'Group ranches in Africa'. *ODI Review*, 2: 2–13.
Paden, John N., 1975 *Religion and political culture in Kano*. University of California Press.
Palmer, T.N., 1986 'Influence of the Atlantic, Pacific and Indian Oceans on Sahel rainfall'. *Nature*, 322: 251–3.
Parker, D.E., Folland, C.K. and Ward, M.N., 1987 'Sea surface temperature anomaly patterns and prediction of seasonal rainfall in the Sahel region of Africa'. *Proceedings of the International Geographical Union Study Group on Recent Climatic Change*, Sheffield.
Parry, D.E. and Trevitt, J.W., 1979 'Mapping Nigeria's vegetation from radar'. *The Geographical Journal*, 145: 265–81.
Paylore, Patricia and Haney, Richard A. Jnr. (eds.), 1976 *Proceedings of the West Africa Conference, Tucson, Arizona, 1976*. University of Arizona.
Piché, V. and Gregory, J., 1977 'Pour une mise en contexte de la famine: le cas du Liptako-Gourma'. In: Dalby, David *et al.* (eds.), 1977: 170–85.
Pitte, J.-R., 1975 'La sécheresse en Mauritanie'. *Annales de Géographie*, 84: 641–64.
Poncet, Yveline, 1973 *Cartes ethno-démographiques du Niger*. Etudes Nigériennes 32. Niamey: Centre Nigérien de Recherches en Sciences Humaines.
   1974 *La sécheresse en Afrique Sahélienne: une étude micro-régionale en République du Niger. La région des Dallols*. Paris: Centre de Développement, Organisation for Economic Co-operation and Development.
Porter, P.W., 1965 'Environmental potentials and economic opportunities – a background for cultural adaptation'. *The American Anthropologist*, 67: 409–20.
Prirard, F., 1966 'Géomorphologie du Manga Nigérien'. *Bulletin, Institut Fondamentale d'Afrique Noire*, Sér. A, 28: 421–5.
Prothero, R. Mansell, 1959 *Migrant labour from Sokoto province, Northern Nigeria*. Kaduna: Government Printer.
   1962 'Some observations on desiccation in North-Western Nigeria'. *Erdkunde*, 16: 111–19.
   (ed.), 1972 *People and land in Africa south of the Sahara*. Oxford University Press.
   1974 'Some perspectives on drought in north-west Nigeria'. *African Affairs*, 73: 162–9.
Pullan, R.A., 1962 *A report on the soil reconnaissance survey of the Nguru-Hadejia-Gumel area*. Bulletin 18, Soil Survey Section, Institute for Agricultural Research, Ahmadu Bello University, Zaria.
   1974 'Farmed parkland in West Africa'. *Savanna*, 3/2: 119–52.
Raeburn, C., 1928 *The Nigerian Sudan. Some notes on water supply and cognate subjects*. Pamphlet 1. Kaduna: Geological Survey of Nigeria.
Rapp, Anders, 1976 'Needs of environmental monitoring for desert encroachment control'. In: Rapp, Anders *et al.* (eds.), 1976: 231–6.
   1978 *A review of desertization in Africa – water, vegetation and man*.

Report no. 1. Stockholm: Secretariat for International Ecology.

Rapp, Anders and Hellden, Ulf, 1977 *Research on environmental monitoring methods for land-use planning in African drylands.* Rapporter och Notiser 42, Naturgeografiska Institution, University of Lund.

Rapp, Anders, Le Houérou, H.N. and Lundholm, B. (eds.), 1976 *Can desert encroachment be stopped?* Ecological Bulletin 24. Stockholm: Secretariat for International Ecology.

Raulin, H., 1964 *Techniques et bases socio-économiques des sociétés rurales Nigériennes.* Etudes Nigériennes 12. Niamey: Centre National de Recherche en Sciences Humaines.

Raynaut, Claude, 1977 'Lessons of a crisis'. In: Dalby, David *et al.* (eds.), 1977: 17–32.

Reining, P. (comp.), 1978 *Handbook on desertification indicators.* Washington, DC: American Association for the Advancement of Science.

  1980 *Challenging desertification in West Africa: insights from Landsat into carrying capacity, cultivation and settlement sites in Upper Volta and Niger.* Papers in International Studies 39, Centre for International Studies, Ohio University.

Renner, G.T., 1926 'A famine zone in Africa: the Sudan'. *The Geographical Review,* 16: 583–96.

Richards, Paul (ed.), 1975 *African environment. Problems and perspectives.* African Environment Special Report 1. London: International African Institute.

  1983 'Ecological change and the politics of African land use'. *African Studies Review,* 26/2: 1–72.

  1985 *Indigenous agricultural revolution. Ecology and food production in West Africa.* London: Hutchinson.

  1986 *Coping with hunger. Hazard and experiment in an African rice-producing system.* London: Allen and Unwin.

  1987 'The politics of famine – some recent literature'. *African Affairs,* 86: 111–16.

Ringrose, Susan and Matheson, Wilma, 1986 'Desertification in Botswana: progress towards a viable monitoring system'. *Desertification Control Bulletin* (UNEP), no.13: 6–11.

Ripley, E.A., 1976(a) 'Comment'. *Quarterly Journal of the Royal Meteorological Society,* 102: 466–7.

  1976(b) 'Drought in the Sahara: insufficient biogeophysical feedback?'. *Science,* 191: 100.

Rippstein, G., Peyre de Fabregues, B. *et al.,* 1972 *Modernisation de la zone pastorale du Niger,* 2 vols. Etude Agrostologique 33. Maison-Alfort: Institut d'Elevage et de Médecine Vétérinaire des Pays Tropicaux.

Roche, C.J.R., 1984 *Cereal banks in Burkina Faso: a case study.* Liverpool Papers in Human Geography, Working Paper 18. Department of Geography, University of Liverpool.

Roder, Wolf, 1976 'Coping with drought in Africa'. In: Paylore, Patricia and Haney, Richard A. Jnr. (eds.), 1976: 36–43.

Romanucci-Ross, L., Moerman, D.E. and Tancredi, L.R. (eds.),
1983 *The anthropology of medicine*. South Hadley, Massachusetts: J.F. Bergin.

Ross, Paul J., 1987 'Land as a right to membership: land tenure dynamics in a peripheral area of the Kano Close-Settled Zone'. In: Watts, Michael (ed.), 1987: 223–47.

Rossignol-Strick, M. and Duzer, D., 1980 'Late Quaternary and West African climate inferred from palynology of Atlantic deep-sea cores'. *Palaeoecology of Africa and the Surrounding Islands* (Balkema), 12: 257–66.

Rouch, Jean, 1956 'Migrations au Ghana'. *Journal de la Société des Africanistes*, 26: 33–196.

Rowling, C.W., 1952 *Report on land tenure in Kano province*. Kaduna: Government Printer.

Russell, Clifford S., 1970 'Losses from natural hazards'. *Land Economics*, 46: 383–93.

Sagua, V.O., Enabor, E.E., Kio, P.R.O., Ojanuga, A.U., Mortimore, M. and Kalu, A.E. (eds.), 1987 *Ecological disasters in Nigeria: drought and desertification*. Proceedings of the National Workshop on Ecological Disasters in Nigeria, Drought and Desertification, Kano, 9–12 December 1985. Lagos: Federal Ministry of Science and Technology.

Salem, B. Ben, 1985 'A strategy on the role of forestry in combatting desertification. Summary of Report of the Expert Consultation on the role of forestry in combatting desertification, Saltillo, Mexico'. International Arid Lands Research and Development Conference – Arid Lands Today and Tomorrow, Tucson, Arizona.

Salifou, André, 1975 'When history repeats itself: the famine of 1931 in Niger'. *African Environment*, 1/2: 22–48.

Sandbrook, J., 1986 'The state and economic stagnation in tropical Africa'. *World Development*, 14/3: 319–32.

Sandford, Stephen, 1976 'Pastoralism under pressure'. *ODI Review*, no.2: 45–68.

1978 'Towards a definition of drought'. In: Hinchley, M.T. (ed.), 1978: 33–40.

1982 'Pastoral strategies and desertification: opportunism and conservatism in dry lands'. In: Spooner, B. and Mann, H.S. (eds.), 1982: 61–80.

1983 *Management of pastoral development in the Third World*. Chichester: Wiley.

Schove, D.T., 1977 'African droughts in the spectrum of time'. In: Dalby, David et al. (eds.), 1977: 38–53.

Seaman, J. et al., 1973 'An enquiry into the drought situation in Upper Volta'. *The Lancet*, 6 October: 774–8.

Seignobos, C., 1979 *Les systèmes de défense végétaux pré-coloniaux, famines, strategies*. Annales de l'Université du Tchad. Paris: L'Harmattan.

Sen, Amartya, 1981 *Poverty and famines: an essay on entitlement*. Oxford: The Clarendon Press.

Sheets, H. and Morris, R., 1974 *Disaster in the desert: failures of international relief in the West African drought*. Washington, DC: Carnegie Endowment for International Peace.

Shenton, B. and Watts, M., 1979 'Capitalism and hunger in northern Nigeria'. *Review of African Political Economy*, 15–16: 53–62.

Sidikou, A.H., 1977 'La stratégie adaptative et ses limites des Zarma du Zarmaganda (République du Niger) face à la sécheresse actuelle (1965–1976)'. In: Gallais, Jean et al., 1977: 141–68.

Simmons, E.B., 1976 *Calorie and protein intakes in three villages of Zaria province, May 1970–July 1971*. Samaru Miscellaneous Paper 55, Institute for Agricultural Research, Ahmadu Bello University, Zaria.

Simpson, James R. and Evangelou, Phylo (eds.), 1983 *Livestock development in Subsaharan Africa. Constraints, prospects, policy*. Boulder, Colorado: Westview Press.

Sirag, Mustafy Ismail, 1980 'Balanites aegyptiaca'. In: Carl Duisberg Ges., 1980: 101–4.

Sivakumar, M.V.K., Virmani, S.M. and Reddy, S.J., n.d. (1984) *Rainfall climatology of West Africa: Niger*. Information Bulletin 5. Andhra Pradesh: International Crop Research Institute for the Semi-arid Tropics.

Smith, M.G., 1955 *The economy of Hausa communities of Zaria*. Colonial Research Series 16. London: Her Majesty's Stationery Office.

Sorokin, P.A., 1942 *Man and society in calamity. The effects of war, revolution, famine, pestilence, upon human mind, behavior, social organization and cultural life*. New York: E.P. Dutton.

Special Sahelian Office, 1973 *Sector review: agriculture*, ST/SSO/9. New York: United Nations Secretariat (Mimeo).

Spink, P.C., 1985 'Letter. Drought in Africa'. *Journal of Meteorology*, 10: 24.

Spooner, Brian, 1982 'Rethinking desertification: the social dimension'. In: Spooner, B. and Mann, H.S. (eds.), 1982: 1–24.

Spooner, B. and Mann, H.S. (eds.), 1982 *Desertification and development: dryland ecology in social perspective*. New York: Academic Press.

Stamp, L.D., 1940 'The southern margins of the Sahara: comments on some recent studies on the question of desiccation in West Africa'. *The Geographical Review*, 30: 297–300.

Stanfield, D.P., 1970 *The Flora of Nigeria: grasses*, 2 vols. Ibadan University Press.

Starns, W.W., 1974 *Land tenure among the Hausa*. Report 104, Land Tenure Center, University of Wisconsin, Madison.

Stebbing, E.P., 1935 'The encroaching Sahara: the threat to the West African colonies'. *The Geographical Journal*, 85: 506–24.

1937(a) 'The threat of the Sahara'. *Journal of the Royal African Society, Supplement* (May).

1937(b) *The forests of West Africa and the Sahara. A study of modern conditions*. Edinburgh: Chambers.

1938(a) 'The man-made desert in Africa. Erosion and drought'. *Journal of the Royal African Society, Supplement* (January).

1938(b) 'Africa and its intermittent rainfall: the role of the savannah forest'. *Journal of the Royal African Society, Supplement* (August).
1953 *The creeping desert in the Sudan and elsewhere in Africa, 15–13 degrees latitude*. Khartoum: McCorquodale.
Stern, R.D., Dennett, M.D. and Larbutt, D.J., 1981 'The start of the rains in West Africa'. *Journal of Climatology*, 1: 59–68.
Stiles, Daniel 1984 'Desertification: a question of linkage'. *Desertification Control Bulletin* (UNEP), no. 11: 1–6.
Stryker, J. Dirk, 1983 'Land use development in the pastoral zone of West Africa'. In: Simpson, James R. and Evangelou, Phylo (eds.), 1983: 175–85.
Susman, P., O'Keefe, P. and Wisner, B., 1983 'Global disasters, a radical interpretation'. In: Hewitt, K. (ed.), 1983: 263–83.
Swami, Kala, 1973 *Moisture conditions in the savanna region of West Africa*. Savanna Research Series 8, Department of Geography, McGill University, Montreal.
Swift, Jeremy, 1975 'Pastoral nomadism as a form of land use: the Twareg of the Adrar n Iforas'. In: Monod, T.(ed.), 1975: 443–54.
1977(a) 'Sahelian pastoralists: underdevelopment, desertification and famine'. *Annual Review of Anthropology*, 6: 457–78.
1977(b) 'Desertification and man in the Sahel'. In: O'Keefe, Phil and Wisner, Ben (eds.), 1977: 171–8.
1982 'The future of African hunter-gatherer and pastoral peoples'. *Development and Change*, 13: 159–81.
Swift, Jeremy and Maliki, Angelo, 1984 *A cooperative development experiment among pastoral herders in Niger*. Pastoral Network Paper 18c. London: Overseas Development Institute.
Swindell, K., 1984 'Farmers, traders and labourers: dry season migration from north-west Nigeria, 1900–33'. *Africa*, 54: 3–19.
Swindell, K., Baba, J.M. and Mortimore, Michael (eds.), in press (1987) *Inequality and development. Some Third World perspectives*. London: Macmillan.
Talbot, M.R. and Williams, M.A.J., 1978 'Erosion of fixed dunes in the Sahel, central Niger'. *Earth Surface Processes*, 3: 107–14.
Tanaka, M., Weare, B.C., Navajo, A.R. and Newell, R.E., 1975 'Recent African rainfall patterns'. *Nature*, 255: 201–3.
Tannehill, I.R., 1947 *Drought. Its causes and effects*. Oxford University Press.
Thomas, M.F. and Whittington, G.W. (eds.), 1969 *Environment and land use in Africa*. London: Methuen.
Tiffen, Mary, 1975 *The enterprising peasant: economic development in Gombe Emirate, NE State, Nigeria, 1900–1968*. Overseas Research Publication 2. London: Her Majesty's Stationery Office.
Timberlanke, Lloyd, 1985 *Africa in crisis. The causes, the cures of environmental bankruptcy*. London: Earthscan (International Institute of Environment and Development).
(ed.), 1986 *The encroaching desert. The consequences of human failure. A report for the Independent Commission on International Humanitarian Issues*. London: Zed Books.

Tolba, M.K., 1986 'Desertification'. *WMO Bulletin*, 35: 17–22.
Toupet, Charles, 1975 *La sédentarisation des nomades en Mauritanie centrale Sahélienne*. Thesis, University of Paris VII. Paris: Librairie Honore Champion.
  1977 'La grande sécheresse en Mauritanie'. In: Dalby, David *et al.* (eds.), 1977: 109–13.
Tourte, R., 1971   'Thémes legers. Thèmes lourdes. Systèmes intensifs. Voies différentes ouvertes au développement agricole du Sénégal'. *Agronomie Tropicale*, 26: 632–71.
Toutain, B., 1977   'Essais de régéneration mécanique de quelques parcours sahéliens dégradés. *Revue d'Elevage et de Médecine Vétérinaire des Pays Tropicaux*, 30/2: 191–8.
Trevallion, B.W., 1966   *Metropolitan Kano Twenty Year Development Plan 1963–83*. Newman Neame, for the Greater Kano Planning Authority.
Trilsbach, A. and Hulme, M., 1984   'Recent rainfall changes in central Sudan and their physical and human implications'. *Transactions, Institute of British Geographers*, NS 9: 280–98.
Turabu, H.M., 1987   'Sand dune fixation techniques as practised in arid and semi-arid areas of Nigeria'. In: Sagua, V.O. *et al.* (eds.), 1987: 285–92.
Turner, Beryl, 1985   'The classification and distribution of fadamas in central northern Nigeria'. *Zeitschrift für Geomorphologie*, 52: 87–113.
UNEP, 1977(a)   United Nations Conference on Desertification, 29 August–9 September 1977. *World Map of Desertification, at a scale of 1:25,000,000*. Nairobi: United Nations Environment Programme.
  1977(b)   United Nations Conference on Desertification, 29 August–9 September 1977. *Status of desertification in the hot arid regions. Climate aridity index map. Experimental world scheme of aridity and drought probability at a scale of 1:25,000,000*. Nairobi: United Nations Environment Programme.
  1977(c)   *Report of the United Nations Conference on Desertification, Nairobi, 29 August–9 September 1977*. Nairobi: United Nations Environment Programme.
  1977(d)   *Report of the Working Group on drought risk insurance and monitoring the human condition, Berlin, 1977*. Nairobi: United Nations Environment Programme.
  1977(e)   United Nations Conference on Desertification, 29 August–9 September 1977. *Desertification: an overview*. Nairobi: United Nations Environment Programme.
  1978   'United Nations Conference on Desertification'. *Desertification Control Bulletin* (UNEP), 1/1.
  1981   *Environment and development in Africa*. UNEP Studies 2. Oxford: Pergamon, for United Nations Environment Programme.
  1983   'Ecology and environment: what do we know about desertification?' *Desertification Control Bulletin* (UNEP), no.9: 2–9.
  1984   *Activities of the United Nations Environment Programme in the combat against desertification. A report prepared by the Desertification*

Branch of UNEP. Nairobi: United Nations Environment Programme.

UNESCO *Arid Zone Research*, 1–17. Paris: United Nations Educational, Scientific and Cultural Organisation.

Union of South Africa, 1923 *Final Report of the Drought Investigation Commission*. Capetown: Government of South Africa.

—— 1951 *Report of the Desert Encroachment Committee*. Pretoria: Government Printer.

USAID, 1972 *Desert encroachment on arable lands: significance, causes and control*. Washington, DC: United States Agency for International Development.

Van Apeldoorn, G.J., 1978(a) *Drought in Nigeria. 1, Context and characteristics. 2, Lessons of the disaster*. Research Report 1, Centre for Social and Economic Research, Ahmadu Bello University, Zaria.

—— (ed.), 1978(b) *The aftermath of the 1972–74 drought in Nigeria*. Proceedings of a conference held at Bagauda, April 1977. Federal Department of Water Resources and Centre for Social and Economic Research, Ahmadu Bello University, Zaria.

—— 1981 *Perspectives on drought and famine in Nigeria*. London: Allen and Unwin.

Van Arcadie, Brian, 1978 'The future of vulnerable societies'. *Development and Change*, 9: 161–74.

Van Dyne, G.M., 1975 'Long term development strategies in relation to environmental management'. In: National Academy of Sciences, 1975.

Van Raay, J.G.T., 1975 *Rural development planning in a savannah region*. Rotterdam University Press.

Van Voorthuizen, E.G., 1978 'Global desertification and range management: an appraisal'. *Journal of Range Management*, 31: 378–80.

Vennetier, P. (ed.), 1972 *La croissance urbaine en Afrique noire et à Madagascar*, 2 vols. Paris: Centre National de la Recherche Scientifique.

Verinumbe, I., 1987 'Utilization of indigenous tree species for the control of desertification in northern Borno'. In: Sagua, V.O. *et al.* (eds.), 1987: 276–84.

Vermeer, D.E., 1981 'Collision of climate, cattle and culture in Mauritania during the 1970s'. *The Geographical Review*, 71: 281–97.

Von Maydell, H.-J., 1977 'The contribution of forestry to regional development in the Sahel'. *Applied Sciences and Development*, 10: 164–74.

Walker, B.H. (ed.), 1979 *Management of semi-arid ecosystems*. Developments in Agricultural and Managed-forest Ecology 7. Lausanne: Elsevier.

Walker, Julia and Rowntree, P.R., 1977 'The effect of soil moisture on circulation and rainfall in a tropical model'. *Quarterly Journal of the Royal Meteorological Society*, 103: 29–46.

Wallace, Tina, 1980 'Agricultural projects and land in Northern Nigeria'. *Review of African Political Economy*, 17: 59–97.

—— 1981 'The challenge of food: Nigeria's approach to agriculture

1975–80'. *Canadian Journal of African Studies*, 15: 239–58.
Walls, James, 1984 'Summons to action'. *Desertification Control Bulletin* (UNEP), no.10: 5–14.
Wane, Clémentine and Kone, Djibril, 1980 'Les petits paysans ne sont pas responsables de la déforestation'. *Afrique Agriculture*, 61: 58–60.
Ware, Helen, 1977 'Desertification and population: sub-Saharan Africa'. In: Glantz, Michael H. (ed.), 1977a: 165–99.
Warren, A. and Maizels, J.K., 1977 *Ecological change and desertification*. Background Document, United Nations Conference on Desertification. Nairobi: United Nations Environment Programme.
Wata, Issoufou, 1979 'Régression de la gommeraie et désertification au Manga (Niger)'. *Environment Africain*, Cahiers d'Etude du Milieu et d'Aménagement du Territoire 37. Dakar: ENDA.
Watson, K.A., 1964 'Fertilizers in Northern Nigeria. Current utilization and recommendations for their use'. *African Soils*, 9: 5–20.
Watson, R.M. and Hemming, C.F., 1983 'Can remote sensing save the nomad?' In: *Remote sensing for rangeland monitoring and management*, pp. 123–33. Remote Sensing Society.
Watts, Michael, 1983(a) *Silent violence. Food, famine and the peasantry in northern Nigeria*. University of California Press.
 1983(b) 'On the poverty of theory: natural hazards research in context'. In: Hewitt, K. (ed.), 1983: 231–62.
 (ed.), 1987 *State, oil and agriculture in Nigeria*. Institute of International Studies, University of California.
Well, Jerome, C., 1974 *Agricultural policy and economic growth in Nigeria, 1962–1968*. Ibadan: Nigerian Institute for Social and Economic Research.
Western, David, 1982 'The environment and ecology of pastoralists in arid savannas'. *Development and Change*, 13: 183–211.
White, Cynthia, 1984 *Herd reconstitution: the role of credit among Wo'daa'be herders in central Niger*. Pastoral Development Network Paper 18d. London: Overseas Development Institute.
White, Gilbert, F. (ed.), 1955 *The future of arid lands*. Publication 43. Washington, DC: American Association for the Advancement of Science.
 (ed.), 1974 *Natural hazards. Local, national, global*. New York: Oxford University Press.
White, R.M., 1986 'Climatic variations'. *WMO Bulletin*, 35: 27–30.
Wickens, G.E. and White, L.P., 1979 'Land use in the southern margins of the Sahara'. In: Walker, B.H. (ed.), 1979: 205–42.
Wigley, T.M., Ingram, M.J. and Farmer, G. (eds.), 1981 *Climate and history*. Cambridge University Press.
Williams, Gavin, 1981 *Inequalities in rural Nigeria*. Development Studies Discussion Paper, University of East Anglia.
Wilson, R. Trevor, 1983 'Goats and sheep in the traditional livestock production systems in semi-arid northern Africa: their importance, productivity and constraints on production'. In: Simpson, James R. and Evangelou, Phylo (eds.), 1983: 91–106.

Winstanley, Derek, 1973(a) 'Recent rainfall trends in Africa, the Middle East and India'. *Nature*, 243: 464–5.
    1973(b) 'Rainfall patterns and general atmospheric circulation'. *Nature*, 245: 190–4.
    1974 'Seasonal rainfall forecasting in West Africa'. *Nature*, 248: 464.
    1985 'Africa in drought – a change of climate?' *Weatherwise*, 38: 74–81.
Wisner, Ben, 1977(a) 'Constriction of a livelihood system: the peasants of Thoraka Division, Meru District, Kenya'. *Economic Geography*, 53: 353–7.
    1977(b) 'Man-made famine in eastern Kenya: the interrelationship of environment and development'. In: O'Keefe, Phil and Wisner, Ben (eds.), 1977: 194–215.
Wisner, Ben and Mbithi, Philip M., 1974 'Drought in eastern Kenya: nutritional status and farmer activity'. In White, Gilbert F. (ed.), 1974: 87–97.
Wood, A.P., 1976 'Farmers' responses to drought in Ethiopia'. In: Hussein, Abdul Mejid (ed.), 1976: 67–88.
    1982 'Spontaneous agricultural resettlement in Ethiopia, 1950–74'. In: Clarke, J.I. and Kosinski, L.A. (eds.), 1982: 157–64.
World Bank, 1979 *Urban growth and economic development in the Sahel.* Staff Working Paper 315. Washington, DC: The World Bank.
    1981 *Accelerated development in subSaharan Africa.* Washington, DC: The World Bank.
World Climate Programme, 1983 *Report of the expert group meeting on the climatic situation and drought in Africa.* WCP 61. Geneva: World Meteorological Organisation.
    1984(a) *World climate applications programme.* WCP 75. Geneva: World Meteorological Organisation.
    1984(b) *Report of the study conference on sensitivity of ecosystems and society to climate change.* WCP 83, Geneva: World Meteorological Organisation.
Worster, D., 1979 *Dust Bowl. The southern plains in the 1930s.* New York: Oxford University Press.
WRECA, 1985 *Groundwater monitoring in Kano State.* Kano: Water Resources and Engineering Construction Agency.
Yayock, J.Y., 1977 'An epidemic of rosette disease and its effect on growth characteristics and yield of groundnuts in Nigeria'. *Oléagineaux*, 32: 113–15.
Zartman, I. (ed.), 1984 *The political economy of Nigeria.* New York: Praeger.
Zelinsky, Wilbur, 1971 'The hypothesis of the mobility transition'. *The Geographical Review*, 61: 219–49.
    1979 'The impasse in migration theory: a sketch map for potential escapes'. In: IUSSP, 1979: 18–20.
Zelinsky, Wilbur, Kosinski, Leszek A. and Prothero, R. Mansell (eds.), 1970 *Geography and a crowding world. A symposium on population pressures upon physical and social resources in the developing lands.* New York: Oxford University Press.

Zonneveld, I.S., 1978 'A critical review of survey methods for range management in the Third World with special emphasis on remote sensing'. In: Hyder, D.N. (ed.), 1978: 510–13.

INDEX

Abalu, G.O.I. 30, 94, 99, 100, 108, 239n24, 240n25, n27, n29
Abdu, P.S. 190
Abdull-Jalil, M.A. 248n7
Acacia albida 159, 163, 164, 201, 204
Acacia senegal 164, 165, 201, 246n14
Achtnich, W. 225
Adamawa 32
Adams, M.E. 226
Adams, W.M. 2
Adansonia digitata see baobab
adaptation 3–4, 136, 189–92, 229–30; A's family 115; agricultural 236n24; desertification 157; farmers 189–90; historical 192–6; M's family 113–15; pastoralists 189–90; poverty 237n27; socio-economic status 192–3; trends 194
Adebayo, S.I. 141, 142
Adefolalu, D.O. 140
Adeoye, K.B. 42, 43, 45, 141
Adepoju, A. 223
administration, rural areas 35
Agboola, S.A. 27, 30
Agnew, C.T. 232n6
Ahmadu Bello University, Department of Biological Sciences 72–3
Ahmed, M.L. 52, 190, 248n10
air photos 238n3, 244n5, n6; ecological change 161; sampling 245n7
albedo 148–9, 243n10
aliens, expulsion of 122
Aliyu, M. 52, 190
Allan, W. 208, 210
almajirai 117; at Dagaceri 84, 103, 133
Amin, S. 197
Andropogon spp. 167, 201–2, 246n16, 249n16
Anglo-French Forestry Commission 13, 184, 185, 244n1
Anyadike, R.N.C. 17
APPER 2, 251n1
Arachis hypogaea see groundnuts
arid zone 7
Arid Zone Afforestation Committee 160, 245n11

aridity: in Africa 6–9; in northern Nigeria 9–11
Aristida spp. 165, 167, 201, 245n14, 246n16
Armstrong, R. 122
aro 102
ashasha hoe 99, 102, 108, 115
asset liquidation 65–6
Atlantic Ocean: pollens 144
Aubréville, A. 14, 16
autonomy in resource use 211–13, 220, 251n5
Ayertey, J. 239n17
Ayoade, J.O. 139
Azadirachta indica see neem trees

Baba, J.M. 101, 196
Babura District 79
Baier, S. 76–7, 187, 192, 210, 215, 224
Baker, R. 6, 221
bak'i see strangers
bakin kasa 99
Bakolori 34
bala'i 80
Balanites aegyptiaca 163, 180, 182, 201, 228
Baldwin, K. 32
Ball, N. 193, 202
baobab 84, 163; leaves 53
bara see begging
Bara, Sudan 204
Barber, W. 152, 244n13
Bargery, G.P. 72, 73, 240n32, 242n14
Barral, H. 165
Barth, H. 73
Basement: groundwater 153, 156
Batagarawa 53
Bayero University: Department of Biological Sciences 239n16, n17; Department of Geography 169; Harmattan Research Group 150; Rural Energy Research Project 241n37
Bechuanaland Protectorate 5,230n5
begging 67, 121, 128, 242n7
Bein, F.L. 190
Bellot, J.M. 225

Index

Bellot-Couderc, B. 225
Bennett, J.W. 3, 196
benniseed *see Sesamum indicum*
Benoit, M. 165
Berg, E. 197, 216
Berkovsky, L. 243n10
Bernus, E. 15, 73, 189, 190, 201, 202, 225, 226, 248n15, 252n14
Berry, L. 6, 15, 17, 190, 200
Bezzaz, F. 234n1
Bille, J.C. 215
Bima Hill 81, 237n36
binne 102
bioclimatic zones of Africa 7, 8
Birch, D.H. 31
Birks, J.S. 117
Birniwa 86; Primary School 239n18; Soils Association 239n20
Biswas, M.R., A.K. 15
Biu 131
Blaikie, P. 247n4
Bonfiglioni, A.M. 166, 190
Bonte, P. 210, 212, 215, 247n1
Borno: circulation 131; groundwater 152–3
Boserup, E. 209
Boudet, G. 201, 202, 205
Bougères, J. 204
Bouquet, C. 189, 234n1
Boutrais, J. 190
Bovill, E.W. 12
Bovin, M. 159
Bowden, M.J. 193
Bradley, P. 147, 201, 219
Bray, T.M. 96
Breman, H. 201, 202
Bremaud, O. 205
Bromfield, A.R. 212
Brooke, C. 187
brousse tigré 201
Brown, K.M. 239n18
Brown, L.R. 15
Brune, S. 212
Bryson, R.A. 243n7
budu see fodder:selling
Budyko ratio 232n7
Bulari dunes 172, 178–9
Bunting, A.H. 137, 243n8
Burkill, H.M. 72
Burkina Faso 202, 204, 220, 227
Burton, I. 4, 11, 191, 214
Burum Gana River 160
bush burning 5, 207, 230n4
bush fallowing 13; *see also* shifting cultivation
Buzaye 120–2, 241n3, n4, n5; *see also* Tuareg

Calatropis procera 165
Caldwell, J.C. 189, 234n1, 250n25
Cameroon 9, 32, 122

Campbell, I. 202
capital needed in farming 107–10
carrying capacity: human population 208; livestock 205
Carter, J.W. 152
cassava 73, 247n24
Catterson, T.M. 228
Cenchrus biflorus 73, 159, 165, 167, 180, 182, 201–2, 246n16, 249n16
census data: Nigeria 238n3
Central Bank of Nigeria 40
Chad 9, 51, 105, 202
Chad Basin 144, 152–3
Chad Formation 153
Chad, Lake 131, 156, 160
Chamard, P.C. 204
Chambers, R. 20, 221
Chapman, M. 197, 198, 248n13
charcoal *see* wood fuel
Charney, J.G. 148
Chasin, B.H. 210, 212
Cheke, R.A. 239n17
Cheri dunes 184
Chevalier, A. 14, 203
Chudeau, R. 232n10
Church, R.J. Harrison 137, 234n1
cin rani see circulation
circulation 196–9; advantages of 134–5, 222–3; Dagaceri 90–7, 115, 130–3; investment 120–1, 132;Kano City 117–22; Kano region 128–30; Wo'daa'be 190
Cisse, A.M. 201, 202
Cissoko, S.-M. 187
Citrullus Lanatus see melon
Clark University 214
Clarke, J.I. 223
Cloudsley-Thompson, J.L. 207
Club du Sahel 227
Cochème, J. 43
Cockrim, E.L. 252n13
Cohen, A. 132
Cohen, R. 84
Collier, F.S. 13, 244n1
Collier, P. 187
colonialism 211–13, 229
Colvin, L.G. 223
Comité d'Etudes 12
Comité d'Information Sahel 212
Commonwealth Development Corporation 29
Copans, J. 193, 212, 213, 214, 234n1
Cossus, L. 204
cotton 27, 28
Courel, M.-F. 204
courts, Shar'ia 35
cowpea 27, 28, 45, 87, 91–7, 105–6
critical population density 208–9

Index

crop mixtures 64
Crop and Weather Reports 53, 90, 91, 235n13
cropping system 105–7
CTFT 201
Currey, B. 229

da 57, 92
Dagace, of Dagaceri 83, 133, 241n38
Dagaceri 82–116, 159; bush fallowing 247n26; kinship in 194; land tenure 225; mobility 130–4; opportunity structure 191; resilience 217; subsistence priority 195; woodcutting 242n18
Dahl, G. 228
Dalby, D. 137, 234n1
Dalziel, J.M. 72–4, 106
Damagaram 130; migrants from 84, 133, *dami* 58
'*dan arba'in* 105, 240n29
Danbatta District 36–40, 64
Dancette, C. 228
Dando, W.A. 248n6
Darfur 201, 248n7
Darkoh, M.B. 225
data reliability: air photos 246n19, n20, n23; family heads 236n20; interviewers 237n37; mobility 122–3; transects of dunes 246n22; village heads 237n28
Daudu, I. 52, 89, 99, 102, 105, 190
Daura 152
Davies, J. 106, 107, 239n17, 240n33, n34
Davy, E.G. 7, 137, 145, 147, 243n4
Dayi 87–8
De Castro, J. 11
De Leeuw, P.N. 28, 226
De Wispelaere, G. 250n20
debts 66–7
DECARP 249n18
deforestation 163; in Bechuanaland 5, 230n5; *see also* desertification
degradation, ecological 4, 13, 18, 199–200, 211; arable soils 168–70; rangelands 249n19; *see also* desertification
Delgado, C.L. 227
Delwaulle, J.C. 249n17
dendal 183; Dagaceri 83; Kaska 174
Dennett, M.D. 138, 143, 145, 147 243n8
Depierre, D. 202
Derrick, J. 36, 189, 234n1
Derrienic, H. 139, 140, 144, 189, 234n12
desert: boundary 15; encroachment (South Africa) 15, (West Africa) 12; end state of desertification 17; from savanna 13
desertification: African perceptions of 188; causes of 199; climatic hypothesis of 200–2; concept of 12–15; definition of 157; and deforestation 163, 202–4; development planning 222; environmental misuse 202; migration 248–9n14; pattern of 15; population pressure 208–11; overcultivation 204–5; overgrazing 205–6; severity of 8; structural hypotheses 211–13; technology for controlling 229, 252n9; will of God 81
Desertification, World Map of 7
desertization 12
desiccation 12, 13
dessèchement 12, 14
determinism 19, 229
development planning 222
Dikwa 237n29
Directorate of Overseas Surveys 245n6
disaster insurance 4
district heads 35
diversification 2–3, 221–5
donkey transport 112–13
Dousse, B. 152, 244n13
Dregne, H.E. 15, 18, 229, 232n14
Dresch, J. 248n8
drought: in A's family 133; adaptive response to 3–4, 136; Danbatta district 36–40; definition of 11–12, 147; farming adaptations 63–5; feedback mechanisms 136, 148–51, 199; hydrological 11, 136, 151–6; impact on assets 120; insurance against 81; land use 155–6; livestock 110; mat-making fluctuates with 111; meteorological 11, 157, 185; migration 223; persistence of 136–44; probability of 46; prostitutes blamed 80, 237n35; resilient instability 215; risk minimisation 30; *see also* Kakaduma
droughts: impact of 193; in Dagaceri, 1980s, 94–5, 133; in M's family 113; in 1970s 9, 36, 42–6, 189; remembered in Danbatta 38
Dry Zone: northern Nigeria 9; West Africa 197; *see also* semi-arid zone
Du Preez, J.W. 152
Duffill, M. 72, 73, 74
dum palm 38, 65, 111–12, 159, 160, 164
Dundas, J. 13, 244n1
dunes: formation 182–4; moving 170–84; rangeland 170–2, 178, 180–2, 185, 206; soils 168; stable 160, 165; village perimeter 170–80, 185; water erosion 249n17
Dupree, H. 190, 191
Durand, J.-H. 141
dust: deposition at Kano 150; suspended 149–51, 243n11
Dutt, R. 187
Dyson-Hudson, N. 6

Eckholm, E. 15, 202, 209, 250n25, n29
ECOWAS 224, 242n19
Ega, L.A. 233n7
Ekrafane 206, 249n16
Elston, J. 243n8
EMASAR 73, 209, 226
Escudero, J.C. 189, 217, 250n33
Essiet, E.U. 169
Ethiopia 252n12
ethno-science 4–6, 229
Etkin, N. 73, 74
Eyre, S.R. 6

fadama soils 33
fallow land 111, 203, 241n38
fallowing 104
family 56; heads 130; livestock 110; meals 61; producing and consuming unit 240n26; size 235n17
famine: entitlement approach 188; foods 38, 67–74, 221; in 1913 13, 74; intensification 39; management 229; relief 38, 79–80, 234n11, 236n22, n25, 251n4; religious and moral framework 80; responses 38–40; social factors 187; social impact 20
famines: chronology 75–7; evaluation 75–81
Famoriyo, S. 233n7
FAO 17, 25, 29, 32, 87, 88, 202, 208, 213, 229
fari see drought
farin kasa 99
farm: distribution, Dagaceri 104, 114; holdings 115; sizes 103, 236n19; trees 221, 228–9
farmed parkland 228, 252n17
Farmer, G. 136, 140, 145, 146
farming systems research 251n8
fartanya hoe 99, 108
fatoma 131, 242n14, n15, n17
Faulkingham, R. 97, 189, 190, 233n6, 238n9
Faure, H. 145
Federal Department of Forestry 167, 246n15
Federal Livestock Department 224
Federal Ministry of Agriculture 51
Federal Office of Statistics 233n3
Federal Surveys 244n2, n6
Ferguson, D.S. 30
fertilizers 100, 239n23
Fété-olé 215
Field, N.J. 245n7
Fisher, H.H. 187
Fisher, M. 239n16
Fishpool, L.D.C. 239n17
Fishwick, R.W. 33
Flohn, H. 145
fodder: collection 133; scarcity 241n38; selling 111, 165

Folland, C.K. 146, 243n10
food: entitlement 4; famine 67–74; production and consumption 86–8
forest reserves 34, 81, 111
forestry 228–9; policy 33
Forrest, G.A. 239n17
Forrest, T.G. 32
fragility: ecosystem 3, 16, 17, 212, 216–17; human system 230n2
Franke, R.W. 210, 212
Franquin, P. 43
Fricke, W. 29, 227
Fuglestad, F. 187
Fulani: farming 227–8; livestock 62–3; migrants 97; nomadic 251n10; sample 56; strangers 120; *see also* Ful'be
Ful'be 85–6, 165, 183–4; *see also* Fulani
Funtua Agricultural Development Project 2
fura 60–1
Furniss, G. 76–7

Gac, J.-Y. 145
gado 102
Gallais, J. 190, 212, 226, 227
Gana, J. 2
Gano, D.Z. 122
Gapp, K.S. 192
Gaptari: dunes 172, 178–9; livestock 183
Garcia, R.V. 189, 192, 214, 217, 250n30, n33
Garduno, M.A. 229
gari 60–1
Garki District 233n6
Gashua 131, 165
Geidam 232n11
gero see millet
gida see households
Gilardi, J.-C. 222
Giles, P.H. 220
Gillet, H. 202
Glantz, M.H. 15, 144, 199, 234n1, 243n5
Gleave, M.B. 209
goat trade 131–3
Goddard, A.D. 32, 191
Goldschieder, C. 248n13
Goldstein, S. 197
Gombe 32, 131
gonar jeji 99
Gould, W.T.S. 197
Gouré 250n21
Gourma 202
grain: exports from Dagaceri 116, 241n43; losses from drought 36, 50, 57–8, 75; measures 58; prices 53–6, 90–1; production 87–8, 114; requirements 87, 238n9; reserves 61–2, 91–7, 115, 219–21; seed 236n25; terms of trade (livestock) 55–6, (mats) 112, 115; yields 88, 166

Index

Grainger, A. 200, 250n31
grasshoppers: at Dagaceri 92, 93, 95–6, 105, 114; at Gumel 239n17
grassland: floristic change 167, 202; regeneration 167; subsystem 165–8, 185, 206
grazing: conditions 95, 166; tenure 225–7, 252n14
Green, L. 197
Greene, M.H. 189
Grégoire, E. 110
Gregory, J. 193
Gregory, S. 140
groundnuts: Bambarra 107; Dagaceri 87, 106; disease 41, 92; effect on soil 212; exports from Niger 234n12; failure of 55; Marketing Board 23, 25, 40; melon, replacing 240n33; prices 29–30; production 23, 28, 40; purchases 26, 40, 106; soil water 45; yield in relation to circulation 132
groundwater: artesian boreholes 32, 152–3; dam construction 155; Danbatta District 36; *fadama* soils 33; Kano State 46; *kwari* soils 164; semi-arid zone 151–5; trend 200; well deepening 234n5
Grove, A.T. 2, 11, 65, 137, 160, 187, 209, 234n10
Guiera senegalensis 49
guinea corn: early maturing 240n34; growth conditions 27; new varieties 96; processing 238n8; production 28, (and consumption) 86–8; response to rainfall 48; soil water 45; yields 91–7; *see also* grain
Gumel: cattle prices 55; crop varieties 107; famine relief 79; grasshoppers 239n17; sesame 107; urban traders 240n33
guna see melon
gurjiya see groundnuts: Bambarra

Haaland, G. 252n14
Hadejia 86: cattle prices 55; circulation 131; famine relief 79–80; mat-making 132
Hammer, T. 204
Hankins, T. 190
Hare, F.K. 136, 146, 151, 199, 205
Haring, K.J. 211, 248n14
Harmattan 150–1
Harris, B. 72
Harrison, M.N. 15
Hart, K. 2, 188
Harvie, D. 6
Hastings, A. 73, 234n10
Hausa: community at Dagaceri 97–8; language in interviews 234n6; sample 56
Hays, H.M. 2, 20, 58, 87, 220, 238n8
Heathcote, R.L. 5, 19, 214

Heijnen, J. 190
Hellden, U. 18, 229
Helleiner, G.K. 30
Hemming, C.F. 229
Hendy, C.R.C. 209
Hergert, G.R. 239n17
Hewitt, K. 11, 214
Hill, A.G. 6
Hill, P. 20, 53, 58, 65, 73, 76–7, 77, 87, 88, 101, 102, 113, 192, 234n10, 252n18
Hills, E.S. 15
Hinds, J. 80
Hjort, A. 227, 228
Hock, J.C. 229
Hogendorn, J.S. 23, 25, 106, 233n1, 240n33
Holling, C.S. 214, 216
Holy, L. 190
Horowitz, M. 159, 190
households: Dagaceri 97–8; heads, 56
Hubert, H. 12
Hugo, G. 224, 247n2
Hulme, M. 137, 147
hunger *see* famine
Huntingdon, E. 12, 187
Hutchinson, J. 72
Hutchinson, P. 143
Hyden, G. 2
hydrological subsystem 19
Hyphaene thebaica see dum palm

Ibadan 132
Ibrahim, F. 151, 203, 204, 249n18
Ibrahim, M.G. 80
IEMVPT 246n14
Ilesanmi, O.O. 141
Illela 52
ILO 2
Ilorin 147
indirect rule 35
inequality, economic 88
Ingessena Hills 190
insect pests 104, 240n29
Institut Géographique National 244n2
insurance 12, 219–21
integrated rural development 2, 97
intensification *see* land use
Inter-Tropical Discontinuity 43–6, 141, 145, 149–50
Ireland, A.W. 141
irrigation 1–2, 33, 207
iyali see family

Jackson, J.K. 15, 227, 228
Jaculus jaculus see jerboa, Tzarza
Jasinski, J. 243n8
Jedrej, M.C. 190
jerboa, Tzarza 95–6
jingina 102

Johnson, D.L. 15, 211, 217, 248n14
Jones, Brynmor 13
Jones, D.G. 152
Jones, G.R.J. 6
Jones, M.J. 203, 212, 213

kaba see mat-making
Kaduna 35, 122
Kakaduma xvii, 20, 115, 116; capital loss 110; Dagaceri 86; evaulation of 78, 193; land selling 195; mobility 123, 127–8, 134
K'ak'alaba 38, 75, 79, 193
Kankiya, Y. 76–7
Kano: rainfall at 89, 140
Kano Close-Settled Zone: compared with Nyala 226; farm sizes 104; farm trees 201, 204, 228; farmlands 13; intensification 209; land (supply) 32, (tenure) 195; manure inputs 239n22; opportunity structure 191; population densities 48; soil management 205
Kano Emirate 35
Kano, Metropolitan: circulation in 117; sample locations 118; urban informal sector 122
Kano River Project 34, 196
Kano State: Agricultural and Rural Development Authority 96, 105, 116; boreholes 244n15; drought-affected areas 20; Drought Relief Committee 80, 234n8, 236n22, 251n4; grazing rights 34; groundnut production 23; groundwater 46, 154–5; markets 53; mobility 32, 119; rainfall 43; Water Resources and Engineering Construction Agency 153–4
Kano–Hadejia River system 153–5
Kanuri, in Kano 120
kanwa see natron
Kanya 133, 134
karatu 117–19, 122, 130
karkara 99
Karnovitz, A. 243n8
Kaska: dunes 172, 173–5; market 173; nursery 245n11
Kassam, A.H. 25, 27, 43, 44, 45, 46, 105, 139, 201, 213, 233n2
Kassas, M. 202
Kates, R.W. 4, 190, 191, 193, 211, 214, 248n14
Katsina 152
Katz, R.W. 138, 243n5
Kazaure: famine relief 79–80; groundwater 46
Keay, R.W.J. 72
Kerri Kerri Formation 152
Keys, A. 192
Khan, M.A. 122

Khartoum 15
Kidson, J.W. 137
King, J.G.M. 32
King, R. 33
Knabe, D.T. 9, 12
Kolawole, A. 230n3
Kolawole, M.I. 108
Komadugu Yobe *see* Yobe River
koranic scholars *see almajirai*
Kordofan 249n18
Kosinski, L.A. 209, 223
Kowal, J.M. 9, 12, 25, 27, 42, 43, 44, 45, 46, 105, 139, 141, 201, 212, 233n2
Kraus, E.B. 138
Krings, T. 206
kuka see baobab
Kukangiwa 241n44
kunu 60–1
Kura, A.G. 190
k'wadago 65, 129
kwari 160, 162, 164, 166, 173, 176, 177, 180, 183, 245n8, 247
Kwubsa market 86, 93, 110, 113, 132
kyauta 102

Labaran, A. 196
labour: costs 114; hiring 102; migration *see* circulation; shortage 101, 107–8
Labouret, H. 197
Lagos 96, 131–5
Lala'a dunes 172, 178–80
Lamb, H.H. 138, 243n12
Lamb, P.J. 138, 146
Lamprey, H.F. 15, 201
land: allocation 133; acquisition 102–3; capability 216; renting 240n28; selling 66, 195; supply 48, 104; tenure 33–4, 195, 225–6; yield 216
Land Resources Division 165, 166, 167, 168, 244n3, 246n16, n17
Land Tenure Law 233n7
land use: changes 105, 161–3; desertification and 185–6; drought and 155–6, 157, 185; ecology of 187; groundwater, effects on 152; intensification 209–10, 225–9, 252n13; opportunistic 206–7; rainfall and 148–50; systems of 207
Land Use Act 233n7
landlord *see fatoma; maigida*
Landsberg, H.E. 139
latitude, correlated with rainfall 44
Laurent, C.K. 32
Lawry, S.W. 227
Laya, D. 19, 190
Le Houérou, H.N. 205, 209, 215, 222
Leach, M. 221
Leftwich, A. 6
Leng, G. 15

Index

Leow, K.S. 141
Leptadenia pyrotechnica 165, 180, 182, 183, 246n23
Lewicki, T. 73
Ligaridi Babba dunes 159, 176–8
Ligaridi Karami dunes 172, 176
livestock: densities 166; dune formation, effects 183; fodder reserves 220–1; holdings 62–3, 110, 115, 121, 166; losses in drought 36, 50–3, 57–9; marketing 235n10; in northern Nigeria 28–9; prices 30; production 2, 227–8; small 228; terms of trade 55–6
Lofchie, M. 1, 226
Longhurst, R. 20, 87–8
longitudinal method 20, 83, 89–98, 193, 239n19
Lough, M. 146
Lovejoy, P.E. 164, 210, 215
Lucas, G.L. 221
Lundholm, B. 214, 222

MAB 228, 232n7
Mabbutt, J.A. 17
Mabogunje, A.L. 2, 197, 198
McCown, R.L. 226, 227
McDonnell, G. 127
McDowell, C.M. 233n7
Macleod, N.H. 149
McTainsh, G.H. 25, 150, 175
Maddox, G.H. 187
Magaria 134
Mageed, Y.A. 229
magirbi hoe 108
Maguzawa 35, 237n34
Mahdi, the 81
Mai Amaro 38
Mai Buhu 38, 75
Mai Tatsine 122
Maiduguri 122, 131
maigida 121; see also households
Maine Soroa 89, 95
Mainguet, M. 15, 204, 229
maitono see jerboa, Tzarza
maiwa 27
Maizels, J.K. 214, 216
malam 130
Malamaduri 235n10
Maley, J. 147, 160
Mali 9, 12, 206
Maliki, A. 221
Mallory, W.H. 187
Malumfashi 233n6
management regimes 19
Manga: area, unsettled 244n1; attitude to Kano 130; circulation 131; community at Dagaceri 97–8; sample 56; women's field work 101, 240n25

Manga Grasslands: boundaries 163, 185, 232n11, 245n10; description 157–60; ecological change 161; location 158; saline lakes 112
Mann, H.S. 228
manure 66, 209, 227–8
manuring 64, 99–100
Maradi 204
market: proposed at Dagaceri 241n44; system 35
Marwa, Mohammadu 122
Mason, B.J. 137
masu cin rani see circulation; migration
masu dan hali 66
masu sarauta 66
mat-making 65, 111–12, 115, 132
Matheson, W. 229
Matlock, W.G. 252n13
Mauny, R. 144
Mauritania 9, 12
Mbithi, P.M. 190
meals 61
Mecca 83, 117, 126
Meek, C.K. 233n7
Meillassoux, C. 196
melon 106–7, 240n33
Mensah, E. 242n7
Mensching, H. 204, 222, 225, 246n18, 249n18
Meteorological Office (United Kingdom) 146
Michon, P. 202
Middle Belt 32
migration: desertification, response to 248–9n14; in- 125–8, 133–4; international 224; investing proceeds 108; out- 97–8, 123–6; policy 223; *see also* circulation
Milas, S. 16, 209, 210
Miles, M.K. 146
millet: cultivation 168–70, (limits) 44–5; grasshoppers 95–6; growth conditions 27; plantings 57, 236n18; processing 238n8; production 28, (and consumption) 86–8; rainfall response 48; yields 91–7; *see also* grain
Ministry of Agriculture: Northern Region 32, 152
Ministry of Agriculture and Natural Resources, Kano State 90, 238n11
Ministry of Land and Survey, Kano State 105
Misau 35
Mischlich, A. 73
MIT 252n13
miya 60
Mkunduge, G.L. 190
mobility *see*: circulation; migration
Mohammed, I. 141, 142
Moock, J.L. 251n8

Index

moral economy 193
Morgan, W.B. 7
Mormoni, Z.L. 190
morphodynamic subsystem 19, 170–84, 185
Morris, R. 9, 234n1
Mortimore, M.J. 20, 25, 27, 28, 29, 32, 36, 48, 105, 117, 127, 135, 188, 191, 197, 204, 209, 220, 238n3, 239n18, n22, 247n3
Mossi 227
Motha, R.P. 138, 140, 162
Mungurun 52
Mustapha, S. 160, 174

Nachtigal, G. 73
naira, value of xxi
National Academy of Sciences 222
National Committee on Arid Zone Afforestation 14
natron 160, 164; trade in 112, 135
natural hazards 191–2, 214
Ndaks, G.D. 52, 190
neem trees 84, 163
Nelson, J.M. 224
Netting, R.McC. 209
Ngelsandi dunes 172, 178–80
Nguru: circulation 131; *egusi* demand 106; grazings 165; rainfall 89
Nicholson, S. 137, 138, 141, 143, 145, 146, 156, 243n12
Niger Agricultural Project 32
Niger Republic: arid zone 15; border with Nigeria 13, 178–80, 224, 244n2, 247n24; cattle exports 51, 224; drought (in 1970s) 9, 19, (in 1980s) 95; farming limits 7, 45; floristic change 202; grazing conditions 166; groundnuts 212, 234n12; migrants (to Kano) 119, 122, (to Nigeria) 127; new crops 105, 107; overcultivation 204; overgrazing 206; price guarrantees 233n5; rainfall 138–44; semi-arid zone 9
Niger–Benue Basin 152
Nigeria: international borders 223, 224, 242n19; rainfall 138–44
Nigerian Grains Board 251n4
nitrogen 233n2
Norman, D.W. 2, 20, 32, 101, 191, 236n23, 251n8
Northern Guinea zone 7, 33
Norton C.C. 149
Nuwanshong, S. 169
Nyala, Sudan 226

Oedaleus senegalensis 95; *see also* grasshoppers
off-farm occupations *see* secondary occupations
Offodile, M.E. 152
Ogallo, L. 138
Ogigirigi, M.A. 252n15
Oguntoyinbo, J.S. 11, 46, 139, 148, 190, 234n7
Ojo, G.O.A. 252n15
O'Keefe, P. 6, 248n9
Okigbo, P. 251n8
Olatunbosun, D. 27
Olofin, E.A. 151
Ologe, K.O. 141
Olowu, J.A.I. 153
Olsson, L. 151, 249n18
Omotosho, J.B. 147
Onitsha 131
Onochie, C.F.A. 72
Onyemelukwe, J.O.C. 131
opportunity structure 191, 222; *see also* adaptation; diversification; secondary occupations
Organisation of African Unity 2
Orlovsky, N. 15
Orogun, E.T. 196
Otegbeye, G.O. 252n15
Otterman, J. 149
overcultivation *see* desertification
over-exploitation 18, 185–6
overgrazing *see* desertification
Owen, J.A. 146, 243n10
Oxby, C. 226

Pacey, A. 20
Paden, J.N. 241n1
Pagot, J. 205
Palmer, T.N. 243n7
Parker, D.E. 146
Parry, D.E. 167, 246n15
peasantry 195–6
pedological subsystem 19
Pennisetum typhoides see millet
Peyre de Fabregues, B. 249n16, 250n20
Piché, V. 193
Pitte, J.-R. 189, 234n1
plant spacing 64
ploughs, ox- 32, 108, 113–15
Poncet, Y. 19, 159
Popov, G.B. 239n17
population: bilocality 198; Dagaceri 84; distribution 31; fertility 208; growth 16, 30; Manga Grasslands 246n19; pressure 208–11; *see also* circulation
Porter, P. 190
Potiskum 152
Poulain, J.F. 228
prices: cattle 55; crops 29–30; data, reliability of 53; farms 113; fodder 111; food 36, 53–6, 90–7, 235nn12–15; labour 102; livestock 29–30, 110, 240n36; mats 111; transport 113; variation with rainfall 91; wood fuel 111
Prirard, F. 160

Index

Prothero, R.Mansell 12, 20, 31, 197, 198, 209, 244n14, 248n13
Pugh, J.C. 7
Pullan, R.A. 28, 73, 160, 204, 239n20

Raeburn, C. 12
rainfall: August 142–4, 242n3; departures from mean 139; dry matter production 166, 215; dynamic processes 145–6; forecasting 136, 144–7, 149; grain yields 166; in 1950–69, 162–3; in 1951–75, 25, 30; in 1970–85, 89; in 1972–73, 42–6; indices 89, 238n9, nn13, n14; monthly distribution 141–3
Niger (1981–84) 133, (and Kano State) 127, 242n8, (and Nigeria) 138–44; prices, effects on 91; probabilities 4, 147, 243n9; reduction 201; sea surface temperatures 146, 243n10; semi-arid zone 9–11; teleconnections 146; trends 3, 144–5; variability 234n4; West Africa 137–8
rainy season 45
Rapp, A. 7, 18, 229
ratchet effect 192
Raulin, H. 108
Raynaut, C. 66, 193, 219
Reining, P. 18, 229, 232n15
remote sensing 229
Renner, G.T. 13, 187
research: costs 22, 233n20; farming systems 2, 6
resettlement 224
resilience 3, 213–17, 230n2; Danbatta District 40; diversity 221–2; farming system 65; interregional dependence 224–5; policies to strengthen 218–19; production systems 20, 116
rice 27, 28
Richards, H.J. 155
Richards, P. 2, 5, 11, 46, 139, 187, 190, 234n7, 248n11
ri'di see Sesamum indicum
Rijiyar-Tsamiyar 52, 248n10
rinderpest 166
Ringrose, S. 229
Ripley, E.A. 148
Rippstein, G. 167
Roche, C.J.R. 220
Roder, W. 190, 191
Roni District 40, 64
rope-making 65
Rosette disease 41, 92
Ross, P. 73, 74, 195
Rouch, J. 197
Rowling, C.W. 233n7
Rowntree, P.R. 149

ruga 85–6
Russell, C.S. 219

Sagua, V. 9
Sahara: advancing 79; boundary fluctuations 144; encroaching 12, 13, 15, 202, 232n11
Saharisation 157
Sahel 7, 33; demographic characteristics 6; desertification trends 17; drought in 1970s 1, 9; economic history 210; economies 216; farming system 107; grain requirement 87; intensification 252n13; rainfall 137, 138, 146, 147, 155–6, 242n1
Sahelo-Sudanian zone: drought 136–7; rainfall 138
Salifou, A. 187
salt making 164
Samaru 147
sana'a 117–19; see also secondary occupations
Sandamu 52
Sandbrook, J. 2
Sandford, S. 147, 206, 250n23
sanfo 58
savanisation 14
Savonnet, G. 189
saye 102
school, at Dagaceri 96; at Birniwa 239n18
Schove, D.J. 137
Seaman, J. 189
secondary occupations 65; see also diversification; fodder : selling; mat-making; opportunity structure; rope-making; salt making; wood fuel: collection, trade
Seignobos, C. 74
semi-arid zone: distribution 7; ecological monitoring 229; groundnuts 41; northern limit 44; political divisions 9–11; regional integration 210; risk 230
Sen, A. 4, 188
Senegal 9, 12, 202
Sesamum indicum 107
Shari'a Law 33
Sheets, H. 9, 234n1
shelter belts 163, 252n15
Shenton, R. 193
shifting cultivation 13, 178, 225, 240n31, 247n26
Sidikou, A.H. 190
Simmons, E.B. 2, 20, 87, 238n7
Sirag, M.I. 201
Sivakumar, M.V.K. 147, 243n9
Smith, M.G. 58, 238n8
social networks 66–7, 221
soil: erosion 13; heating 100; management 99–100; moisture 149, 152, 243n10;

nutrients 212–13; properties 169, 246n17, n18; subsystem 168–70, 185, 205; suitable for groundnuts 25; water 45
Sokoto, Caliphate 233n9; capitalist farming 196; close-settled zone 32, 191; migration 32; rainfall and wells 12
Sokoto–Rima Basin 152
Sorghum bicolor see guinea corn
Sorghum dura 237n29
Sorokin, P.A. 192
South Chad Irrigation Project 230n3
Southern Guinea zone 33
Spink, P.C. 250n30
Spitz, P. 192
Spooner, B. 222
Stamp, L.D. 13
Stanfield, D.P. 72
Starns, W.W. 233n7
Stebbing, E.P. 13, 14
Stern, R.D. 141
Stiles, D. 16
Stone, P.H. 148
storage of food 12, 61, 67, 88, 219–21
strangers 117–22
Strategic Grain Reserve 220, 251n3
Stryker, J.D. 206, 210
Sudan Republic 15; *ashasha* 240n35; deforestation 203; Gezira scheme 132; overcultivation 204
Sudan zone 12–13, 33; farming systems 107; rainfall 138, 147
Sudano-Guinean zone 138
Sudano-Sahelian region 6, 17
Sud-Tamesna 250n20
Sukumaland 190
sungumi hoe 99
survey areas, location 47
survey villages 56
Swami, K. 232n7
Swift, J. 221, 225, 226, 248n8
Swindell, K. 197

tafki 99, 159, 160, 168, 170
taki see manure
talakawa 66
Talbot, M.R. 249n17
Tanaka, M. 137, 143
Tannehill, I.R. 11
tax: assessment records 235n9, 238n3; defaulting 55; remissions 96
Ténéré dunes 159
termite granaries 73, 74, 237n29, n32
Thorbahn, P.F. 97, 189, 233n6, 238n9
Tiffen, M. 32
Timberlake, L.C. 202, 211, 219, 232n14
tiya 87

tobacco 27
Tolba, M.K. 16, 209, 250n26
Toupet, C. 189, 212
Tourte, R. 213
Toutain, B. 202
tractor hire 108
transhumance 29, 32, 224
Trevallion, B.W. 233n4
Trevitt, J.W. 167, 246n15
Trilsbach, A. 137
Tuareg 212, 215; *see also* Buzaye
tudu 160, 162, 163, 164, 183; soils 33
Tulatura market 174
Turabu, H.M. 252n15
Turner, B. 33
tuwo 60–1

UNEP 6, 7, 12, 16, 17, 185, 218, 219, 232n7, 249n18, 250n26, 251n1, n7; Desertification Branch 232n14
UNESCO 15
Ungogo District 234n11
Union of South Africa 11, 15
United Nations: Conference on Desertification 12, 15, 218, 250n26, 251n1, n7; Special Sahelian Office 238n9; *see also* UNEP
Universal Primary Education 96
urban informal sector 122, 222–3
US Monthly Climatic Data 239n14
USAID 14–15
Usambara Mountains 190
Uzoma, J.U. 153

Van Apeldoorn, G.J. 20, 42, 43, 45, 50, 51, 52, 53, 54, 55, 66, 190, 219, 220, 222, 224, 235n12, 237n33, 238n11, 248n10, 251n6
Van Arcadie, B. 219
Van Dyne, G.M. 252n11
Van Raay, J.G.T. 29, 227
Van Voorthuizen, E.G. 228
vegetation subsystem 19
Verinumbe, I. 228
Vermeer, D.E. 189
Vigna unguiculata see cowpea
village heads 47–8, 78–9, 222, 234n8, n11
Von Maydell, H.-J. 228
vulnerability: ecosystem 212; human system 3, 193, 230n2

Walker, J. 149
Walker, P.H. 150
Wallace, T. 2, 196
Ware, H. 210
Warren, A. 214, 216
Wata, I. 165, 201

Index

water table 13; *see also* groundwater
Watson, K.A. 213
Watson, R.M. 229
Watts, M. 2, 66, 187, 190, 192, 193, 196, 198, 234n10, 235n14, n15, 247n3, 248n12
weeding 102, 109
welfare perceptions 195
Wells, J.C. 23, 32
Western, D. 215
White, C. 190
White, G.F. 4, 15, 191, 192, 214
White, H.P. 209
White, L.P. 165, 225, 250n22
White, R.M. 252n9
WHO Malaria Control Project 233n6
Wickens, G.E. 165, 221, 225, 250n22
Wigley, T.M.L. 136, 140, 145, 146
Wild, A. 203, 212, 213
Williams, G. 67
Williams, M.A.J. 249n17
Wilson, J. 25, 117
Wilson, R.T. 228
winds: dune formation 247n25
Winstanley, D. 144, 145, 146, 243n6
Wisner, B. 6, 190, 193, 241n42, 248n9
Wood, A.P. 192, 223, 247n3
wood fuel: charcoal 73, 232n15; (Sudan) 15; collection 133; deforestation 204; farm forestry 228; Kano region 241n37; offtake 28; requirements, Niger 252n16; rights 34; trade 65, 111, 236n26
woodland: subsystem 163–5, 185, 203; tenure 227; tree mortality 201
World Bank 1, 2, 97, 222, 225
World Climate Programme 138, 145, 146, 149, 242n1
World Meteorological Organisation 7, 231n6
Worster, D. 252n15

'Yan Buhu 38
'Yar Gusau 38, 75
Yayock, J.Y. 41
yields: major crops 48–50, 57–8
Yobe River 131, 160, 165, 207, 244n3
Yola 122
Yoruba language 131
yunwa see famine

zamani 81
Zaria 32, 87
Zartman, I. 187
Zelinsky, W. 199, 209, 248n13
Zinder 89, 133
Zonneveld, I.S. 229